Smart Innovation, Systems and Technologies

Volume 268

Series Editors

Robert J. Howlett, Bournemouth University and KES International, Shoreham-by-Sea, UK

Lakhmi C. Jain, KES International, Shoreham-by-Sea, UK

The Smart Innovation, Systems and Technologies book series encompasses the topics of knowledge, intelligence, innovation and sustainability. The aim of the series is to make available a platform for the publication of books on all aspects of single and multi-disciplinary research on these themes in order to make the latest results available in a readily-accessible form. Volumes on interdisciplinary research combining two or more of these areas is particularly sought.

The series covers systems and paradigms that employ knowledge and intelligence in a broad sense. Its scope is systems having embedded knowledge and intelligence, which may be applied to the solution of world problems in industry, the environment and the community. It also focusses on the knowledge-transfer methodologies and innovation strategies employed to make this happen effectively. The combination of intelligent systems tools and a broad range of applications introduces a need for a synergy of disciplines from science, technology, business and the humanities. The series will include conference proceedings, edited collections, monographs, handbooks, reference books, and other relevant types of book in areas of science and technology where smart systems and technologies can offer innovative solutions.

High quality content is an essential feature for all book proposals accepted for the series. It is expected that editors of all accepted volumes will ensure that contributions are subjected to an appropriate level of reviewing process and adhere to KES quality principles.

Indexed by SCOPUS, EI Compendex, INSPEC, WTI Frankfurt eG, zbMATH, Japanese Science and Technology Agency (JST), SCImago, DBLP.

All books published in the series are submitted for consideration in Web of Science.

More information about this series at https://link.springer.com/bookseries/8767

Jie-Fang Zhang · Chien-Ming Chen ·
Shu-Chuan Chu · Roumen Kountchev
Editors

Advances in Intelligent Systems and Computing

Proceedings of the 7th Euro-China
Conference on Intelligent Data Analysis
and Applications, May 29–31, 2021,
Hangzhou, China

Editors
Jie-Fang Zhang
Institute of Intelligent Media Technology
Communication University of Zhejiang
Hangzhou City, China

Shu-Chuan Chu
College of Computer Science
and Engineering
Shandong University of Science
and Technology
Qingdao, Shandong, China

Chien-Ming Chen
College of Computer Science
and Engineering
Shandong University of Science
and Technology
Qingdao, Shandong, China

Roumen Kountchev
Department of Radio Communications
and Video Technologies
Technical University of Sofia
Sofia, Bulgaria

ISSN 2190-3018 ISSN 2190-3026 (electronic)
Smart Innovation, Systems and Technologies
ISBN 978-981-16-8050-2 ISBN 978-981-16-8048-9 (eBook)
https://doi.org/10.1007/978-981-16-8048-9

© The Editor(s) (if applicable) and The Author(s), under exclusive license to Springer Nature Singapore Pte Ltd. 2022

This work is subject to copyright. All rights are solely and exclusively licensed by the Publisher, whether the whole or part of the material is concerned, specifically the rights of translation, reprinting, reuse of illustrations, recitation, broadcasting, reproduction on microfilms or in any other physical way, and transmission or information storage and retrieval, electronic adaptation, computer software, or by similar or dissimilar methodology now known or hereafter developed.

The use of general descriptive names, registered names, trademarks, service marks, etc. in this publication does not imply, even in the absence of a specific statement, that such names are exempt from the relevant protective laws and regulations and therefore free for general use.

The publisher, the authors and the editors are safe to assume that the advice and information in this book are believed to be true and accurate at the date of publication. Neither the publisher nor the authors or the editors give a warranty, expressed or implied, with respect to the material contained herein or for any errors or omissions that may have been made. The publisher remains neutral with regard to jurisdictional claims in published maps and institutional affiliations.

This Springer imprint is published by the registered company Springer Nature Singapore Pte Ltd.
The registered company address is: 152 Beach Road, #21-01/04 Gateway East, Singapore 189721, Singapore

Preface

This volume contains the proceedings of the 7th Euro-China Conference on Intelligent Data Analysis and Applications (ECC 2021), which is hosted by Communication University of Zhejiang and Shandong University of Science and Technology and was held in Hangzhou, China, during May 29–31, 2021. ECC 2021 is technically co-sponsored by Springer, Communication University of Zhejiang, Shandong University of Science and Technology, and Taiwan Association for Web Intelligence Consortium. It aims to bring together researchers, engineers, and policymakers to discuss the related techniques, to exchange research ideas, and to make friends.

37 regular papers were accepted in this proceeding. We would like to thank the authors for their tremendous contributions. We would also express our sincere appreciation to the reviewers, program committee members, and the local committee members for making this conference successful. Finally, we would like to express our special thanks for the great help from Communication University of Zhejiang for locally organizing the conference.

Hangzhou City, China	Jie-Fang Zhang
Qingdao, China	Chien-Ming Chen
Qingdao, China	Shu-Chuan Chu
Sofia, Bulgaria	Roumen Kountchev

Contents

Intelligent Data Analysis

An Integrated System for Unbiased Parkinson's Disease Detection from Handwritten Drawings 3
Liaqat Ali, Ce Zhu, Hengling Zhao, Zhonghao Zhang, and Yipeng Liu

Improving Supply Chain Agility to Reduce Enterprise Risk Based on QFD-MADM 15
Ru-Yue Yu, Chih-Hung Hsu, and An-Ching Sun

A CNN-Based Method for AAPL Stock Price Trend Prediction Using Historical Data and Technical Indicators 25
Yuxiao Gong, Jimmy Ming-Tai Wu, Zhongcui Li, Shuo Liu, Lingyun Sun, and Chien-Ming Chen

Research on Supply Chain Agility Based on Bullwhip Effect 35
Ming-Ge Li, Chih-Hung Hsub, and Jun-Wei Liuc

Using Agility to Reduce the Bullwhip Effect of Supply Chains 43
Jun-Yi Zeng, Chih-Hung Hsu, and Xiu Chen

Mining Frequency-Utility Patterns from a Big Data Environment 53
Ranran Li, Jimmy Ming-Tai Wu, Min Wei, Ke Wang, and Qian Teng

The Association Between Related-Party Transactions and Tax Planning 63
Wen-Jye Hung, Tsui-Lin Chiang, Ya-Min Wang, and Qi Luo

Mind-Media System: A Consumer-Grade Brain-Computer Interface System for Media Applications 77
Chang Liu, Yijie Zhou, and Dingguo Yu

Artificial Intelligence

A Cooperative Evolution Framework Based on Fish Migration Optimization 85
Wenqi Li, Shu-Chuan Chu, and Jeng-Shyang Pan

Improving K-Means with Harris Hawks Optimization Algorithm 95
Li-Gang Zhang, Xingsi Xue, and Shu-Chuan Chu

Weighted Multi-task Sparse Representation Classifier for 3D Face Recognition 105
Linlin Tang, Zhangyan Li, Tao Qian, Shuhan Qi, Yang Liu, Jiajia Zhang, Shuaijie Shi, Churan Liu, and Jingyong Su

Multiple Kernel Clustering with Direct Consensus Graph Learning 117
Yanlong Wang and Zhenwen Ren

Density Peaks Clustering Algorithm Based on K Nearest Neighbors 129
Shihao Yin, Runxiu Wu, Peiwu Li, Baohong Liu, and Xuefeng Fu

Exponential Fine-Tuning Harmony Search Algorithm 145
Lipu Zhang and Xuewen Shen

Computational Intelligences

A Location Gradient Induced Sorting Approach for Multi-objective Optimization 157
Lingping Kong, Václav Snášel, Swagatam Das, and Jeng-Shyang Pan

Production Line Balance Optimization Based on Improved Imperial Competition Algorithm 167
Xue-Hua Yang and Chih-Hung Hsu

Research on the Key Resilience Indexes of Logistics Enterprises in Response to Supply Chain Disruption Risks 175
Xu-He, Chih-Hung Hsu, and Xian-Tuo Xiao

Optimization of Resource Service Composition in Cloud Manufacture Based on Improved Genetic and Ant Colony Algorithm 183
Wang Zhengcheng

Real-Time Multi-person Multi-camera Tracking Based on Improved Matching Cascade 199
Yundong Guo, Xinjie Wang, Hao Luo, Huijie Pu, Zhenyu Liu, and Jianrong Tan

Design and Optimization of the Seat of the Elderly Scooter Based on Solar Energy 211
Ya-Zheng Zhao and Yi-Jui Chiu

Research on Digital Intelligence Enabled Omnimedia
Communication System and Implementation Path in 5G Era 221
Weilong Chen and Jing Zhang

Analysis of Interaction Patterns of Intercountry Cooperation
and Conflict Events Based on Complex Networks 229
Xiao-Mei Mo and Ding-Guo Yu

Video, Image, and Others

Application of the Novel Parallel QUasi-Affine TRansformation
Evolution in WSN Coverage Optimization 241
Jeng-Shyang Pan, Geng-Chen Li, Jianpo Li, Min Gao,
and Shu-Chuan Chu

Parameters Extraction of Solar Cell Using an Improved
QUasi-Affine TRansformation Evolution (QUATRE) Algorithm 253
Jeng-Shyang Pan, Ai-Qing Tian, Tien-Szu Pan, and Shu-Chuan Chu

QUATRE Algorithm for 5G Heterogeneous Network Downlink
Power Allocation Problem 265
Pei-Cheng Song, Shu-Chuan Chu, Anhui Liang, and Jeng-Shyang Pan

Reversible Image Watermarking Based on Deep Learning 275
Jianchuan He, Linlin Tang, Jiawei Chen, Tao Qian, Shuhan Qi,
Yang Liu, and Jiajia Zhang

Image Encryption with Logistic Chaotic Model Using C-QUATRE
Algorithm 285
Xiao-Xue Sun, Jeng-Shyang Pan, Tsu-Yang Wu, Lingping Kong,
and Shu-Chuan Chu

Visualization of Population Convergence Results by Sammon
Mapping in Multi-objective Optimization 295
Václav Snášel, Lingping Kong, and Jeng-Shyang Pan

A Novel Binary QUasi-Affine TRansformation Evolution
(QUATRE) Algorithm and Its Application for Feature Selection 305
Fei-Fei Liu, Shu-Chuan Chu, Xiaopeng Wang, and Jeng-Shyang Pan

Networks and Security

On the Security of a Lightweight Three-Factor-Based User
Authentication Protocol for Wireless Sensor Networks 319
Shuangshuang Liu, Zhiyuan Lee, Lili Chen, Tsu-Yang Wu,
and Chien-Ming Chen

Design and Optimization of the Seat of New Energy Electric Vehicle ... 327
Ya-Zheng Zhao, Yi-Jui Chiu, and Shu-Hao Zhao

Cryptanalysis of an Authentication Protocol for IoT-Enabled Devices in Distributed Cloud Computing Environment 339
Zhen Li, Lei Yang, Tsu-Yang Wu, and Chien-Ming Chen

Copyright Storage Method of Dance Short Video Based on Blockchain 349
Yang Yang and Dingguo Yu

Comments on a Secure and Efficient Three-Factor Authentication Protocol Using Honey List for WSN 359
Xuanang Lee, Lei Yang, Zhenzhou Zhang, Tsu-Yang Wu, and Chien-Ming Chen

Integrating Autonomous Decentralized Communication and Edge Computing for Real-Time Control in IoT System 367
Masaya Harada, Zhaoyang Du, Celimuge Wu, Tsutomu Yoshinaga, Wugedele Bao, and Yusheng Ji

Meta-Graph-Based Embedding for Recommendation over Heterogeneous Information Networks 377
Shiyuan Shuai, Xuewen Shen, Jun Wu, and Zhiqi Xu

Comments on a Secure AKA Scheme for Multi-server Environments 391
Qian Meng, Zhiyuan Lee, Tsu-Yang Wu, Chien-Ming Chen, and Kuan-Han Lu

Author Index 399

About the Editors

Jie-Fang Zhang received the Ph.D. degree in the Institute of Applied Mathematics and Mechanics, Shanghai University, China. Currently, he is a Professor at the Institute of Intelligent Media Technology, Communication University of Zhejiang, China. In the past, he was a professor at Zhejiang Normal University, Jinhua City, Zhejiang Province, China. His research interests include optics communication and Intelligent media technology.

Chien-Ming Chen received the Ph.D. degree from the National Tsing Hua University, Taiwan. He is currently an Associate Professor with the College of Computer Science and Engineering, Shandong University of Science and Technology, Shandong, China. He has published more than 80 SCI-Indexed International journals. He serves as an Associate Editor for IEEE ACCESS and an Executive Editor for the International Journal of Information Computer Security. He is a Senior Member of IEEE. His current research interests include network security, IoT, blockchain and big data.

Shu-Chuan Chu received the Ph.D. degree in 2004 from the School of Computer Science, Engineering and Mathematics, Flinders University of South Australia. She joined Flinders in December 2009 after 9 years at the Cheng Shiu University, Taiwan. She has been the Research Fellow in the College of Science and Engineering of Flinders University, Australia from 2009. Currently, she is the Research Fellow and Ph.D. Supervisor in the College of Computer Science and Engineering of Shandong University of Science and Technology. Her research interests are mainly in Evolutionary Computation, Swarm Intelligence, Wireless Sensor Networks.

Prof. Dr. Roumen Kountchev D.Sc. is with the Faculty of Telecommunications, Dept. of Radio Communications and Video Technologies - Technical University of Sofia, Bulgaria. He has over 400 papers published in magazines and conference proceedings, 21 books, 50 book chapters; 21 patents. A member of Euro Mediterranean Academy of Arts and Sciences; President of Bulgarian Association for Pattern Recognition (member of IAPR); Editorial board member of IJBST Journal Group;

Editorial board member of: Intern. J. of Reasoning-based Intelligent Systems; Intern. J. Broad Research in Artificial Intelligence and Neuroscience, Intern. J. of Computational Vision and Robotics; Intern. J of Intelligent Decision Technologies; Intern. J. of Bio-Medical Informatics and e-Health; Advances in Artificial Intelligence and Machine Learning; KES Focus Group on Intelligent Decision Technologies; Senior Member of Intern. Engineering and Technology Institute. Editor of books in Springer SIST series.

Intelligent Data Analysis

An Integrated System for Unbiased Parkinson's Disease Detection from Handwritten Drawings

Liaqat Ali, Ce Zhu, Hengling Zhao, Zhonghao Zhang, and Yipeng Liu

Abstract Current Parkinson's disease (PD) diagnosis relies on a series of hospital-based clinical examinations. To enable early PD detection at home, recognition from hand-written drawings is one way for automated PD detection system. However, existing methods have two main problems i.e., biasedness and lack of generalization in independent testing. The biasedness problem is due to two factors. The first factor is subject overlap between training and testing datasets caused by conventional validation methods. The second factor is imbalanced classes. In this paper, to avoid biasedness in the constructed models we utilize a balanced handwritten images. To avoid biasedness due to subject overlap, we use a more robust cross validation scheme i.e., leave one subject drawings out. In order to develop a decision support system to generalize to unseen data, we use several feature driven systems, and integrate F-score based feature selection model with those systems. Experimental results show that integration of F-score based model with Gaussian Naive Bayes model is a good candidate for PD detection based on hand-written drawings. It yields PD detection accuracy of 71.21% on main dataset and 63.04% on another dataset during independent testing.

L. Ali (✉) · C. Zhu · H. Zhao · Z. Zhang · Y. Liu
School of Information and Communication Engineering, University of Electrnic Science and Technology of China (UESTC), Chengdu 611731, China
e-mail: eczhu@uestc.edu.cn

H. Zhao
e-mail: zhaohengling@126.com

Z. Zhang
e-mail: zhonghaozhang@yeah.net

Y. Liu
e-mail: yipengliu@uestc.edu.cn

© The Author(s), under exclusive license to Springer Nature Singapore Pte Ltd. 2022
J.-F. Zhang et al. (eds.), *Advances in Intelligent Systems and Computing*, Smart Innovation, Systems and Technologies 268, https://doi.org/10.1007/978-981-16-8048-9_1

1 Introduction

Parkinson's disease (PD) is known to a rampant neurological disease worldwide [1]. The disease mainly affects elder people with age equal to or greater than 60 years [2]. People with Parkinsonism (PWP) show different symptoms like poor balance, tremors, bradykinesia, impaired voice, etc. [1]. Lack of gold standard medical test has made diagnosis of PD a challenging task [3]. However, literature has shown that many PWP have distorted hand writing. Based on these findings, numerous researchers have proposed and developed automated methods for screening of PD using handwritten exams.

Recently, Drotar et al. recorded 8 different hand writing exams from 75 individuals (38 healthy individuals and 37 PD patients) [4]. They achieved 81.3% of PD detection accuracy by constructing three well-known machine learning models i.e., Adaptive Boosting (Adaboost) ensemble model, k-nearest neighbours (KNN) and support vector machine (SVM). Based on micrographia, Pereira et al. proposed various machine learning based systems to facilitate automated diagnosis of PD [3, 5]. They conducted their first study on the spiral drawings that were recorded from 55 subjects [5]. By developing three machine learning models i.e., SVM, optimum path forest (OPF) and Naive Bayes (NB), they obtained highest PD detection accuracy of 78.9%. Pereira et al. extended their study by establishing an imbalanced dataset that contained spiral and meander drawings of 92 subjects [3]. The dataset was named HandPD and is publicly available at [6]. From the HandPD dataset, Pereira et al. extracted a set of 9 statistical features by applying image processing and computer vision methods and obtained PD detection accuracy of 67%. Recently, Ali et al. highlighted a potential problem of biasedness in machine learning models that is caused by imbalance classes in the HandPD data [7]. To balance the training process, they applied random undersampling method and developed unbiased machine learning models. Their developed unbiased method resulted in accuracy of 76.44%, specificity of 81.94% and sensitivity of 70.94%. The main problem in these methods is the use of conventional validation methods and imbalanced data which results in biasedness. Additionally, the reported results are based on train-test validation or cross validation on the available data. Hence, a test for generalization of the developed model on new data was not provided.

In this paper, we attempt to resolve two main problems (biasedness and lack of generalization) that are present in PD detection. The first problem is biasedness in the developed machine learning models which is due to two factors i.e. imbalanced data and subject overlap. As the data has replicated drawing samples for each subject, thus, if one or more samples of a subject are in the testing dataset and other samples of the same subject are in the training dataset, an overlap will occur. Hence, machine learning models developed with such type of subject overlap will result in biased PD detection accuracy. To avoid this kind of biasedness, we use a more robust cross validation scheme i.e., Leave One Subject Drawings Out (LOSDO). Another kind of baisedness could be due to the imbalanced classes in the data, as discussed and demonstrated in [7]. To avoid such kind of biasedness, in this paper we utilize a balanced dataset namely NewHandPD dataset. The second problem or challenge is the lack of generalization in the developed models to unseen data during independent testing. To mitigate this issue, we develop different decision support systems (DSSs) by integrating F-score model with numerous machine learning models. The main objective is to explore a feature driven DSS that will mitigate the above discussed problems. Based on the experimental results, it was pointed out that integration of F-score model with GNB model yields a good DSS for PD detection based on hand-written drawings.

2 Dataset

The HandPD and NewHandPD datasets utilized in this study were collected online from [3, 6]. For the HandPD dataset, hand-written images from 92 individuals were collected having 74 PD patients and 18 healthy individuals. For the NewHandPD dataset, hand-written images of 66 individuals (35 healthy individuals and 31 affected with PD) were considered. Following the approach of [5], we utilized spiral images in this study for both the datasets.

During the feature extraction process, each hand-written drawing was segmented out using image segmentation methods and from each drawing, the hand-written trace (HT) was separated from the exam template (ET) through automated method as can be seen in Fig. 1. Based on comparative analysis between HT and ET, a set of 9 statistical features were extracted. For in-depth discussion on the feature extraction process, readers are referred to [7]. Details about the extracted set of features and their rankings are reported in Table 1.

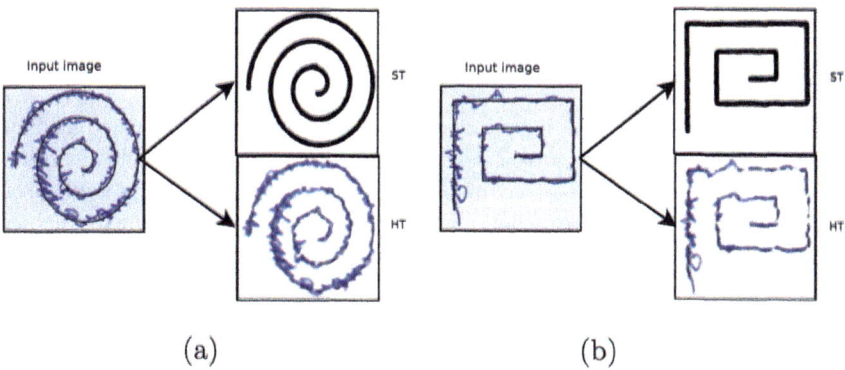

Fig. 1 a Denotes extraction of HT and ET for spiral image and **b** denotes extraction of HT and ET for meander image

Table 1 Extracted features and their ranking, R: rank of a feature

R	Description	Formulas or details
4	Root mean square (RMS) of the difference between HT and ET radius	$RMS = \sqrt{\frac{1}{n}\sum_{i=1}^{n}(r_{HT}^{i} - r_{ET}^{i})^2}$ where n = the number of sampled points, r_{HT}^{i} = HT radius of i-th sample point and r_{ET}^{i} = ET radius considering the i-th sample point
3	The maximum difference between ET and HT radius	$d_{max} = \arg\max_i\{|r_{HT}^{i} - r_{ET}^{i}|\}$
7	The minimum difference between ET and HT radius	$d_{min} = \arg\min_i\{|r_{HT}^{i} - r_{ET}^{i}|\}$
9	The standard deviation of the difference between ET and HT radius	–
1	Mean Relative Tremor (MRT)	$MRT = \frac{1}{n-d}\sum_{i=d}^{n}\left|r_{ET}^{i} - r_{ET}^{i-d+1}\right|$ where d is the displacement of the sample points used to compute the radius difference
8	The Maximum ET	Computed based on the relative tremor $\left|r_{ET}^{i} - r_{ET}^{i-d+1}\right|$
5	The Minimum ET	Computed based on the relative tremor $\left|r_{ET}^{i} - r_{ET}^{i-d+1}\right|$
2	The standard derivation of ET values	Computed based on the relative tremor $\left|r_{ET}^{i} - r_{ET}^{i-d+1}\right|$
6	The number of times the difference between HT and ET radius changes from negative to positive, or vice-versa	–

3 Methods

Recent research highlighted the problem of biasedness in the constructed machine learning models caused by imbalance classes in the HandPD data [7]. It was shown that this biasedness can be overcome by exploiting some resampling methods or using a balanced dataset for model training and development. Additionally, subject overlap between training and testing data also cause biasedness. In this paper, we use LOSDO validation approach and a balanced data for model development to avoid biasedness. Moreover, lack of generalization of unbiased models developed under small scale data is another issue that is rampant in the previously developed automated methods for PD detection based on hand-written images. However, the generalization issue was not taken into account in previous studies. These problems have still left some challenges open which need to be tackled.

To tackle these challenges, in this paper we construct feature driven DSSs. We integrate F-score based feature selection model with numerous machine learning models namely Logistic Regression (LR), Gaussian Naive Bayes (GNB), Linear Discriminant Analysis (LDA) and SVM. The extracted set of features from hand-written drawings are given to the feature ranking model i.e. F-score model. It evaluates the discrimination of two sets of real numbers [8]. If $I_j, j = 1, 2, 3, \ldots, n$ represents the samples of the NewHandPD dataset, m_+ denotes samples of the healthy group and m_- denotes the samples of the PD group, then the F-score of kth feature is defined as:

$$F(k) = \frac{(\bar{I}_k^{(+)} - \bar{I}_k)^2 + (\bar{I}_k^{(-)} - \bar{I}_k)^2}{\frac{1}{(m_+-1)}\sum_{j=1}^{m_+}(\bar{I}_{j,k}^{(+)} - \bar{I}_k^{(+)})^2 + \frac{1}{(m_--1)}\sum_{j=1}^{m_-1}(\bar{I}_{j,k}^{(-)} - \bar{I}_k^{(-)})^2} \qquad (1)$$

where $\bar{I}_k, \bar{I}_k^+, \bar{I}_k^-$ are the average of the kth feature of the whole, positive and negative datasets, respectively. Moreover, $\bar{I}_{j,k}^{(-)}$ is the kth feature of the jth negative sample. And $\bar{I}_{j,k}^{(+)}$ is the kth feature of the jth positive sample [9]. It is important to note that a feature with higher F-score value imply the higher discriminative power while a feature with lower F-score value means lower discriminative power. After features ranking, based on their importance different subsets of features are constructed. The newly constructed subsets of features are given to numerous machine learning models resulting in different DSSs. The subsets of features are given to machine learning model for evaluating the classification performance of each subset of features. An optimal subset of features for each DSS is selected based on the classification results. The block diagram of the developed feature driven DSSs is depicted in Fig. 2.

In order to develop or construct a DSS with good configuration, the usual strategy is trail and error [11]. However, this method result in high complexity. The alternative solution is automatic design of machine learning models through genetic algorithm (GA) [11]. Motivated by effective automated designing of machine learning models through GA in [10, 11], we also utilize GA for hyperparameters optimization of the machine learning models during model development. During the model development stage, we model the cross validation accuracy as the fitness function for the GA. Based on the cross validation performance, optimal subset of features and optimal hyperparameters are selected. After the model development phase, the generalization of the selected features and the optimized hyperparameters is evaluated using a new dataset.

In summary, we use leave one subject drawings out (LOSDO) cross validation approach on the NewHandPD dataset which is a balanced dataset. During the cross validation, drawings of a subject are taken out for validation and the data of remaining subjects is considered for training. In second iteration, drawings of the second subject are taken out for validation and the model is trained on the data of the remaining subjects. The same process is repeated until the drawings of all the subjects are validated. Finally, the LOSDO cross validation accuracy is evaluated. Based on the cross validation accuracy, model's hyperparameters are optimized through GA.

Different from previous studies, we further carry out a secondary level validation by testing the performance of the optimally developed models on another dataset i.e. HandPD dataset. Additionally, for evaluation purposes, we utilize five different

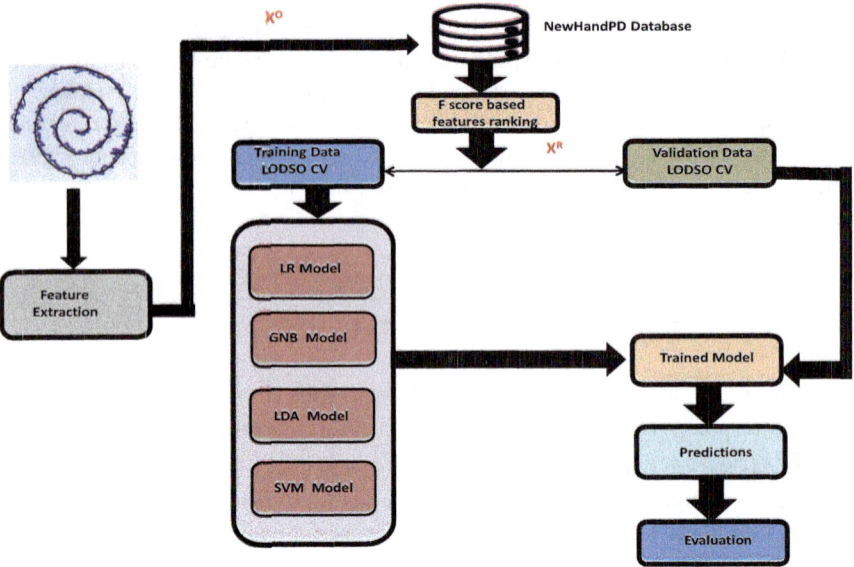

Fig. 2 Block diagram of feature driven decision support systems for PD detection based on handwritten images

evaluation metrics namely classification accuracy, area under curve (AUC), Mathews correlation coefficient (MCC), specificity and sensitivity.

4 Experiments and Discussion

All the experiments were performed using Python and Google Colab. The code can be obtained from (https://github.com/LiaqatAli007/An-Optimal-Feature-Driven-DSS-for-PD-Detection-based-on-Hand-Written-Drawings/blob/main/EuroChinaPaper.ipynb).

4.1 Model's Development Using Cross Validation

During the model development stage, we develop the feature driven DSSs on the main dataset i.e. NewHandPD dataset. The dataset is divided into N equal parts where N denotes the total number of subjects present in the dataset. After the data partitioning, LOSDO cross validation is carried out. The main purpose of this experiment is to obtain optimal subset of features and its integration with the configured machine

Table 2 Results of different feature driven DSSs during cross validation or model development phase. N: size of selected features

DSS	ACC(%)	Sen.(%)	Spec.(%)	MCC	N
F-GNB	71.21	96.77	48.57	0.508	2
F-LR	89.39	77.41	100.0	0.803	2
F-SVM	71.21	90.32	54.28	0.472	7
F-LDA	69.69	83.87	57.14	0.421	2

learning model. The results of the model development in terms of classification accuracy, sensitivity, specificity and MCC are reported in Table 2.

4.2 Evaluation of Generalization of the Developed Feature Driven Decision Support Systems Using Independent Testing

In this experiment, we evaluate the generalization capabilities of the optimally selected features and optimally configured machine learning models during the model development stage. For this purpose, we use another dataset namely HandPD dataset. We train the developed DSSs using the NewHandPD dataset under the optimal features and hyperparameters configuration. The trained models are (independently) tested during the secondary level validation. The performance of different feature driven DDSs during independent testing is reported in Table 3.

Table 3 Results of different feature driven DSSs during independent testing phase

DSS	ACC (%)	Sen. (%)	Spec. (%)	MCC	N
F-GNB	63.04	70.27	33.33	0.031	2
F-LR	19.56	0.000	100.0	–	2
F-SVM	70.65	87.83	0.000	–0.162	7
F-LDA	80.43	100.0	0.000	–	2

4.3 Analysis of the Obtained Results

For more in-depth analysis of the developed DSSs, ROC plots and AUC are taken into account for the cross validation or model development phase and the secondary level validation or independent testing phase. The ROC plots for the developed DSSs under different machine learning models are given in Fig. 3a–h. It is obvious that AUC of 0.730 and 0.645 is obtained for the cross validation and independent testing phases of GNB based DSS, respectively. From the figures (c) and (d), it is obvious that AUC of 0.864 and 0.535 is obtained for the cross validation and independent testing phases of LR based DSS, respectively. The ROC plots for the DSS under the SVM model are depicted in Fig. 3e and f. It is evident from the figures that AUC of 0.803 and 0.602 are obtained for the cross validation and independent testing phases, respectively. Finally, Fig. 3g and h depict the ROC plots for the LDA based DSS. The figures show that AUC of 0.763 and 0.465 are produced for the cross validation and independent testing phases, respectively.

Critical analysis of the obtained results for the developed feature driven DSSs revealed that for small scale datasets under the cross validation approach on the available data, outstanding results can be obtained by integrating feature selection methods with machine learning models. However, obtaining similar results during the independent testing phase is a challenging task. This could be mainly due to the fact that small scale datasets does not represent complete distribution of the data for a given problem. That is why many researchers preferred to report the cross validation results on the available data in their published studies [12–15].

This study pointed out another important fact regarding the use of evaluation metrics for checking the performance of developed models. Many previous methods have only used classification accuracy for evaluation purposes, which is not a pertinent way of a model's evaluation. For example, consider the results produced by the DSS developed under LDA model. The DSS yields highest accuracy of 80.43% during the independent testing phase. However, when we look at the sensitivity of 100% and specificity of 0%, we came to know that the model is a complete failure in terms of predictions. As it predicts every subject's data applied at its input to be a PD patient. That is why the sensitivity is maximum i.e., 100% and specificity is minimum i.e., 0%. Evidently, such a high accuracy is of no use. On the other hand, the DSS developed under the GNB model yields classification accuracy of 71.21%, sensitivity of 96.77%, specificity of 48.57% and MCC of 0.508 during the model development phase and accuracy of 63.04%, sensitivity of 70.27%, specificity of 33.33% and MCC of 0.031 during independent testing phase. Hence, the results reveal that the DSS developed under GNB model yields good performance and generalization.

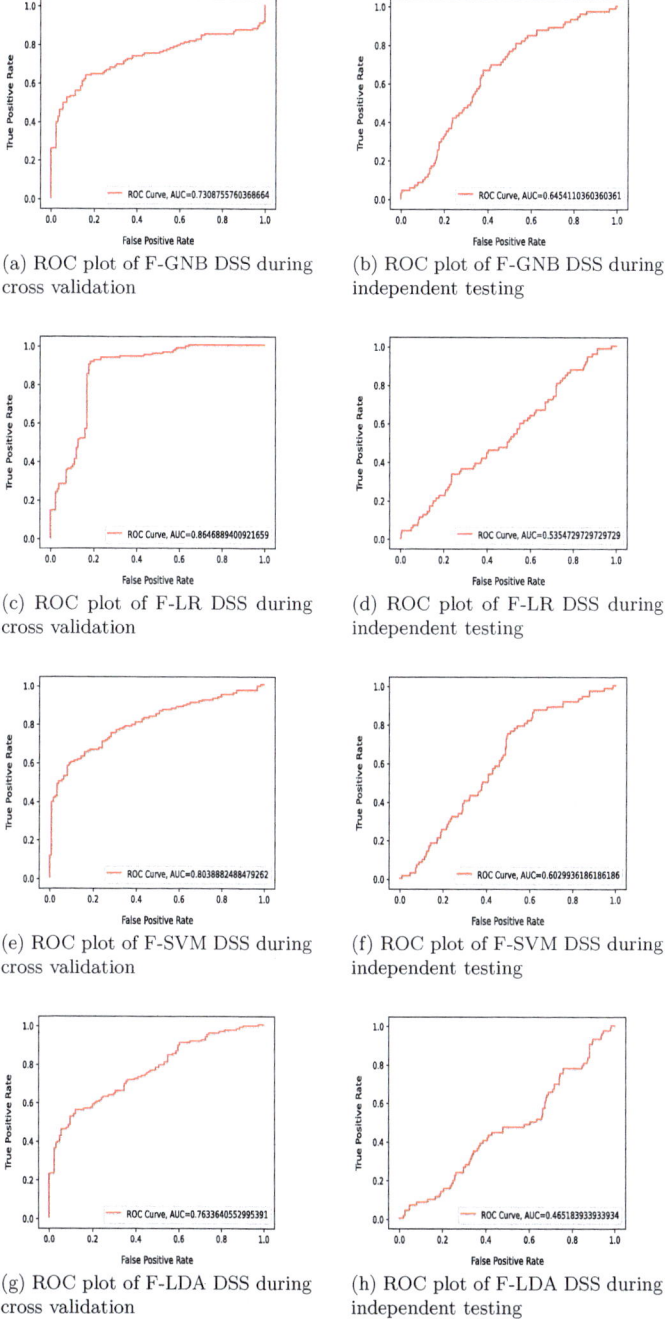

(a) ROC plot of F-GNB DSS during cross validation
(b) ROC plot of F-GNB DSS during independent testing
(c) ROC plot of F-LR DSS during cross validation
(d) ROC plot of F-LR DSS during independent testing
(e) ROC plot of F-SVM DSS during cross validation
(f) ROC plot of F-SVM DSS during independent testing
(g) ROC plot of F-LDA DSS during cross validation
(h) ROC plot of F-LDA DSS during independent testing

Fig. 3 ROC plots of different feature driven DSSs

5 Conclusion

In this paper, problems in the automated PD detection based on hand-written drawing images were considered. Unbiased DSSs were developed by using balanced dataset for model development and LOSDO cross validation to avoid subject overlap during the training and testing phases of the cross validation approach. Based on integration of F-score based statistical model with four different well-known machine learning models, four different DSSs were developed. Different from the methodologies adopted in the previous studies, to ensure generalization of the developed DSSs, the feature driven DSSs were developed using one dataset i.e. NewHandPD dataset. During model development phase, we utilized cross validation on the NewHandPD dataset in order to find out optimal subset of features and optimally configures machine learning models. To check the generalization capabilities of the developed DSSs, in the next phase i.e. independent testing phase, the machine learning models were tested using another dataset i.e. HandPD dataset. Analysis of the obtained results showed that integration of F-score model with GNB model yields an good DSS for PD detection.

References

1. Ali, L., Zhu, C., Zhou, M., Liu, Y.: Early diagnosis of Parkinson's disease from multiple voice recordings by simultaneous sample and feature selection. Expert Syst. Appl. **138**, 22–28 (2019)
2. Van Den Eeden, S.K., Tanner, C.M., Bernstein, A.L., Fross, R.D., Leimpeter, A., Bloch, D.A., Nelson, L.M.: Incidence of Parkinson's disease: variation by age, gender, and race/ethnicity. Am. J. Epidemiol. **157**(11), 1015–1022 (2003)
3. Pereira, C.R., Pereira, D.R., Silva, Francisco, A., Joo, P., Masieiro, S.A.T., Weber, C.H., Joo P.P.: A new computer vision-based approach to aid the diagnosis of Parkinson's disease. Comput. Methods Progr. Biomed. **136**, 79–88 (2016)
4. Drotr, P., Mekyska, J., Rektorov, I., Masarov, L., Smkal, Z., Faundez-Zanuy, M.: Evaluation of handwriting kinematics and pressure for differential diagnosis of Parkinson's disease. Artif. Intell. Med. **67**, 39–46 (2016)
5. Pereira, C.R., Pereira, D.R., Silva, F.A., Masieiro, J.P., Weber S.A.T., Hook, C., Papa, J.P.: A step towards the automated diagnosis of parkinson's disease: Analyzing handwriting movements. In: IEEE 28th International Symposium on Computer-based Medical Systems, pp. 171–176 (2015)
6. Pereira, C.R., Pereira, D.R., Silva, F.A., Masieiro, J.P., Weber, S.A.T., Hook, C., Papa, J.P.: HandPD Dataset (2016). http://wwwp.fc.unesp.br/~papa/pub/datasets/Handpd/, Accessed: 15-Jan-2019
7. Ali, L., Zhu, C., Golilarz, N.A., Javeed, A., Zhou, M., Liu, Y.: Reliable Parkinson's disease detection by analyzing handwritten drawings: construction of an unbiased cascaded learning system based on feature selection and adaptive boosting model. IEEE Access **7**, 116480–116489 (2019)
8. Chen, Y.-W., Lin, C.-J.: Combining SVMs with various feature selection strategies. In: Feature Extraction, pp. 315–324 (2006)
9. Song, Q.J., Jiang, H.Y., Liu, J.: Feature selection based on FDA and F-score for multi-class classification. Expert Syst. Appl. **81**, 22–27 (2017)

10. Ali, L., Zhu, C., Zhang, Z., Liu, Y.: Automated detection of Parkinson's disease based on multiple types of sustained phonations using linear discriminant analysis and genetically optimized neural network. IEEE J. Transl. Eng. Health Med. **7**, 1–10 (2019)
11. Kapanova, K.G., Dimov, I., Sellier, J.M.: A genetic approach to automatic neural network architecture optimization. Neural Comput. Appl. **29**(5), 1481–1492 (2018)
12. Tsanas, A., Little, M.A., McSharry, P.E., Spielman, J., Ramig, L.O.: Novel speech signal processing algorithms for high-accuracy classification of Parkinson's disease. IEEE Trans. Biomed. Eng. **59**(5), 1264–1271 (2012)
13. Little, M.A., McSharry, P.E., Hunter, E.J., Spielman, J., Ramig, L.O., et al.: Suitability of dysphonia measurements for telemonitoring of Parkinson's disease. IEEE Trans. Biomed. Eng. **56**(4), 1015–1022 (2009)
14. Zuo, W.-L., Wang, Z.-Y., Liu, T., Chen, H.-L.: Effective detection of Parkinson's disease using an adaptive fuzzy k-nearest neighbor approach. Biomed. Signal Process. Control **8**(4), 364–373 (2013)
15. Kebin, W., Zhang, D., Guangming, L., Guo, Z.: Learning acoustic features to detect Parkinson's disease. Neurocomputing **318**, 102–108 (2018)

Improving Supply Chain Agility to Reduce Enterprise Risk Based on QFD-MADM

Ru-Yue Yu, Chih-Hung Hsu, and An-Ching Sun

Abstract In the context of economic globalization, many enterprises will encounter more and more risk barriers in the process of development, from small losses to large failures. Therefore, the importance of enterprise agility is becoming increasingly prominent. Taking enterprise A as an example, this study adopts the combination of quality function deployment and multi-attribute decision-making method (FDM, AHP, DEMATAL and GRA) to establish the house of quality model and carry out quantitative measurement analysis. The empirical results show that in the risk environment, when considering the improvement of enterprise agility, the enhancement of production and sales ability and strategic decision-making can be prioritized, so as to reduce the enterprise key risks.

1 Introduction

With the rapid development of economic globalization and information technology, the internal and external environment of manufacturing enterprises is also changing constantly, which brings with it all kinds of risks. In addition, the diversified demands of customers and the explosive growth of big data also bring environmental uncertainties to enterprises. Therefore, enterprises must improve their ability to deal with the uncertainties, namely, supply chain agility, so as to make rapid response to the market to form a core competitive advantage.

In order to better cope with the market uncertainty, domestic and foreign scholars began to focus on the exploration of improving the agility of supply chain. Supply

---

R.-Y. Yu (✉) · C.-H. Hsu · A.-C. Sun
Institute of Industrial Engineering, School of Transportation, Fujian University of Technology, No. 3 Xueyuan Road, University Town, Minhou, Fuzhou City, Fujian Province, China
e-mail: yu_ruyue@163.com

C.-H. Hsu
e-mail: chhsu886@fjut.edu.cn

A.-C. Sun
e-mail: 747776736@qq.com

chain agility can quickly respond to market changes and respond quickly to potential and actual disruptions in the supply chain [1]. Considering that the idea of Quality Function Deployment (QFD) can help enterprises improve customer satisfaction and participate in market competition. Therefore, on the basis of the predecessors, combined with the instance, this study will supply chain agility combined with enterprise risk, and puts forward a kind of based on QFD integration architecture combined with multi-criteria decision-making method for the enterprise to evaluate risk factors and agility of supply chain principles, find out the key risk factors and agility of supply chain principles, help enterprises to put forward the preventive measures, and improve supply chain agility and reduce risk.

The main contents of this study include: according to the existing domestic and foreign studies, the main enterprise risk factors and supply chain agility criteria are sorted out, and questionnaires are distributed to experts. Based on an example, the relationship between risk factors and the relationship between risk factors and supply chain agility criteria is analyzed by using QFD-multi-attribute decision-making (QFD-MADM). According to the results of data analysis, this study put forward reasonable risk prevention plan for enterprises.

2 Research Reviewed

In the current market, any enterprise operating has a complex supply chain and is faced with many risks and its adverse effects may be huge [2]. Due to the risk of supply chain becoming more common, the enterprise planning is often upset, thus the enterprise's overall performance has also been affected, as a result, many scholars began to analyze and define the enterprise risk, such as credit risk, environment risk, social risk and risk of energy, and actively explore the effective method to risk identification and assessment, make recommendations for the enterprise managers. Naudé and Johanna [3] divided corporate risks into five categories, including financial, operational, external, market, reputational, and identified key risk factors through data analysis [3]. Congjun Rao et al. [4] designed a two-stage compound supplier selection mechanism based on multi-attribute auction and supply chain risk management, and the risk types include technology risk, information risk, management risk, economic risk, environmental risk, societal risk and ethical risk [4].

Therefore, on the basis of the above studies and in combination with other literatures, this study identifies and summarizes enterprise risk classification, as shown in Table 1.

Due to the diversity and uncertainty of enterprise risks, some scholars have proposed to improve supply chain agility to reduce risks. For example, Braunscheidel and Suresh [1] mitigated supply chain risks by improving supply chain agility. Agility is a key driving force for organizations to survive and thrive in a dynamic market environment [5]. In the face of today's changing market environment, agile capability plays an increasingly important role in improving the competitiveness of enterprises. Some scholars focus on identifying the key factors of supply chain agility. Mandal

Table 1 Risk factors classification

Risk categories	Risk factors
The external environment	Risk of supply and demand changes Industrial climate index Exchange rate fluctuations/tax changes Seasonal adjustment and tidal current fluctuation Floods, earthquakes, typhoons Disease
Enterprise product supply	Product safety and quality Production capacity is insufficient Risk of waste discharge Customized design concepts Delivery process is damaged or delayed Poor traffic regulations
Supplier material supply	Cooperation risk, breach of commitment Key supplier failure bankruptcy Limited green suppliers Supplier competence and reliability Suppliers are dependent and production is delayed The supplier has delayed the delivery of the goods
Human resource dimensions	External human attack, error Insurrection, war terrorism Man-made accident The labor dispute led to a strike Child labor, forced labor Omit supervision during homework Employee's illegal operation Information of the staff was leaked
Enterprise interior and equipment dimensions	Information equipment failure Inadequate information security and leakage The information system was compromised Insufficient information method concepts and tools Availability and accuracy of information Risk factors for information transfer IT and information sharing risks

and Saravanan [6] used partial least squares analysis to show that entrepreneurship orientation, environment orientation, supply chain orientation, technology orientation, market orientation and learning orientation can affect the agile and resilient development of tourism supply chain [6]. Kim and Chai [7] found that the innovation capability of suppliers positively affects information sharing, which in turn affects the agility of supply chain [7]. Supply chain agility is influenced by a variety of factors. This study sorted out and summarized the known literature to obtain the supply chain agility criteria, as shown in Table 2.

QFD is considered to be a useful tool for product development, which can effectively translate customer requirements into corresponding engineering characteristics of various stages of product development and production (such as planning, product design, production process development and manufacturing) [8]. With the continuous development of QFD theory, some scholars combine QFD with multi-criteria decision-making tool, and they think it is an effective and useful decision-making tool. Haq and Boddu [9] improve supply chain agility by using fuzzy QFD method, combined with Analytic Hierarchy Process (AHP) and TOPSIS. They illustrate the proposed approach by using the food processing industry as an example [9]. Fargnoli and Haber [10] combine QFD with Network Analysis (ANP) to evaluate the interaction between product and service elements to elicit customer needs and expectations and validate the effectiveness of this approach through a case study of a biomedical company [10]. Therefore, it can be inferred that the combination of QFD and MCDM method will be a successful decision-making tool. This study will use this tool to evaluate the relationship between enterprise risk and supply chain agility and provide suggestions for enterprise managers.

MCDM is a method that can simultaneously consider multiple criteria and process evaluation information given by decision-makers. Due to the shortcomings of the traditional QFD computing framework, Ocampo et al. [11] used the MADM tool integrated in MCDM to solve some of the computational problems of the traditional QFD for sustainable product design [11]. The hybrid method of QFD-MCDM has been widely used in various fields of system evaluation. For example, Liu [12] proposed a fuzzy MCDM method to select the best prototype product in consideration of engineering characteristics and factors involved in product development [12]. Dursun and Karsak [13] proposed a fuzzy MCDM method based on QFD concept for supplier selection in consideration of the influence of internal dependencies among evaluation standards [13]. Li et al. [14] proposed a MCDM method based on the combination of QFD and TOPSIS in an intuitionistic fuzzy environment to help users evaluate and select knowledge management systems from customer criteria and system criteria [14]. Song et al. [15], in order to solve the problems existing in the existing QFD framework, combined MCDM methods such as improved maximum consensus (MC) method, step-wise weight assessment ratio analysis (SWARA) method and statistical distance (SD) method with QFD, and applied this framework to the selection of logistics service providers [15].

Therefore, in order to realize sustainable supply chain, improve supply chain agility and reduce supply chain risk, this study proposed an integrated framework of

Table 2 Supply chain agility classification

Supply chain agility categories	Supply chain agility guidelines
Cooperative competition	Integration of supply chain partners Long-term cooperation and strengthening of trust relationship with partners Work with partners to develop core competitiveness Leverage the capabilities of your partners With partners to improve product quality, care for social benefits Select the partner with good performance and basic ability
Information technology	Information data integration Data accuracy Using information technology Electronic shipment of finished products to facilitate control of shipment operations Adopt e-business Enhance order processing speed and accuracy
Market supply	Improve market sensitivity/avoid shortage in the case of game behavior Meeting changing requirements/responding to changing enterprise environment requirements Forecast and analyze potential market demand/forcsight
Customer relationship	Reduce lead time/lead time control Rapid launch of new products/increase the frequency of new product introduction to market Improve customer service Use order-driven rather than forecast-driven Provide customized products Rapid customer response
Manufacturing technology capability	Develop flexible production technology Producing flexible products/reducing the complexity of product design process Set up a virtual enterprise Reduce facility setup and switching times to increase production Raise awareness of technology and information technology
Organization and team management	Process integration Cost minimization Establish a reward system Reducing uncertainty Integrate agility into the supply chain strategic environment Enhance brand value Production and sales capability

QFD-FDM-AHP-DEMATAL-GRA through the combination of QFD and MCDM, so as to build sustainable risk resistance ability and agility.

3 Research Methods

Based on the theory of QFD, this study combined fuzzy Delphi method (FDM), analytic hierarchy process (AHP), decision-making trial and evaluation laboratory (DEMATAL) and grey relation analysis (GRA) to establish the house of quality (HoQ) model, as shown in Fig. 1. Among them, the literature survey method was used to find out the initial risk factors and the initial criteria of supply chain agility, the expert questionnaire method was used as the data source, and the above multi-criteria decision-making method was used to analyze the relationship between risk factors, risk factors and the criteria of supply chain agility, as well as their weights.

According to the above initial risk factors and agility criteria and the results of expert questionnaire, the importance of initial risk factors and agility criteria was calculated by FDM, and the threshold value was set, and 19 key risk factors ① and 20 key agility criteria ② were screened out. Use DEMATAL to calculate the interaction matrix of key risk factors ⑦. According to the results of the questionnaire, the interaction matrix of agility criteria and the interaction matrix between risk factors ④ and agility criteria ③ were obtained by using the arithmetic mean to solve the

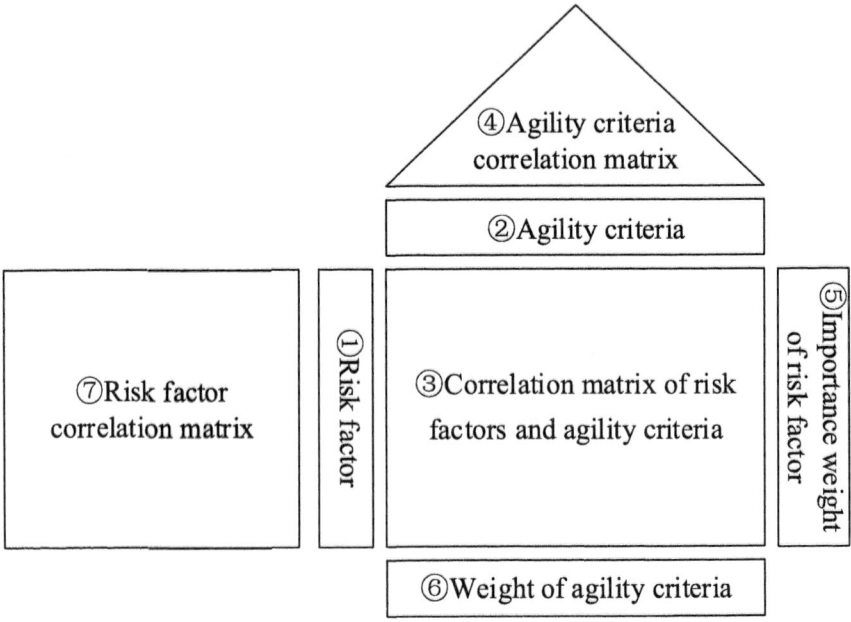

Fig. 1 HoQ model

Improving Supply Chain Agility to Reduce ...

fuzziness. Considering the interaction between the two, the integration weight of risk factors can be obtained ⑤. Finally, GRA is used to calculate the grey relational degree ⑥ of agility criteria and to sort them.

4 The Empirical Research

The case company in this study is A Electronic Assembly (Xiamen) Co., Ltd., whose production divisions are all over China and shoulder different production objectives. The main business of A Xiamen Branch is the design and manufacture of electrical, electronic and optical fiber connectors and cables. The target markets are communication and information processing markets, and the products are widely used in 3C, aerospace, military, automobile and railway.

In this study, expert questionnaire method was adopted. A total of 4 kinds of questionnaires were sent out for 4 times, 8 were sent out each time and 8 were collected with a recovery rate of 100%. The respondents of the questionnaire were all professionals of the manufacturing enterprise with sufficient experience and knowledge. The analysis steps of this study are as follows:

Step 1: Identify key risk factors and supply chain agility guidelines

FDM is used to analyze the importance of initial risk factors and supply chain agility criteria G, and the threshold value S is set. The G value and S value are compared to select the key risk factors and supply chain agility criteria.

Step 2: Determine the weights of key risk factors

Under the results of Step 1, AHP was used to calculate the weight value of each key factor. The greater the weight value, the higher the importance. Among them, the top-three risk factors are successively: the company's financial processing and control process, management policy error 'R11'; product safety and quality 'R1'; the company's market share decreased by 'R12'.

Step 3: Determine the relationship matrix of risk factors

Use DEMATAL to consider the mutual influence relationship between risk factors. In the analysis process, the original relation matrix, the normalized direct relation matrix and the comprehensive influence matrix are obtained, respectively. Among them, the comprehensive impact matrix is used as the risk factor relational matrix on the left side of the QFD.

After that, a causal map of the risk factors was drawn. The greater the value of centrality, the stronger the relationship between this factor and other factors. When the cause degree is positive, it means that the factor is an influencing factor. The greater the value, the greater the influence on other factors. When the cause degree is negative, it means that the factor is the factor affected, and the smaller the value is, the easier it is to be affected. The results show that the highest centrality is the reputation and ability of the managers and leaders 'R10', and the highest reason is the intrusion of the information system 'R5'.

Step 4: Determine the relationship matrix between risk factors and supply chain agility criteria, the relationship matrix of supply chain agility criteria, and the grey correlation degree of supply chain agility criteria

After deconfuzzing the questionnaire data, the relationship matrices of risk factors and supply chain agility criteria and the relationship matrices of supply chain agility criteria were obtained, respectively. The risk factor weight in Step 2 and the impact matrix in Step 3 were sorted to get the importance weights. The top three, in terms of weight, are: inadequate supply liquidity and poor finance, resulting in interruption or delay of goods 'R9'; long product lead time 'R19'; the data information system has been compromised 'R5'.

GRA was used to analyze the grey correlation degree of supply chain agility criteria, and the ranking was carried out. Following the above steps, the top-three agility criteria are, respectively, quick decision-making, strategic flexibility 'A11', improve customer service level 'A9' and production and sales capacity 'A1'.

5 Conclusions

In order to help enterprises improve supply chain agility to achieve the goal of risk reduction, this study proposes an integrated architecture of QFD combined with multi-criteria decision-making method and identifies enterprise risk factors and supply chain agility criteria. Based on a practical case, the paper analyzes the interrelationship between risk factors, the interrelationship between risk factors and supply chain agility criteria, and the final grey correlation degree, and puts forward the measures to prevent risk and improve agility for enterprise managers. At the same time, the accuracy of the result and the feasibility of the method are also illustrated.

Although the proposed framework can be applied to other industries, different industries have their unique risks and agility, and the evaluation indexes need to be re-determined. The framework can also be connected with other tools, such as the Internet, Internet of Things, and Big Data, to help monitor, plan and optimize supply chains in real time. The evaluation framework also has some limitations. First of all, this study only evaluates risk and agility, and other variables can be added in the future to build a more comprehensive and deeper SSC. In addition, due to time and cost constraints, this study only takes one enterprise in the manufacturing industry as the research object, and the number of enterprises can be increased in the future to make the evaluation results more applicable.

Acknowledgments This paper was supported by Natural Science Foundation of Fujian Province of China (Grant No. 2019J01790).

References

1. Braunscheidel, M.J., Suresh, N.C.: The organizational antecedents of a firm's supply chain agility for risk mitigation and response. J. Oper. Manag. **27**(2), 119–140 (2009)
2. Cano-Olivos, P., Hernández-Zitlalpopoca, R., Sánchez-Partida, D., et al.: Risk analysis of the supply chain of a tools manufacturer in Puebla, Mexico. J. Contingencies Cris. Manag. **27**(4), 406–413 (2019)
3. Badenhorst-Weiss, J.A., Naudé, R.T.: The challenges behind producing a bottle of wine: supply chain risks. J. Trans. Supply Chain Manag. **14**(1), 1–15 (2020)
4. Rao, C., Xiao, X., Goh, M., Zheng, J., Wen, J.: Compound mechanism design of supplier selection based on multi-attribute auction and risk management of supply chain. Comput. Ind. Eng. **105**, 63–75 (2017)
5. Gligor, D., Bozkurt, S., Gölgeci, I., et al.: Does supply chain agility create customer value and satisfaction for loyal B2B business and B2C end-customers? Int. J. Phys. Distrib. Logist. Manag. **50**(7/8), 721–743 (2020)
6. Mandal, S., Saravanan, D.: Exploring the influence of strategic orientations on tourism supply chain agility and resilience: an empirical investigation. Tour. Plan. Dev. **16**(6), 612–636 (2019)
7. Kim M., Chai, S.: The impact of supplier innovativeness, information sharing and strategic sourcing on improving supply chain agility: global supply chain perspective. Int. J. Prod. Econ. **187**, 42–52 (2017)
8. Xu, J., Xu, X., Xie, S.Q.: A comprehensive review on recent developments in quality function deployment. Int. J. Prod. Qual. Manag. **6**(4), 457–494 (2010)
9. Haq, A.N., Boddu, V.: Analysis of enablers for the implementation of leagile supply chain management using an integrated fuzzy QFD approach. J. Intell. Manufact. **28**(1), 1–12 (2017)
10. Fargnoli, M., Haber, N.: A practical ANP-QFD methodology for dealing with requirements' inner dependency in PSS development-ScienceDirect. Comput. Ind. Eng. **127**, 536–548 (2019)
11. Ocampo, L.A., Labrador, J.J.T., Jumao-As, A.M.B., et al.: Integrated multiphase sustainable product design with a hybrid quality function deployment-multi-attribute decision-making (QFD-MADM) framework. Sustain. Prod. Consump. **24**, 62–78 (2020)
12. Liu, H.T.: Product design and selection using fuzzy QFD and fuzzy MCDM approaches. Appl. Math. Model. **35**(1), 482–496 (2011)
13. Dursun, M., Karsak, E.E.: A QFD-based fuzzy MCDM approach for supplier selection. Appl. Math. Model. **37**(8), 5864–5875 (2013)
14. Li, M., Jin, L., Wang, J.: A new MCDM method combining QFD with TOPSIS for knowledge management system selection from the user's perspective in intuitionistic fuzzy environment. Appl. Soft Comput. **21**, 28–37 (2014)
15. Song, C., Wang, J.Q., Li, J.B.: New framework for quality function deployment using linguistic Z-numbers. Mathematics **8**(2) (2020)

A CNN-Based Method for AAPL Stock Price Trend Prediction Using Historical Data and Technical Indicators

Yuxiao Gong, Jimmy Ming-Tai Wu, Zhongcui Li, Shuo Liu, Lingyun Sun, and Chien-Ming Chen

Abstract The stock price is a non-stationary time series, so it is challenging to predict the stock price. Some statistics and machine learning research hope to solve this problem, but these methods require complex feature engineering. Deep learning without feature extraction has brought a breakthrough for this. This paper uses the convolutional neural network (CNN) to establish a three-category prediction model based on historical stock prices and technical analysis indicators to predict stock price trends. Experiments conducted on AAPL show that adding technical indicators can improve the performance of the CNN prediction model.

1 Introduction

Stock forecasting has always been an appealing issue in the financial field. Many investors and researchers are committed to mining potential patterns from historical stock data to predict future stock price changes. However, stock forecasting is a challenging task. The stock market is a giant chaotic system with many people participating and a lot of variables. Many factors affect stock prices, such as policies, the company's development and investor psychology. Due to incomplete information, it is difficult to predict stock trends.

Y. Gong · J. Ming-Tai Wu · Z. Li · S. Liu · L. Sun · C.-M. Chen (✉)
College of Computer Science and Engineering, Shandong University of Science and Technology, Qingdao, China
e-mail: gongyuxiao@sdust.edu.cn

J. Ming-Tai Wu
e-mail: wmt@wmt35.idv.tw

Z. Li
e-mail: 17685458562@163.com

S. Liu
e-mail: 407209281@qq.com

L. Sun
e-mail: 2442303052@qq.com

© The Author(s), under exclusive license to Springer Nature Singapore Pte Ltd. 2022
J.-F. Zhang et al. (eds.), *Advances in Intelligent Systems and Computing*, Smart Innovation, Systems and Technologies 268, https://doi.org/10.1007/978-981-16-8048-9_3

Many researchers applied statistical time series models to stock forecasting problems such as Auto Regressive Moving Average (ARMA) models [4] and Autoregressive Integrated Moving Average (ARIMA) models [5]. However, stock forecasting is a non-linear problem that contains many uncertain factors, and the establishment of a linear model cannot fit the stock price trend well. Due to the limitations of linear models, many researchers study the non-linear models such Autoregressive Conditional Heteroskedasticity (ARCH) models [2] and Generalized Autoregressive Conditional Heteroskedasticity (GARCH) models [3, 7]. Some traditional machine learning methods, such as support vector machines (SVM) [14] and artificial neural networks (ANN) [24] have also been applied to stock forecasting. However, these non-linear methods all depend on the quality of the input features. These methods usually require complex feature engineering, which requires solid professional knowledge.

Deep learning is gradually applied in speech recognition [1], text classification [25, 26], image recognition [10], autonomous driving [16] and other fields [6, 17, 19–21]. It has achieved the most advanced performance. Successful applications in these fields motivate researchers to apply deep learning to stock prediction tasks. Moreover, deep learning such as convolutional neural networks can extract features from raw data, so it is an advanced tool for stock prediction. Thus, this paper aims to use a two-dimensional CNN to predict the price trend of AAPL on the next trading day, using historical stock prices and 64 types of technical indicators as the input of CNN.

2 Related Work

In the past few decades, deep learning has made progress in financial time series forecasting. Nelson et al. used LSTM to predict whether the stock price will rise in 15 min [15]. Based on historical price data and online financial news, Huynh et al. put forward a new framework using Bidirectional Gated Recurrent Unit (BGRU) to predict future stock price fluctuations [11].

Technical indicators are valuable tools for the technical analysis of stocks. Vargas et al. use financial news headlines and five technical indicators as input to the deep learning model to forecast the daily price movement of stock prices [18]. Based on stock basic trading data and technical analysis index, Gao et al. used principal component analysis (PCA) to reduce data dimension and then used LSTM to forecast the next day's closing price of stock [9]. Kim et al. used different representations of the same feature (images and time series) to build a stock prediction model based on the feature fusion of CNN and LSTM. But they only considered the impact of historical price data on stock prices [13]. Wu et al. proposed a graph-based CNN model that uses both historical prices and leading indicators to predict future trends of stocks

3 Method

In this paper, one-day interval historic price data and technical analysis indicators are used to forecast the next trading day's price trend.

3.1 Data and related Processing

Here, we introduce the experimental data and data preprocessing methods in detail.

3.1.1 Data

Technical analysis indicators are an effective tool for traders to analyze stock trends. Therefore, this paper combines technical analysis indicators with historical price data to predict stock trends.

The historical price data was collected from NASDAQ[1] website, ranging from September 28, 2018, to August 30, 2019, including 232 trading days. This paper uses one-day interval stock price data and technical indicators of AAPL. The AAPL historical price data is shown in Fig. 1. The raw data of AAPL contains six attributes ($date, close, volume, open, high$ and low). Technical indicators can reflect past market conditions, and they have the possibility to predict future market trends. TA-Lib[2] library is used to generate technical analysis indicators. Figure 2 shows technical analysis indicators used in this paper (John et al. [12]).

3.1.2 Data Preprocessing

In this paper, the price trend of the next trading day is used as the label of the data set. Firstly, Eq. 1 is used to calculate the c_t. The c_t is the rate of increase or decrease of the closing price of the trading day t compared to the closing price of the previous trading day $t-1$, $close_t$ is the closing price of day t.

---

[1] https://www.nasdaq.com.

[2] http://ta-lib.org.

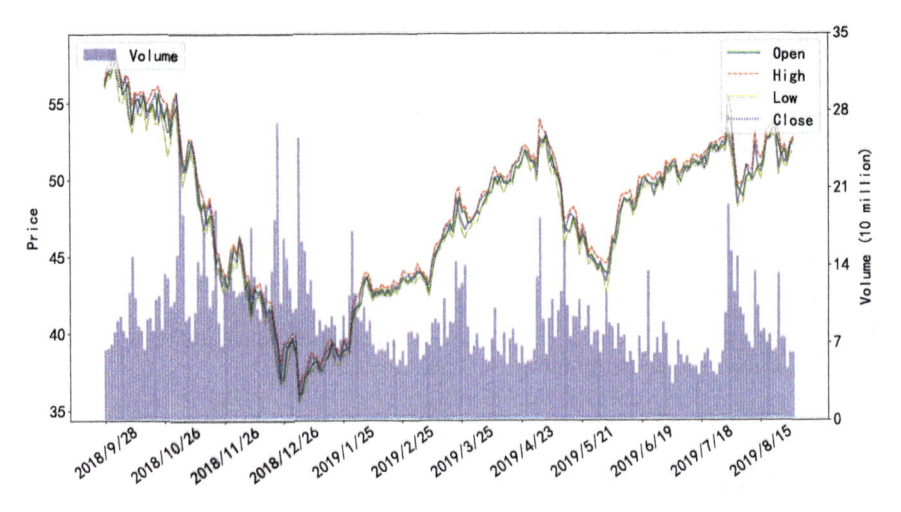

Fig. 1 AAPL historical price data

$$c_t = \frac{close_t - close_{t-1}}{close_{t-1}} * 100 \tag{1}$$

Secondly, after calculating the daily fluctuations, use Eq. 2 to add labels to the data set.

$$y_t = \begin{cases} 1 & \text{if } c_t > 0.5 \\ -1 & \text{if } c_t < -0.5 \\ 0 & \text{otherwise} \end{cases} \tag{2}$$

The threshold is set to 0.5, which represents the range of price fluctuations that can be tolerated. That is to say, if the increase or decrease is greater than 0.5, the stock price is considered to rise and mark it as 1; if the increase or decrease is less than −0.5, the stock price is considered to be down and mark it as −1; otherwise, the stock price is considered unchanged and mark it as 0. Then use one-hot encoding for labels. Finally, a sliding window with a width of 30 was used to generate experimental samples. The sliding window slides forward one step at a time. Each sample is reshaped into a three-dimensional image with 1 channel, which is used as the input of CNN.

The dimensions and numerical values of variables are different (take AAPL as an example, the opening price is $56.435 while the volume is 90241760 on January 2, 2020.). If the original data is used to predict stock movements directly, the variables of different scales will have very different effects on the prediction results. Therefore, it is necessary to standardize the data to eliminate the dimensional relationship between variables. This paper uses the standardized method proposed by Wu et al. [22], the function is show in Eq. 3:

$$x_{t,new}^T = \frac{x_t^T - mean^T}{max^T - min^T} \tag{3}$$

Category	Technical Analysis Indicators
Overlap Studies Functions	BBANDS - Bollinger Bands DEMA - Double Exponential Moving Average EMA - Exponential Moving Average HMA - Hull Moving Average HT_TRENDLINE - Hilbert Transform - Instantaneous Trendline KAMA - Kaufman Adaptive Moving Average MA - Moving average MAMA - MESA Adaptive Moving Average MAVP - Moving average with variable period MIDPOINT - MidPoint over period MIDPRICE - Midpoint Price over period T3 - Triple Exponential Moving Average TRIMA - Triangular Moving Average WMA - Weighted Moving Average
Momentum Indicator Functions	ADX - Average Directional Movement Index ADXR- Average Directional Movement Index Rating APO - Absolute Price Oscillator AROON - Aroon AROONOSC - Aroon Oscillator BOP - Balance Of Power CCI - Commodity Channel Index CMO - Chande Momentum Oscillator DX - Directional Movement Index MACD - Moving Average Convergence/Divergence MFI - Money Flow Index MINUS_DI - Minus Directional Indicator MINUS_DM - Minus Directional Movement MOM - Momentum PLUS_DI - Plus Directional Indicator PLUS_DM - Plus Directional Movement PPO - Percentage Price Oscillator ROC - Rate of change ROCP - Rate of change Percentage ROCR - Rate of change ratio ROCR100 - Rate of change ratio 100 scale RSI - Relative Strength Index STOCH - Stochastic STOCHF - Stochastic Fast STOCHRSI - Stochastic Relative Strength Index TRIX - 1-day Rate-Of-Change (ROC) of a Triple Smooth EMA ULTOSC - Ultimate Oscillator WILLR - Williams' %R
Volume Indicators	Chaikin Accumulation/Distribution Line ADOSC - Chaikin A/D Oscillator OBV - On Balance Volume
Volatility Indicator Functions	ATR - Average True Range NATR - Normalized Average True Range TRANGE - True Range
Price Transform Functions	TYPPRICE - Typical Price WCLPRICE - Weighted Close Price
Cycle Indicator Functions	HT_DCPERIOD - Hilbert Transform - Dominant Cycle Period HT_DCPHASE - Hilbert Transform - Dominant Cycle Phase HT_PHASOR - Hilbert Transform - Phasor Components HT_SINE - Hilbert Transform - SineWave HT_TRENDMODE - Hilbert Transform - Trend vs Cycle Mode
Statistic Functions	BETA - Beta CORREL - Pearson's Correlation Coefficient LINEARREG - Linear Regression LINEARREG_ANGLE - Linear Regression Angle LINEARREG_INTERCEPT - Linear Regression Intercept LINEARREG_SLOPE - Linear Regression Slope STDDEV - Standard Deviation TSF - Time Series Forecast VAR - Variance

Fig. 2 Technical analysis indicators used in this paper

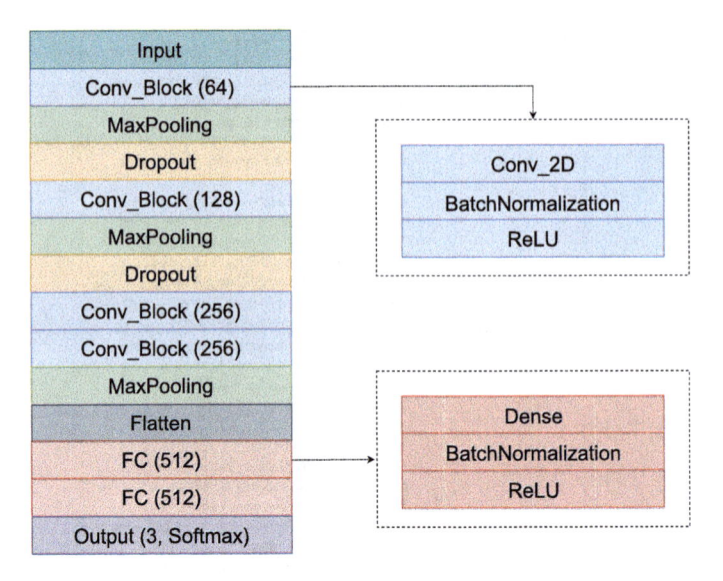

Fig. 3 AAPL historical price data

where x_t^T is the value of price related variable ($open, high, low...$) or technical index (EMA_{12}, EMA_{26}...) at time t, $mean^T$, max^T and min^T are the mean, maximum and minimum of a given period T, $x_{t,new}^T$ is the standardized variable value. In this paper, T is set to 120.

3.2 Model Architecture

Convolutional Neural Network (CNN) is an effective learning model for solving time series classification problems. This paper uses a CNN-based network model to predict the trends of AAPL. The architecture of the model is shown in Fig. 3. The model has 4 convolutional blocks with 64, 128, 256 and 256 filters, respectively. The convolution block includes a convolutional layer, a batch normalization layer and a Rectified Linear Units (ReLU) layer. The max pooling layer is used to reduce the size of the feature map. After the convolutional layer extracts the features, the fully connected layer maps the learned feature representation to the sample label space. To avoid overfitting, early stop and dropout are also added.

(a) With technical indicators.   (b) Without technical indicators.

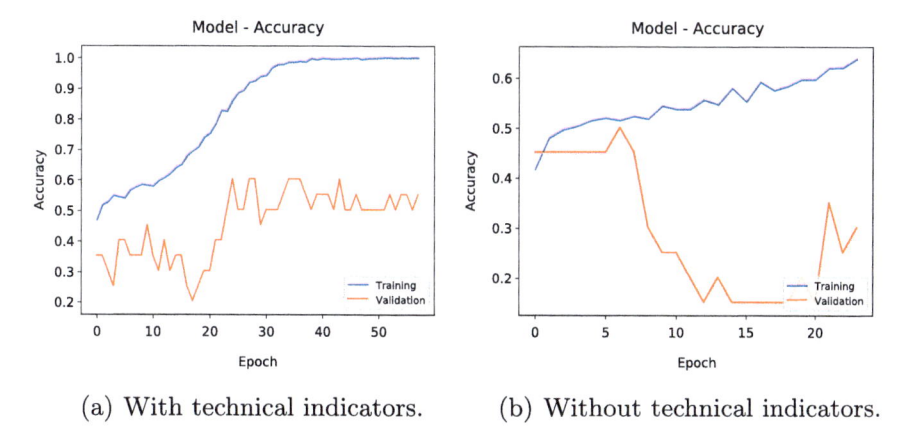

Fig. 4 The accuracy of the CNN prediction model on AAPL stocks

4 Experimental Results

In this section, to analyze the impact of technical indicators on stock price trend prediction, the CNN model with historical data as input is used as the baseline. The experiment uses the data of AAPL in the past 30 trading days to predict the price rise and fall of the next trading day. The first 80% of the samples are used for model training, the remaining 10% are used for verification and 10% are used for testing. The model was built using Keras [8]. Early stopping is set to 20 epochs, which means that if there is no improvement in the loss of the validation set for 20 epochs, the training is stopped. Save the model with the least loss in the validation set for testing. Figure 4 shows the accuracy of the experiment. Figure 4a is the accuracy of the model using historical prices and technical indicators and Fig. 4b is the accuracy of the model using historical prices. Finally, on the test set, the accuracy of the model using only historical prices is 0.428571, while the accuracy of the model using historical prices and technical indicators can reach 0.47619. The experimental results show that by adding technical indicators, the model can be effectively prevented from overfitting and the prediction accuracy can be improved.

5 Conclusion

In this paper, the CNN model is used for AAPL stock price trend prediction using historical data and technical indicators. In addition, experiments on AAPL have proved that adding technical indicators can effectively improve the performance of the CNN stock price trend prediction model. More experiments will be done to study the impact of technical indicators on the performance of other deep learning models in the future. In addition, it is also a meaningful direction to find out which combination of technical indicators is most effective, which can reduce model input and shorten running time.

References

1. Afouras, T., Chung, J.S., Senior, A., Vinyals, O., Zisserman, A.: Deep audio-visual speech recognition. IEEE Trans. Pattern Anal. Mach. Intell. (2018)
2. Alam, M.Z., Siddikee, M.N., Masukujjaman, M.: Forecasting volatility of stock indices with arch model. Int. J. Financ. Res. **4**(2), 126 (2013)
3. Alberg, D., Shalit, H., Yosef, R.: Estimating stock market volatility using asymmetric garch models. Appl. Financ. Econ. **18**(15), 1201–1208 (2008)
4. Anaghi, M.F., Norouzi, Y.: A model for stock price forecasting based on arma systems. In: 2012 2nd International Conference on Advances in Computational Tools for Engineering Applications (ACTEA), pp. 265–268. IEEE (2012)
5. Ariyo, A.A., Adewumi, A.O., Ayo, C.K.: Stock price prediction using the arima model. In: 2014 UKSim-AMSS 16th International Conference on Computer Modelling and Simulation, pp. 106–112. IEEE (2014)
6. Chen, C.M., Chen, L., Gan, W., Qiu, L., Ding, W.: Discovering high utility-occupancy patterns from uncertain data. Inf. Sci. **546**, 1208–1229 (2021)
7. Franses, P.H., Van Dijk, D.: Forecasting stock market volatility using (non-linear) garch models. J. Forecast. **15**(3), 229–235 (1996)
8. François, C.: Keras. https://github.com/keras-team/keras (2015)
9. Gao, T., Chai, Y.: Improving stock closing price prediction using recurrent neural network and technical indicators. Neural Comput. **30**(10), 2833–2854 (2018)
10. He, K., Zhang, X., Ren, S., Sun, J.: Deep residual learning for image recognition. In: Proceedings of the IEEE Conference on Computer Vision and Pattern Recognition, pp. 770–778 (2016)
11. Huynh, H.D., Dang, L.M., Duong, D.: A new model for stock price movements prediction using deep neural network. In: Proceedings of the Eighth International Symposium on Information and Communication Technology, pp. 57–62 (2017)
12. John, B., Brian, C.: Ta-lib documentation (2020). https://github.com/mrjbq7/ta-lib/blob/master/docs/doc_index.md
13. Kim, T., Kim, H.Y.: Forecasting stock prices with a feature fusion lstm-cnn model using different representations of the same data. PLoS ONE **14**(2), e0212320 (2019)
14. Lin, Y., Guo, H., Hu, J.: An svm-based approach for stock market trend prediction. In: The 2013 International Joint Conference on Neural Networks (IJCNN), pp. 1–7. IEEE (2013)
15. Nelson, D.M., Pereira, A.C., de Oliveira, R.A.: Stock market's price movement prediction with lstm neural networks. In: 2017 International Joint Conference on Neural Networks (IJCNN), pp. 1419–1426. IEEE (2017)
16. Sallab, A.E., Abdou, M., Perot, E., Yogamani, S.: Deep reinforcement learning framework for autonomous driving. Electron. Imag. **2017**(19), 70–76 (2017)
17. Tseng, K.K., Zhang, R., Chen, C.M., Hassan, M.M.: Dnetunet: a semi-supervised cnn of medical image segmentation for super-computing ai service. J. Supercomput. **77**(4), 3594–3615 (2021)
18. Vargas, M.R., dos Anjos, C.E., Bichara, G.L., Evsukoff, A.G.: Deep leaning for stock market prediction using technical indicators and financial news articles. In: 2018 International Joint Conference on Neural Networks (IJCNN), pp. 1–8. IEEE (2018)
19. Wang, E.K., Wang, F., Kumari, S., Yeh, J.H., Chen, C.M.: Intelligent monitor for typhoon in iot system of smart city. J. Supercomput. **77**(3), 3024–3043 (2021)

20. Wang, K., Chen, C.M., Hossain, M.S., Muhammad, G., Kumar, S., Kumari, S.: Transfer reinforcement learning-based road object detection in next generation iot domain. Comput. Netw. 108078 (2021)
21. Wang, K., Xu, P., Chen, C.M., Kumari, S., Shojafar, M., Alazab, M.: Neural architecture search for robust networks in 6g-enabled massive iot domain. IEEE Internet Things J. (2020)
22. Wu, J.M.T., Li, Z., Lin, J.C.W., Pirouz, M.: A new convolution neural network model for stock price prediction. In: International Conference on Genetic and Evolutionary Computing, pp. 581–585. Springer (2019)
23. Wu, J.M.T., Li, Z., Srivastava, G., Tasi, M.H., Lin, J.C.W.: A graph-based convolutional neural network stock price prediction with leading indicators. Softw.: Pract. Experi. **51**(3), 628–644 (2021)
24. Yetis, Y., Kaplan, H., Jamshidi, M.: Stock market prediction by using artificial neural network. In: 2014 World Automation Congress (WAC), pp. 718–722. IEEE (2014)
25. Zhao, Z., Zhou, H., Li, C., Tang, J., Zeng, Q.: Deepemlan: Deep embedding learning for attributed networks. Inf. Sci. **543**, 382–397 (2021)
26. Zhao, Z., Zhou, H., Qi, L., Chang, L., Zhou, M.: Inductive representation learning via cnn for partially-unseen attributed networks. IEEE Trans. Netw. Sci. Eng. **8**(1), 695–706 (2021)

Research on Supply Chain Agility Based on Bullwhip Effect

Ming-Ge Li, Chih-Hung Hsub, and Jun-Wei Liuc

Abstract The competition in the twenty-first century is not the competition between enterprises, but the competition in the supply chain. "Bullwhip effect" is a common phenomenon in supply chain management. With the development of economy and the intensification of competition, the structure of supply chain is increasingly complex. Therefore, how to eliminate and effectively prevent the harm of "Bullwhip effect" has become a problem that people pay special attention to. The objective of this study is to provide a new way of using agility to prevent bullwhip effect. Through the combination and application of a variety of analytical methods, the key criteria of agility are ranked, so that enterprises can control these factors in priority, and formulate strategies in advance, so as to improve the agility of enterprises to prevent the negative impact brought by the bullwhip effect, and find the optimal solution to the "bullwhip effect." Taking a Guangdong Yinglian Technology Company as an example, this study found that the best way to reduce the "bullwhip effect" of the supply chain is to improve customer satisfaction, and secondly to improve employees' work capabilities.

1 Introduction

In today's rapidly changing environment, the company should constantly reinvent itself by establishing a strong internal and external network of relationships [1]. Companies without integrated business processes and weak partnerships are unlikely

---

M.-G. Li (✉) · C.-H. Hsub · J.-W. Liuc
Institute of Industrial Engineering, School of Transportation, Fujian University of Technology, No. 3 Xueyuan Road, University Town, Minhou, Fuzhou City, Fujian Province, China
e-mail: changtuili@126.com

C.-H. Hsub
e-mail: chhsu886@fjut.edu.cn

J.-W. Liuc
e-mail: 1787528227@qq.com

to successfully enter the highly competitive global markets [2]. Therefore, more and more companies recognize the importance of supply chain agility (SCA).

Therefore, companies must quickly and accurately seize opportunities and dynamically adjust their internal resources and capabilities to be able to flexibly respond to market changes to cope with the increasingly complex and volatile market environment, and agility will also be used as a strategic resource. High-tech enterprises can survive in a highly competitive business environment. However, for many years, most companies have not paid attention to the transformation of economic model, and are still "plagiarism" and "staking the race", making them into a crisis of survival. How to make companies overcome inertia and inertia while undergoing economic transformation and market changes Maintaining agility in the process has become a very important research topic.

The bullwhip effect will distort the information in the supply chain, and the degree of distortion will be amplified step by step, so that each node in the supply chain must maintain a high inventory level. There will be a huge deviation between this and the actual demand, so that the company has excess inventory and costs. Rising, the economy, market, environment, and society will be adversely affected, and even lead to chaos in supply, production, and sales. Therefore, the study of the bullwhip effect is a beneficial and necessary supplement to the entire supply chain management, and it is an indispensable part of the supply chain management.

In a fiercely competitive environment, the negative impact of the bullwhip effect on supply chain management, as well as the individual needs of customers, the short product life cycle, uncertain market demand, and the rapid development of technology, are faced by the development of enterprises. Many problems require that the entire supply chain be able to quickly adjust and respond to market changes. Therefore, it is imperative to improve the agility of the supply chain, and it is also an important way to improve corporate performance, gain competitive advantage, and survive. Aiming at how companies use the advantages of agility to suppress the occurrence of the bullwhip effect, this study analyzes the causes of the bullwhip effect, and uses agility to flexibly respond to market changes to improve the competitiveness of the company, and it is for the supply chain managers of the company to make decisions. for reference.

After reviewing many articles, I found that there are almost no articles linking the bullwhip effect with agility. Therefore, this article attempts to improve the company's agility by analyzing the causes of the bullwhip effect. Thereby reducing the negative impact of the company when it encounters fluctuations.

2 Research Reviewed

This chapter mainly focuses on the theoretical issues involved in this research, collects and analyzes literature data, including the influencing factors of the bullwhip effect, and studies on the criteria for improving agility.

At the beginning of the twentieth century, managers at Proctor and Gamble noticed fluctuations in the production of diapers, even though the consumption rate was fairly stable throughout the year. Managers call this phenomenon the bullwhip effect of Disney et al. [3]. The bullwhip effect is a phenomenon in which the variability of demand orders placed by downstream participants in the supply chain increases as the orders move upstream.

The bullwhip effect describes information distortion as the infiltration of orderly information upstream. Lee Padmanabhan, and Whang first gave the definition of the bullwhip effect and identified four main reasons for the bullwhip effect: demand forecast updates, order batches, price fluctuations, rationing and shortage games [4]. Based on previous academic literature, Lee et al. put forth the main factors contributing to the bullwhip effect which are as follows:

(1) Demand forecast updating
(2) Order batching
(3) Price flfluctuation
(4) Rationing and shortage gaming
(5) Lead times.

The bullwhip effect is one of the most widely studied phenomena in supply chain management. Based on the analysis of key factors, this study evaluated the importance of the bullwhip effect factor in the supply chain. And through the analysis of related theories and the cause of the bull-surface effect by scholars in recent years,

In recent years, some researches on the bullwhip effect at home and abroad are shown in Table 1.

Table 1 Research on bullwhip effect by foreign scholars in recent years

Scholar	Research content
An-Yuan Chang et al. [1]	The bullwhip effect in the retail supply chain is discussed. Based on the analysis of key factors, the importance of each influencing factor of the bullwhip effect in the supply chain and the causal relationship between these factors are evaluated to provide more information for supply chain managers. Effective analysis of the cause of the bullwhip effect
Tesfay [2]	It is found that the bullwhip effect may be caused by the cooperation of the internal coordination of the supplier and the internal coordination of the organization, and a coordination method called super hybrid coordination is also proposed to control the influence of the internal coordination of the organization on the bullwhip effect
Ma et al. [5]	A multi-channel supply chain including manufacturers, dual-channel retailers, and online retailers has been constructed. Research has pointed out that the bullwhip effect is affected by the speed of price adjustments and the sensitivity of price discounts, especially in a dual-cycle or chaotic state. The bullwhip effect of the chain has increased significantly, so controlling the sensitivity of price discounts is useful for supply chain node companies

Agility was first proposed by American scholars in the early 1990s. It is a new type of strategic thinking specifically proposed for the manufacturing sector. It can improve the ability of the enterprise's internal manufacturing system to respond to rapid changes in the external environment. In short, agility is adaptation Ability to change. However, with the development of science and technology and the popularization of network information technology, the current thinking on agility is not limited to the manufacturing field. Improving the agility of the entire supply chain has become a key to the survival and development of enterprises in the fierce market competition.

By collecting relevant literature on agility research in recent years, the following 63 principles for improving agility have been sorted out, and they are divided into market supply, customer relationship, cooperative competition, information technology, organization and team management, and technology according to their characteristics. Six blocks of ability.

In recent years, some researches on agility at home and abroad are shown in Table 2.

Since the bullwhip effect is endogenous to the structure of the traditional supply chain, it cannot be fundamentally resolved from the perspective of information perfection and simplified decision-making. The fact is that the uncoordinated target interests of the members of the supply chain and the lack of effective incentives and supervision mechanisms make it impossible to improve information and simplify decision-making under the traditional supply chain structure.

The supply chain structure can be divided into three aspects: organizational structure, information structure, and decision-making structure. These three aspects complement each other. To overcome the bullwhip effect, it is necessary to improve the three aspects of the supply chain structure at the same time, so as to achieve the

Table 2 Research on agility by domestic and foreign scholars in recent years

Scholar	Research content
Dubey et al. [6]	It is pointed out that information sharing and supply chain connectivity resources will affect the visibility of the supply chain, but under the regulation of the top management's commitment, the agility, adaptability and consistency of the supply chain will be enhanced, so that the enterprise can obtain a sustainable competitive advantage
Hwang and Kim [7]	The role of logistics strategy (flexibility, collaboration and differentiation) as a prerequisite for supply chain agility (SCA) and financial performance is discussed. Research shows that logistics strategy directly or indirectly affects financial performance through SCA, which provides empirical evidence for the core role of SCA in helping companies quickly perceive and respond to the ever-changing environment
Roscoe et al. [8]	The study found that the connectivity of internal and external processes has a separate and common positive impact on supply chain agility, while complexity has a moderate impact under certain circumstances, and extended life cycle theory to supply chain research

coordination of the interests and objectives of all principals and agents within the supply chain, the integration of information, and the improvement of contracts.

In order to coordinate the interests of enterprises within the supply chain, a feasible method is to cooperate between enterprises and establish a partnership similar to an alliance. It is also an effective way to improve the internal information and decision-making structure of the supply chain and promote the integration of information. Effective cooperation and partnership depend on the establishment of trust between enterprises, and trust is established through long-term cooperative games between enterprises. Because trust is not easy to obtain, and because traditional competition theories make companies accustomed to regard other members in the supply chain as competitors, it is generally difficult to maintain cooperation and partnership within the supply chain, especially among members in the same supply chain role. The cooperative relationship between. This is because the contractual relationship between them is loose, often in a state of direct competition, it is difficult to provide incentives for cooperation, and it is also difficult to establish an effective cooperation supervision mechanism.

It should be pointed out that not only the bullwhip effect, but also many other supply chain efficiency losses can also be found in the supply chain structure, including the organizational structure and information structure. The principal–agent relationship within the supply chain is inevitable. Even in the integrated enterprise, there is also a principal–agent relationship. To completely solve the problem of efficiency loss, it is necessary to establish an effective incentive and supervision mechanism, and improve internal and external contracts. Especially in the aspect of perfecting external contracts, there is still a lot of work to be studied.

3 Research Methods

Based on the quality function deployment (QFD), this research constructs a house of quality (HOQ) to build the correlation between the bullwhip effect and agility. First collect and sort relevant documents, organize key group interviews, focus on the key risk factors of the case company, import the interview results into the House of Quality, and perform calculation and analysis on the key risk factors. Faced with risks, companies must establish or strengthen the agility of the supply chain. In this study, a series of agility indicators were obtained through literature discussion and collation, and HOQ was used to combine the cause of bullwhip and agility indicators. Finally, using fuzzy Delphi, fuzzy interpretation structure model, and other methods, the results of the House of Quality are obtained. As the picture shows in Fig. 1.

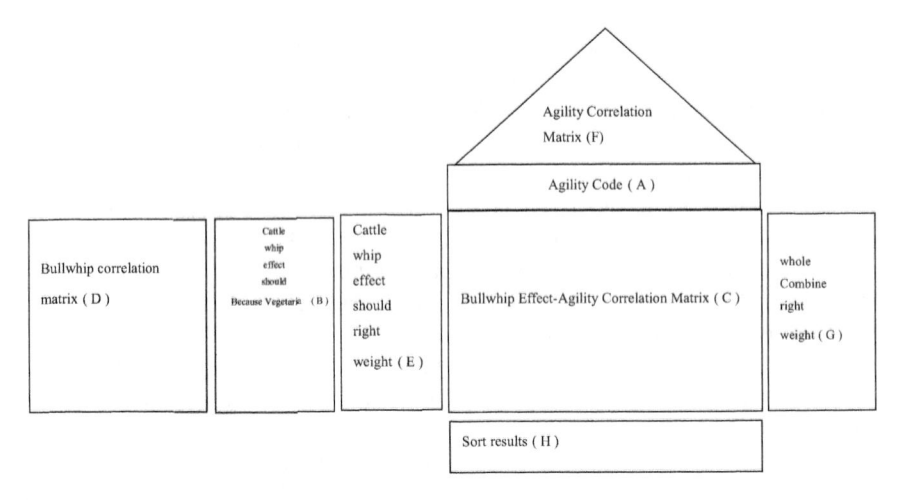

Fig. 1 QFD

4 The Empirical Research

A supply chain company in Guangdong is engaged in road cargo transportation, NVOCC, intra-provincial express business, inter-provincial express business, international express business, agent international express (except postal enterprise franchise business), special cargo transportation (refrigerated preservation, container) And third-party pharmaceutical modern logistics business (accepting pharmaceutical production and operation companies to entrust the storage and distribution of drugs, except for specially managed drugs) and other logistics companies, have a very mature logistics operation system. This study conducted a three-stage questionnaire survey. The purpose is to find the factors that cause the bullwhip effect in logistics companies, and how to suppress the bullwhip effect by improving corporate agility. So that companies can take measures to deal with risks in advance, respond to market changes faster and better, and improve their competitiveness.

1. This research has sorted out 36 factors that cause the bullwhip effect by searching relevant domestic and foreign literature in recent years. Through the fuzzy Delphi method (FDM) to integrate 10 expert questionnaires, the extreme values are excluded first, and then the most conservative cognitive values and the most optimistic cognitive values assessed by the experts are not within the range of twice the standard deviation (Excluded from the calculation). Then use the formula to calculate the triangular fuzzy number and the consensus value. The higher the value, the higher the consensus of the experts. Finally, the threshold is set at 6.65.
2. After using the fuzzy Delphi method to filter out the key factors of the bullwhip effect, the formula in the fuzzy interpretation structure model is used to calculate the evaluation values in the five recovered expert questionnaires to establish a

direct relationship matrix for the bullwhip effect, and to calculate the bullwhip effect. The direct relationship matrix transforms the binary variables, and the fuzzy interpretation structure matrix is obtained.

3. After using the fuzzy interpretation structure model to find out the relationship between the bullwhip effect factors, the network analytic hierarchy process is used to design the third stage questionnaire using the fuzzy interpretation structure matrix to obtain the value of the degree of interaction. Through the collected questionnaire data, the weight of the bullwhip effect is calculated and the super matrix is established.

4. The data collected from the five third-stage expert questionnaires were calculated and tested for consistency to determine that the C.R values were all less than 0.1, and a paired comparison matrix was established using formulas.

In this study, the relationship between the bullwhip effect factor and the agility criteria and the relationship between the agility criteria were analyzed by the gray correlation analysis method, and scores of 0, 1, 3, and 9 were performed. If the degree of correlation between the two is mildly relevant, please fill in "1" in the form; if it is moderately relevant, please fill in "3" in the form; if it is highly relevant, please fill in the form "9"; if there is no correlation between the two, fill in "0".

5 Conclusions

Faced with the individual needs of customers, short product life cycles, uncertain market demands, and rapid technological changes, companies must be flexible to respond to market changes in order to survive in a highly competitive market environment. The bullwhip effect is Most companies are faced with problems. The purpose of this research is to find solutions to improve the agility of companies to deal with and alleviate the impact of the bullwhip effect, and take Guangtou Bangda Supply Chain Technology Co., Ltd. as an example to conduct research. The research results of this study can provide references for the decision-making of the logistics enterprise managers.

Acknowledgments This paper was supported by Natural Science Foundation of Fujian Province of China (Grant No. 2019J01790).

References

1. Chang, A.-Y., Yu, H.-F., Yang, T.-H.: A study of the key factors contributing to the bullwhip effect in the supply chain of the retail industry. J. Ind. Prod. Eng. **30**(7) (2013)
2. Tesfay, Y.Y.: Central applications of Tesfay coordination in transportation supply chain: case of airline industry. J. Traffic Transp. Eng. (Engl. Ed.) **6**(05), 514–525 (2019)

3. Disney, S.M., Farasyn, I., Lambrecht, M., et al.: Taming the bullwhip effect whilst watching customer service in a single supply chain echelon [J]. Eur. J. Oper. Res. **173**(1), 151–172 (2006)
4. Lee, H.L., Padmanabhan, V., Whang, S.: Information distortion in a supply chain: the bullwhip effect [J]. Manage. sci. **43**(4), 546–558 (1997)
5. Ma, J., Lou, W., Tian, Y.: Bullwhip effect and complexity analysis in a multi-channel supply chain considering price game with discount sensitivity. Int. J. Prod. Res. **57**(17) (2019)
6. Dubey, R., Altay, N., Gunasekaran, A., Blome, C., Papadopoulos, T., Childe, S.J.: Supply chain agility, adaptability and alignment. Int. J. Oper. Prod. Manag. **38**(1) (2018)
7. Hwang T., Kim, S.T.: Balancing in-house and outsourced logistics services: effects on supply chain agility and firm performance. Serv. Bus. **13**(3) (2019)
8. Roscoe, S., Eckstein, D., Blome, C., Goellner, M.: Determining how internal and external process connectivity affect supply chain agility: a life-cycle theory perspective. Prod. Plan. Control. **31**(1) (2020)

Using Agility to Reduce the Bullwhip Effect of Supply Chains

Jun-Yi Zeng, Chih-Hung Hsu, and Xiu Chen

Abstract With the intensification of market competition and the improvement of environmental awareness, sustainable development has become the theme of the development of the global society in the twenty-first century. More and more people are paying attention to environmental friendliness and sustainable development of society. As an enterprise is a link in the supply chain, it is impossible to avoid the problem of bullwhip effect common in the supply chain. Agility can effectively improve the adverse impact of bullwhip effect and improve sustainability. The main purpose of this study is to explore the use of the key criteria of agility to quickly respond to the problem of the bullwhip effect in the supply chain to achieve the purpose of sustainability. Dragon net environmental protection company in Fujian as an example, the Quality Function Deployment (QFD) as a basic framework, through literature catalogued, it is concluded that the bullwhip effect factors and agile force rules, the Fuzzy Delphi Method (FDM) screening of bullwhip effect of the key factors and the key principles of agile force, then using Fuzzy Interpretive Structure Modeling (FISM) analysis of bullwhip effect between the key factors influence each other relations, again by Analytic Network Process (ANP) to explore the correlation between the bullwhip effect factors and weight size, finally by the Grey Relation Analysis (GRA) to get the final sorting result. The research of this paper is of great significance to the sustainable development of environmental protection enterprises and can provide the basis for managers to make decisions.

J.-Y. Zeng (✉) · C.-H. Hsu · X. Chen
Institute of Industrial Engineering, School of Transportation, Fujian University of Technology, No. 3 Xueyuan Road, University Town, Minhou, Fuzhou City, Fujian Province, China
e-mail: 1178968775@qq.com

C.-H. Hsu
e-mail: chhsu886@fjut.edu.cn

X. Chen
e-mail: 2431870512@qq.com

1 Introduction

In sustainable development has become the needs of customers and enterprises, social development needs of environment, sustainable supply chain of thought by Drum wright in 1994 first: enterprise should not merely pursue the economic efficiency, should also pay more attention to its social responsibility, attaches great importance to the enterprise the organization, such as purchasing, manufacturing, and consumption in the process of the social benefit [1]. Since then, more and more scholars have begun to pay attention to the research on the environment and social benefits in supply chain management, and enterprises have begun to strengthen the cooperation of supply chain partners to improve the awareness of social responsibility in the whole supply chain.

Supply chain management (SCM) is considered as a strategy and method that can improve supply chain and enterprise performance. A key and significant problem is the lack of information about the correct quantity required and setting the exact order quantity, which becomes one of the most critical decisions in SCM. Lack or variation in information can increase the deviation between orders and actual requirements. This difference becomes significant when orders are passed up the supply chain. This phenomenon of demand variation difference and fluctuation is also known as the bullwhip effect [2]. The bullwhip effect will lead to inefficiencies in the supply chain, such as excessive inventory investment, poor customer service, loss of revenue, misguided capacity planning, ineffective transportation, and missed production plans [3]. Then, in order to improve the competitiveness of enterprises, supply chain management needs to reduce the negative impact of the bullwhip effect.

Christopher (2000) pointed out that the agility of supply chain can reduce the negative impact of market changes, and then improve enterprise performance [4]. Therefore, this paper suggests that enhanced supply chain agility may be used to reduce the negative impact of the bullwhip effect. Based on this assumption, you then try to determine which agility can be effective.

QFD is a flexible system that can transform customer requirements (CRS) into company design requirements (DRS) at all stages from product design, manufacturing to sales [5]. With the spread of QFD, its scope gradually expands to manufacturing, service and supply chain [6]. Therefore, from the perspective of supply chain management, this paper uses QFD to reduce the impact of bullwhip effect and improve sustainability by improving agility.

The purpose of this study is to identify the supply chain bullwhip effect factor of environmental protection enterprises, and with an integration of QFD method through agility to reduce the adverse impact of the bullwhip effect factors, analysis of bullwhip effect and agility of the relationship between the two groups of variables, and ultimately for the environmental protection enterprise determine the priority of the rule of agile force, to help the implementation of supply chain management.

2 Literature Review

2.1 Supply Chain Bullwhip Effect

Bullwhip effect is one of the most studied phenomena in supply chain management literature and one of the main sources of inefficiency in current supply chain. Hussain et al. [7] used simulation and Taguchi design experiments to study the impact of capacity constraints and safety stocks on the bullwhip effect in a two-layer supply chain model [7]. Mahbub et al. [8] established a multi-product, multi-stage, and multi-cycle supply chain model and set the objective function to maximize profits, thus achieving the goal of minimizing the bullwhip effect of the supply chain [8]. Pastore et al. [9] proposed that the bullwhip effect is affected by unknown parameters and adopted analytical approximation method to obtain the influence degree of unknown parameters on the bullwhip effect [9]. This paper summarizes the causes of the bullwhip effect by referring to the previous research results and combining with the actual enterprise investigation. See the fourth section for details.

The relevant research contents of bullwhip effect are shown in Table 1.

Table 1 Studies of scholars related to the bullwhip effect

Scholars	The research content
Hussain et al. [7]	This study illustrates the impact of capacity constraints and safety stocks on the bullwhip effect in a two-tier supply chain model. The nonlinear effects of capacity constraints in supply chain are studied by simulation and Taguchi design experiments. This study provides a practical method for supply chain managers to solve the problem of tradeoff capability and safety stock at different levels in a timely manner, and to improve their capability and safety stock and make more reasonable decisions
Mahbub et al. [8]	The study is for each entity to find the optimal prediction of sales volume, so as to achieve the maximization of profit, and reduce the negative impact of bullwhip effect in supply chain, through the establishment of multi-product, multi-stage, multiple cycles of supply chain model, set up to realize the profit maximization of the objective function, the model is applied in the supply chain network, for instance, So as to minimize the bullwhip effect of supply chain
Pastore et al. [9]	This paper presents a two-stage single product supply chain in which the bullwhip effect is affected by unknown parameters and the estimated updating frequency of parameters. In order to improve the low accuracy of the existing model, an analytical approximation method is adopted to study the influence of unknown parameters on the bullwhip effect when the process parameters are unknown and the prediction is updated but the parameter estimation remains unchanged

2.2 Supply Chain Agility

Agility is the ability to effectively respond to changes in user needs, efficiency is the relationship between flexibility and performance, and agility determines the level of enterprise operation efficiency. Sangari et al. [10] obtained 12 types of influencing factors that can quickly realize the agility of supply chain by applying the practical classification method, and ranked the factors [10]. Wu et al. [11] creatively proposed a new multi-standard decision-making structure to deal with complex problems in supply chain and the interrelationship between various aspects and attributes to improve agility [11]. Mahbub et al. [8] aimed to explore the mediating factors of the gap between information processing theory and supply chain agility, analyze 60 cases, and use qualitative analysis method to analyze the relationship between information processing and supply chain agility. The mediation configuration and agile adaptation types in supply chain that define three kinds of adaptability are proposed [8]. Dubey et al. [12] pointed out that under the adjustment effect of top management commitment, the agility, adaptability, and consistency of supply chain will be enhanced, so that enterprises can obtain sustainable competitive advantages [12]. Roscoe et al. [13] found that the connectivity of internal and external processes has a separate and common positive impact on supply chain agility, while complexity has a moderate impact in specific circumstances [13]. As for the research on supply chain agility, most scholars focus on how to improve supply chain agility, but few further study on the positive effect of agility on sustainable supply chain.

The relevant research content of agility is shown in Table 2.

3 Methodology

3.1 Methods the Framework

Based on QFD as a framework, through literature catalogued, it is concluded that the bullwhip effect factors and agile force rules, the FDM screening of bullwhip effect of the key factors and the key principles of agile force, then using FISM analysis of bullwhip effect between the key factors influence each other relations, again by the ANP to explore the correlation between the bullwhip effect factors and weight size, finally by the GRA to get the final sorting result. Figure 1. shows the detailed structure of the House of Quality (HoQ), and the details on how to build the HoQ are shown below.

Table 2 Author studies on agility

Scholars	The research content
Sangari et al. [10]	The main purpose of this study is to create a practical taxonomy. A survey of supply chain managers working in automotive companies was conducted to identify and verify 12 categories of factors that contribute to the rapid realization of supply chain agility. Then, according to the reasonable range of each factor, the results are ranked from large to small, and an integrated framework is obtained, which provides a way of thinking for the development of supply chain agility
Wu et al. [11]	Based on the case study in supply chain network, this study creatively proposes a new multi-standard decision structure to deal with the complex problems in supply chain and the interrelationships among various aspects and attributes. The fuzzy Delphi method is used to screen out the unnecessary factors, and then the fuzzy set theory is combined with decision test and evaluation to evaluate the core competitive advantage in supply chain
Mahbub et al. [8]	The purpose of this study is to explore the mediating factors that contribute to the gap between information processing theory and supply chain agility. Based on the analysis of 60 cases, the mediating configuration and agile adaptation types in supply chain are proposed to define three kinds of adaptations around the link between information processing and supply chain agility using qualitative analysis methods
Dubey et al. [12]	It is pointed out that under the adjustment effect of top management commitment, the agility, adaptability and consistency of supply chain will be enhanced, so that enterprises can obtain sustainable competitive advantage
Roscoe et al. [13]	It is found that internal and external process connectivity has separate and collective positive effects on supply chain agility, while complexity has moderate effects in specific situations

4 Empirical Research

This paper is a major research and development production base for environmental pollution control equipment and other environmental products such as the design of dust removal devices and flue gas desulfurization devices, including the electric dust removal control system and other technical equipment, technical equipment, technical level, product sales, and enterprise strength to achieve the current international advanced level and domestic leading position. The proposed method is used to build the quality house of the enterprise, and eight supply chain experts from different departments are invited to fill in three stages of the questionnaire to make a comprehensive decision. The resulting data is presented in the QFD framework.

4.1 Questionnaire Design and Collation

The first stage: Identify key factors and criteria and Screening key factors and criteria.

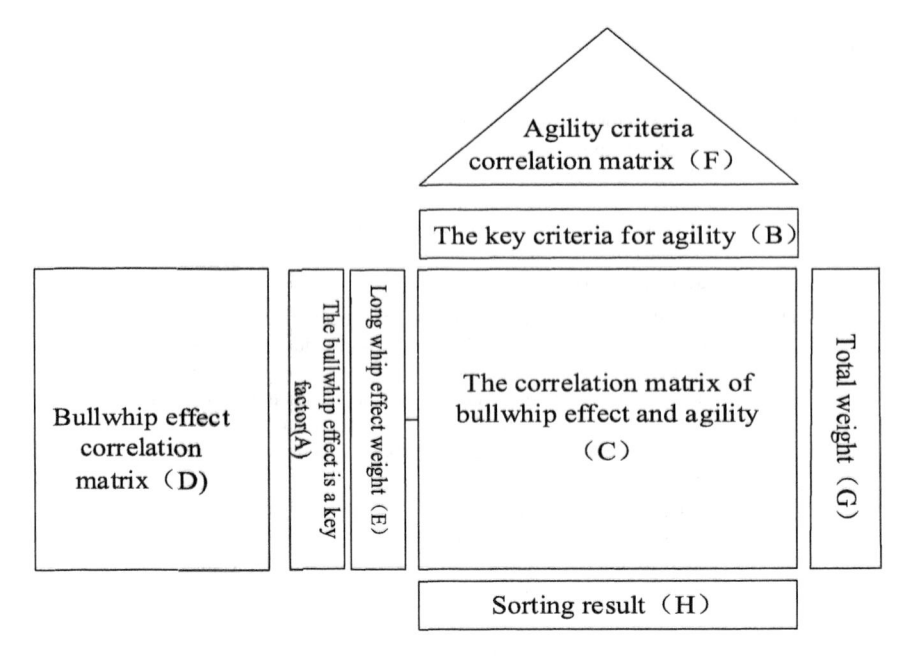

Fig. 1 Structure of the House of Quality

Through literature search and industry expert consultation, the factors of bullwhip effect and agility criteria are determined.

Will find out the rule of bullwhip effect factors and agile force questionnaire design, the first stage to the environmental protection enterprises of experts, the expert scoring extreme value selection, using the FDM to calculate the values, and set the threshold value, select bullwhip effect key factors (B) and agile force key criteria (A). Therefore, 12 key factors of bullwhip effect and 20 key criteria of agility are listed. The calculation results are shown in Table 3 and Table 4, which are respectively the data content of (B) and (A) in HoQ.

The second stage: Find the relevance of key factors.

In order to find out the correlation between the bullwhip effect factors, based on the screening results of the first stage, a questionnaire was designed for the second stage, and experts were invited to evaluate the 12 key factors using a 0-1-2-3 scale. The FISM was used to calculate the evaluation value. After setting the threshold value, the matrix {0, 1} is converted, namely, the Fuzzy Interpretative Structure Modeling matrix. At the same time, the evaluation criteria of 0-1-3-9 scale were sent to experts to fill in, recovered and sorted, and the correlation matrix (F) of agility and the correlation matrix (C) of bullwhip effect and agility were directly obtained.

The third stage: Obtain the weight value of the factors of bullwhip effect, establish a super matrix and the integrated weight of bullwhip effect is obtained.

The correlation of bullwhip effect factors was used as the basis for the design of the third stage questionnaire, which was assessed in pairs on a scale of 1–9. The

Table 3 The factors of bullwhip effect screened by Fuzzy Delphi Method

Number	Factor	Gi	Rank
B1	Multi-level organization of supply chain	6.79	1
B2	Demand forecasting	6.27	2
B3	Information asymmetry	5.93	3
B4	Periodicity of demand	5.83	4
B5	Factory capacity constraints	5.72	5
B6	Price fluctuations	5.53	6
B7	Shortage problem	5.49	7
B8	Order batch size	5.48	8
B9	Changes in demand	5.37	9
B10	Batch ordering strategy	5.32	10
B11	Sales time	5.22	11
B12	Lack of synchronization	5.06	12

ANP is used to verify whether the results meet the principle of consistency, and the consistency is less than 0.1. The original matrix of bullwhip effect factors filled in by 8 experts was used to calculate the feature vectors by ANP, and the weight Wm of the feature vectors in each row was calculated. After sorting out, the weight value of bullwhip effect factors (E) was obtained.

According to the calculation steps of the weight value of bullwhip effect in (2), the weight value of the 8 experts under the influence of various relevant factors and conditions is calculated, which is (D) after integration.

The weight of the bullwhip effect factor (E) is multiplied by the super matrix (D) to obtain the integrated weight of the bullwhip effect (G).

4.2 Grey Relation Analysis

First: Calculate the Grey Relation Analysis matrix.

The agile force correlation matrix (F), the bullwhip effect, and the agile force correlation matrix (C) are multiplied by the matrix, and then normalized. The grey correlation coefficient matrix is obtained by using the GRA.

Second: The grey correlation sorting result is obtained.

The integrated weight of bullwhip effect (G) is multiplied by the grey correlation coefficient matrix to obtain the grey correlation degree of agile force criterion, and the ranking result is (H).

The steps and methods for establishing the HoQ in this paper are shown in Fig. 1. The symbols from A to H correspond to the letters in Fig. 1. This paper converts the data generated by analysis into HoQ, which is also the final result. The grey ranking of agility criteria indicates the first three factors, with " A16-developing

Table 4 Agility criteria filtered by Fuzzy Delphi Method

Number	Principle of agility	Gi	Rank
A1	Improve the competitiveness of the market and the environment	7.717	1
A2	Long-term cooperation with partners and strengthening trust	7.215	2
A3	A partner who selects performance and basic skills	7.057	3
A4	Improve market sensitivity	6.959	4
A5	Order production rather than forecast production	6.888	5
A6	The information of the market in the supply chain is transparent and transparent, which can respond quickly to customer needs	6.869	6
A7	Data accuracy	6.817	7
A8	Cost minimization	6.765	8
A9	Shorten the prompt response of advance period	6.638	9
A10	Improve the development ability of innovative products	6.630	10
A11	Integration of supply chain partners	6.618	11
A12	Managing the core competitiveness of the enterprise	6.578	12
A13	Information data integration	6.504	13
A14	Shorten delivery time	6.496	14
A15	Introduce appropriate information technology and incorporate new hardware and software and new products	6.482	15
A16	Development with knowledge as a driver	6.451	16
A17	E-commerce	6.451	17
A18	Continuously cultivate multi-skill and flexible employees and improve employee working ability	6.418	18
A19	Develop potential customers for the future as a chance for the new market	6.358	19
A20	Improve technology awareness and information technology	6.287	20

knowledge-driven enterprises" as the key criterion, followed by "A10-improving the R&D capability of innovative products," and then "A11-integration of supply chain partners."

5 Conclusion

Through the empirical study of one of the leading environmental protection enterprises in China, the main findings are as follows: the top three bullwhip effect factors are price fluctuation, multi-level organization of supply chain, order batch size. The top three Agile criteria are developing knowledge-driven enterprises, enhancing R&D capabilities for innovative products, and integrating supply chain partners.

This study has some contributions in theory and practice. First, a primitive QFD method is designed to rank the supply chain agility criteria of environmental companies to reduce the impact of the bullwhip effect. This study is the first of its kind and fills the literature gap discussed in Sect. 2. Second, the ranking scheme has been further recognized by environmental industry insiders and supply chain companies. Even in the short term, organizations can counter the negative effects of the bullwhip effect through better strategies and operations. Therefore, this study also has a certain practical significance for the majority of environmental protection industry companies.

Acknowledgments This paper was supported by Natural Science Foundation of Fujian Province of China (Grant No.2019J01790).

References

1. Beamon, B.M.: Environmental and sustainability ethics in supply chain. Manag. Sci. Eng. Ethics **11**(2), 221–234 (2005)
2. Zhou, W., Piramuthu, S.: Technology regulation policy for business ethics: an example of RFID in supply chain management. J. Bus. Ethics **116**(2), 327–340 (2013)
3. Turrisi, M., Bruccoleri, M., Cannella, S.: Impact of reverse logistics on supply chain performance. Int. J. Phys. Distrib. Logist. Manag. **43**(7), 564–585 (2013)
4. Christopher, M.: The agile supply chain: competing in volatile markets. Ind. Mark. Manage. **29**(1), 37–44 (2000)
5. Bottani, E., Rizzi, A.: Strategic management of logistics service: a fuzzy QFD approach. Int. J. Prod. Econ. **103**, 585–599 (2006)
6. Lam, J.S.L., Dai, J.: Developing supply chain security design of logistics service providers: an analytical network process-quality function deployment approach. Int. J. Phys. Distrib. Logist. Manage. **45**(7), 674–690 (2015)
7. Hussain, M., Khan, M., Sabir, H.: Analysis of capacity constraints on the backlog bullwhip effect in the two-tier supply chain: a Taguchi approach. Int. J. Logist. Res. Appl. **19**(1) (2016)
8. Mahbub, N., Hasin, A.A., Aziz, R.A., Sharin, A.: Maximisation of total supply chain profit and minimisation of bullwhip effect in a multi-echelon supply chain network using particle swarm optimisation and genetic algorithm. Inderscience Publ. **11**(2/3) (2017)
9. Pastore, E., Alfieri, A., Zotteri, G., Boylan, J.E.: The impact of demand parameter uncertainty on the bullwhip effect. Eur. J. Oper. Res. **283**(1) (2020)
10. Sangari, M.S., Razmi, J., Gunasekaran, A.: Critical factors for achieving supply chain agility: towards a comprehensive taxonomy. Inderscience Publ. **23**(3) (2016)
11. Wu, K.J., Tseng, M.L., Chiu, A.S., Lim, M.K.: Achieving competitive advantage through supply chain agility under uncertainty: a novel multi-criteria decision-making structure. Int. J. Prod. Econ. (2016)
12. Dubey, R., Altay, N., Gunasekaran, A., Blome, C., Papadopoulos, T., Childe, S.J.: Supply chain agility, adaptability and alignment. Int. J. Oper. Prod. Manag. **38**(1) (2018)
13. Roscoe, S., Eckstein, D., Blome, C., Goellner, M.: Determining how internal and external process connectivity affect supply chain agility: a life-cycle theory perspective. Prod. Plan. Control **31**(1) (2020)

Mining Frequency-Utility Patterns from a Big Data Environment

Ranran Li, Jimmy Ming-Tai Wu, Min Wei, Ke Wang, and Qian Teng

Abstract Both frequent itemset mining (FIM) and high utility itemset mining (HUIM), which have been widely discussed before, consider the frequency or utility of the project separately to find the information required by users. However, this information is not complete as merchants need to find the commodities that occur frequently and have higher profits. To tap into multiple user preferences, the concept of skyline was proposed, which considered both the frequency and utility of the item to better support user decision-making. Nevertheless, in real life, the generation of data from multiple sources leads to a large volume of data, and the previously proposed algorithm is not enough to handle large-scale data sets on a single machine. In this paper, a skyline mining framework based on MapReduce is proposed. The algorithm adopts distributed method to process large data sets in parallel, which solves the problem that a single machine cannot process a large number of distributed data. Experimental results show that the MR-Skyline algorithm is superior to the most advanced algorithms in large data sets.

R. Li · J. Ming-Tai Wu (✉) · M. Wei · K. Wang · Q. Teng
College of Computer Science and Engineering, Shandong University of Science and Technology, Shandong, China
e-mail: wmt@wmt35.idv.tw

R. Li
e-mail: 734181156@qq.com

M. Wei
e-mail: 17685541390@163.com

K. Wang
e-mail: 870848546@qq.com

Q. Teng
e-mail: qrape@foxmail.com

1 Introduction

Frequent pattern mining and high utility itemset mining (HUIM) have important research value in data mining field. FIM [1, 2, 8] aims to find frequently occurring itemsets from the database and then extract valuable information, while HUIM [3, 10] reveals high-profit itemsets from the database to help users make decisions. Both of these patterns require a user-defined threshold to be set when the itemset is no less than this user-defined threshold to be considered a frequent or highly utility itemset, and both patterns extract information from a single dimension. The Skyline concept combines the two dimensions of frequency and utility to find more valuable and complete information.

All along, the FIM has been widely researched, such as in the analysis of the market, but it ignores those items that are not frequent but can obtain high profits of goods, to solve this problem, the concept of high utility itemset mining (HUIM) is put forward, which calculates the utility of the item in the database by considering both the unit profit (external utility) and quantity (internal utility) of the item. Simultaneously, high utility itemset mining does not have the feature of downward closure, so the algorithm proposed for frequent itemset mining cannot be directly used for high utility itemset mining. To maintain this feature, Liu et al. proposed a TWU model [10] and used to find HUIs from the database. Subsequently, many algorithms [4, 6, 12, 16] have been proposed for mining high utility itemsets, and different pruning strategies [5, 13] have been adopted for pruning search space. However, whether it is FIM or HUIM, the size of the search space and the size of the generated candidate set largely depends on the size of the user-defined threshold, and setting an appropriate threshold is a challenging and difficult task in itself.

The process of mining skyline points does not require the user to set thresholds, and it considers both the frequency and utility of the itemset, obtaining more valuable information from multiple dimensions, which have high usage value in real life. For example, hotels that can satisfy travelers in terms of distance and price. Skyline points are a set of points that are not dominated by other points, called non-dominated points, also known as the Skyline Frequent Utility Patterns (SFUPs). The previous SFUPs mining algorithm [9] are all performed on a machine. However, as the amount of information increases, a single machine cannot handle large-scale data sets.

Hadoop is a distributed architecture proposed by Google Apache. Its key components are HDFS and MapReduce. The full name of HDFS is Hadoop Distributed File System, which is mainly used to store the data in nodes, and it is characterized by its high reliability. MapReduce is an abstract general-enough programming model whose main principle is divide-and-conquer and merge sort recursively. MapReduce can be roughly divided into two phases: the Map phase and the Reduce phase, and a Map must be followed by a Reduce. Mapper and Reducer are the projects of Map and Reduce respectively. The key-value pairs generated by Mapper are transmitted to the Reducer, which merge those with the same key together to generate another merged key-value pair. The final or intermediate results are then output to HDFS. As far as we know, there is currently no algorithm that can mine skyline points in big data. Therefore, in this paper, we propose a parallel mining of skyline algorithm MR-Skyline based on MapReduce framework. The main contributions of this paper are as follows:

1. The MR-Skyline proposed is conducted on the Hadoop platform, so it has the general advantages of Hadoop, such as high reliability, which can be embodied in bit storage.
2. MR-Skyline adopts the MapReduce framework, which can divide the original database into several independent partitions without modifying the original database data, and process these data with HDFS. Therefore, it can perform the task of parallel mining on the large data set.
3. Due to the superiority of utility-max function [9], it can find the maximum utility under the itemset with the same occurrence frequency, so it will continue to be used.

The remaining structure of this paper is as follows: Sect. 2 reviews the related work of skyline. Section 3 introduces the definition of mining skyline points. Section 4 presents the algorithm. Experimental results and summary are presented in Sect. 5 and Sect. 6, respectively.

2 Related Work

In this section, we will review some of the research related to skyline.

2.1 Skyline

Unilateral consideration of frequency or utility cannot meet the needs of users. Yeh et al. [15] first proposed a two-stage algorithm to find those itemsets that both have high efficiency and frequency by setting two minimum thresholds (minimum utility threshold and minimum support threshold). Next, Podpecan et al. [11] proposed an improved algorithm to improve mining performance, but the algorithm still requires two user-defined thresholds, however, setting a threshold of an appropriate size is a challenging task in itself. Goyal et al. [7] proposed a SKYLINE algorithm based on UP-tree structure, which solved the above problem of setting two thresholds. Therefore, it did not need to set thresholds to mine a group of skyline points that were non-dominated by other points. Even so, the algorithm is not efficient because it produces a large number of candidate sets when dealing with large data. Lin et al. [9] proposed two efficient algorithms, namely, SKYFUP-D and SKYFUP-B. This algorithm is generated by DFS and BFS, respectively, and it is based on a data structure that is more efficient than UP-tree, namely the Utility-List structure. All the above algorithms consider frequency and utility at the same time. Wu et al. [14] proposed a new Skyline Quantity-Utility Pattern (SQUP). In addition, by

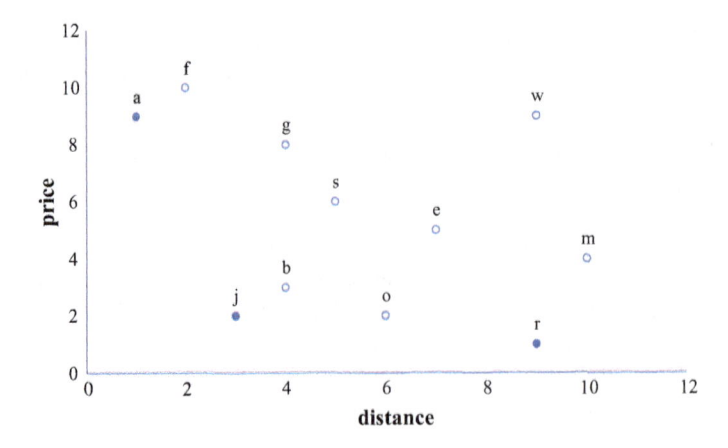

Fig. 1 Example of skyline points

considering the quantity and utility of items, two efficient algorithms, SQU-MINER and SKYQUP, are proposed to mine SQUPs.

The concept of skyline is a group of points that are not governed by other dimensions points. That is to say, when comparing multiple dimensions, the skyline point is superior to other points or equally important, or at least one dimension is superior to other points, this group of the non-dominated points as the final result to return, which is of great significance for the study of large-scale data sets. Tourists, for example, may consider two factors at the same time, such as distance from the hotel to the city center or scenic spot and the price of the booking hotel. Generally, the price of the hotel near the city center is higher than that of the hotel far away from the city center. Therefore, it is necessary to find a hotel that is close to the city center and has a suitable price. As shown in Fig. 1, the horizontal axis represents the location of the hotel from the city center, the vertical axis represents the price of the hotel, the points in the graph represent hotels, while points a, j, and r are a set of skyline points that we found, which are superior to the other points in both distance and price. They are a set of non-dominant points.

3 Preliminary and Problem Statement

3.1 Preliminaries

Let $I = \{i_1, i_2, ..., i_m\}$ be a set of items in a transaction database D, which contains n transactions, and $D = \{T_1, T_2, ..., T_n\}$. Each transaction T_q in D consists of several different items such that $T_q = \{i_1, i_2, ..., i_k\}$. Thus, we can have that $i_j \in I$, and $T_q \subseteq D$. Table 1 shows the running examples in the original database, it contains

Mining Frequency-Utility Patterns from a Big Data Environment

Table 1 A transaction database

TID	Items: quantities
T_1	B:2,C:1,D:2
T_2	A:4,D:1,E:1
T_3	A:2,B:4,C:2,D:1
T_4	B:5,C:2,D:1,E:2
T_5	A:3,B:4,D:1
T_6	D:2,E:1
T_7	B:2,C:1,D:1

Table 2 A profit table

Item	Profit
A	2
B	1
C	2
D	4
E	5

the internal utility (purchase quantity) of the item i_j in each transaction T_q. Table 2 shows the unit profit $pr(i_j)$ of each item i_j.

Definition 1 The occurrences frequency of itemset X in the database D is called $f(X)$ and defined it as:

$$f(X) = |\{T_q | X \subseteq T_q \wedge T_q \in D\}|. \tag{1}$$

For example, in Table 1, f(D) = 7, f(BCD) = 4.

Definition 2 The utility of an item i_j in a transaction T_q is denoted as $u(i_j, T_q)$ and defined as:

$$u(i_j, T_q) = q(i_j, T_q) \times pr(i_j). \tag{2}$$

For example, in transaction T_1, the utility of the item (D) can be calculated as $u(D, T_1) = q(D, T_1) \times pr(D) = 2 \times 4 = 8$.

Definition 3 The utility of itemset X in a transaction database D, can be denoted as $u(X)$ and defined as:

$$u(X) = \sum_{X \subseteq T_q \wedge T_q \in D} u(X, T_q). \tag{3}$$

Definition 4 The utility of a transaction in a transaction database D is denoted as $tu(T_q)$ and defined as:

$$tu(T_q) = \sum_{i_j \subseteq T_q} u(i_j, T_q). \tag{4}$$

Definition 5 The transaction-weighted utility of an itemset X in a transaction database D is denoted as $twu(X)$ and defined as:

$$twu(X) = \sum_{X \subseteq T_q \wedge T_q \in D} tu(T_q). \tag{5}$$

For the running example, the *twu* of each item is as follows: $twu(A) = 45$, $twu(B) = 71$, $twu(C) = 59$, $twu(D) = 103$, $twu(E) = 53$.

In order to consider both the frequency and utility of the itemsets, skyline frequency-utility pattern mining (SFUPM) is defined as follows:

Definition 6 For itenset X and itemset Y, if $f(X) \geq f(Y)$ and $u(X) > u(Y)$ or $f(X) > f(Y)$ and $u(X) \geq u(Y)$, we can say itemset Y is dominated by itemset X which is denoted as $X \succ Y$.

Definition 7 When considering both frequency and utility, the itemset X is said to be SFUPM iff the itemset X is not dominated by other points.

3.2 Problem Statement

Given a large-scale database and a profit table, based on the above definitions, our problem is to find skyline frequency-utility patterns (SFUPs) from a MapReduce-based framework.

4 Proposed a MapReduce Framework

In this section, we propose a MR-Skyline algorithm based on the Hadoop platform to mine SFUPs. The input data are the transaction database DB. After processing in the calculating phase, sorting phase, and mining phase, we finally output the mined skyline points. The utility-max (*utilmax*) function is used to prune the dataset of each partition to reduce the search space. The proposed framework is shown in Fig. 2. The description of each stage is as follows:

4.1 Calculating Phase

The original database as the input data are divided into several independent partition sub-databases and sent to each Mapper for processing, and a pair in the form of <key,

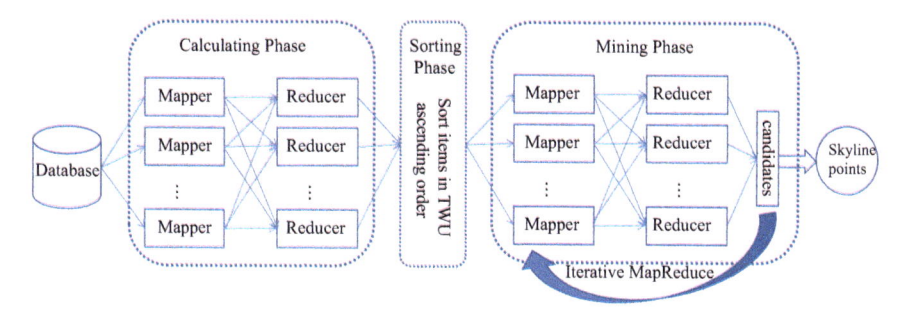

Fig. 2 The MapReduce framework

value> is generated. The Reducer merges and calculates the pairs with the same key, and obtains the occurrence frequency $f(X)$ of each item X and its transaction weighted utility $TWU(X)$.

4.2 Sorting Phase

In the previous stage, the *TWU* of each item is obtained. In this phase, we sort the items according to the *TWU* value of each item, finally, the output is sorted by *twu-ascending* order for each item. Such as *A, E, C, B, D* in the running example.

4.3 Mining Phase

During this phase, the algorithm finds the skyline points in parallel through several iterations. According to the sorting results obtained in the previous stage, each transaction in the partitioned sub-database is sorted in *twu-ascending* order. Each partitioned sub-database is processed by the MapReduce framework. The *utilmax* is used to determine whether this item and its extension are potential skyline points, and a candidate set is generated, in which the points dominated by other points are deleted. This information is then sent to the Mapper, where iterative processing continues until no candidate set is generated and then a set of skyline points we need are found.

5 Experimental Results

In this section, our experiment evaluates the running time of SQUP algorithm under data sets of different sizes. The experimental data set is copied on the basis of the real data set Foodmart, and the data sets of 0.2, 0.4, 0.6, 0.8, 1.0GB are respectively

Fig. 3 Runtimes under different database sizes

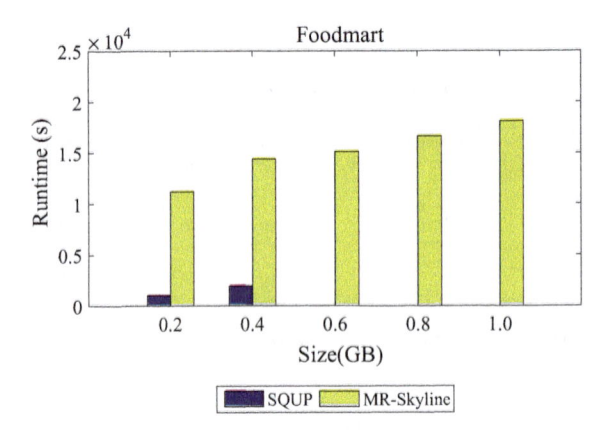

formed. The experimental result is shown in Fig. 3. As can be seen from the figure, with the increase of the size of dataset, SQUP algorithm on a single machine failed to get experimental results due to memory leakage, while the designed algorithm can still be executed.

From the experimental result, we can see that the algorithm proposed in this paper is superior to the most advanced algorithms, especially in processing with large data sets. This is because when the data volume is large, the algorithm is executed on a Hadoop-based platform and the data set is divided into multiple partitions to process the data in parallel.

6 Conclusion and Future Work

In this paper, we propose a new MapReduce-based framework for mining skyline in large data sets. A new MR-Skyline algorithm is proposed for parallel mining of skyline points from big data sets. It runs under the Hadoop platform, so it inherits the general advantages of Hadoop, such as high reliability and high fault tolerance of data processing. Our proposed algorithm is superior to the most advanced algorithms in processing large data sets. In the future, we will continue to explore how to improve our algorithms based on the MapReduce framework, as well as how to efficiently mine skyline points on the Spark platform, since the most advanced algorithms currently do not work with large data sets. Spark and the MapReduce framework are important topics for future research.

Acknowledgements This research is supported by Shandong Provincial Natural Science Foundation (ZR201911150391).

References

1. Agrawal, R., Imieliński, T., Swami, A.: Mining association rules between sets of items in large databases. In: Proceedings of the 1993 ACM SIGMOD International Conference on Management of Data, pp. 207–216 (1993)
2. Agrawal, R., Srikant, R., et al.: Fast algorithms for mining association rules. In: Proceedings of the 20th International Conference Very Large Data Bases, VLDB, vol. 1215, pp. 487–499. Citeseer (1994)
3. Chan, R., Yang, Q., Shen, Y.D.: Mining high utility itemsets. In: Third IEEE International Conference on Data Mining, pp. 19–19. IEEE Computer Society (2003)
4. Chen, C.M., Chen, L., Gan, W., Qiu, L., Ding, W.: Discovering high utility-occupancy patterns from uncertain data. Inf. Sci. **546**, 1208–1229 (2021)
5. Fournier-Viger, P., Wu, C.W., Zida, S., Tseng, V.S.: Fhm: Faster high-utility itemset mining using estimated utility co-occurrence pruning. In: 21st International Symposium on Methodologies for Intelligent Systems (ISMIS 2014) (2014)
6. Gan, W., Lin, J.C.W., Fournier-Viger, P., Chao, H.C., Hong, T., Fujita, H.: A survey of incremental high-utility itemset mining. In: Data Mining and Knowledge Discovery, Wiley Interdisciplinary Reviews (2018)
7. Goyal, V., Sureka, A., Patel, D.: Efficient skyline itemsets mining. In: Proceedings of the Eighth International C* Conference on Computer Science and Software Engineering, pp. 119–124 (2015)
8. Han, J., Pei, J., Yin, Y.: Mining frequent patterns without candidate generation. ACM SIGMOD Rec. **29**(2), 1–12 (2000)
9. Lin, J.C.W., Yang, L., Fournier-Viger, P., Hong, T.P.: Mining of skyline patterns by considering both frequent and utility constraints. Eng. Appl. Artif. Intell. **77**, 229–238 (2019)
10. Liu, Y., Liao, W.k., Choudhary, A.: A two-phase algorithm for fast discovery of high utility itemsets. In: Pacific-Asia Conference on Knowledge Discovery and Data Mining, pp. 689–695. Springer (2005)
11. Podpecan, V., Lavrac, N., Kononenko, I.: A fast algorithm for mining utility-frequent itemsets. Constr.-Based Mining Learn. 9 (2007)
12. Wu, C.W., Shie, B.E., Tseng, V.S., Yu, P.S.: Mining top-k high utility itemsets. In: Proceedings of the 18th ACM SIGKDD International Conference on Knowledge Discovery and Data Mining, pp. 78–86 (2012)
13. Wu, J.M.T., Lin, J.C.W., Tamrakar, A.: High-utility itemset mining with effective pruning strategies. ACM Trans. Knowl. Discov. Data (TKDD) **13**(6), 1–22 (2019)
14. Wu, J.M.T., Teng, Q., Srivastava, G., Pirouz, M., Lin, J.C.W.: Efficient mining of non-dominated high quantity-utility patterns. In: 2020 International Conference on Data Mining Workshops (ICDMW), pp. 690–695. IEEE (2020)
15. Yeh, J.S., Li, Y.C., Chang, C.C.: Two-phase algorithms for a novel utility-frequent mining model. In: Pacific-Asia Conference on Knowledge Discovery and Data Mining, pp. 433–444. Springer (2007)
16. Yu, P., Tseng, V., Shie, B., Wu, C.: An efficient algorithm for high utility itemset mining. In: Proceedings of the 16th ACM SIGKDD KDD-2010, pp. 253–262 (2010)

The Association Between Related-Party Transactions and Tax Planning

Wen-Jye Hung, Tsui-Lin Chiang, Ya-Min Wang, and Qi Luo

Abstract Due to the emerging of anti-tax avoidance, the act of companies arranging tax avoidance is being examined. Related party transaction is complicated and nontransparent; therefore, companies often use it to undergo inappropriate interest arrangement to ease tax burden. This research examines the related party transactions of China public listed companies from 2007 to 2018 to study the relationship between related party transaction and tax planning. This research divides transactions into five types, total, sales, purchase, non-operating, and contingent. Research shows higher tax planning when companies adopt contingent transaction with other related enterprise. But the other research illustrates lower tax planning when companies adopt sales transaction with other related enterprise.

1 Introduction

Related party transactions are a necessary part in the financial activities. For example, affiliated party transaction may be the suppliers and buyers in the same industrial chain based on their business needs. To manipulate related party transactions under the greedy desire, inevitably, investors perhaps doubt the motivations are just to feather his own nest for window dressing financial statement [1]. They, scholars, investors, and regulators, are concerned whether full disclosure of the information of related party transactions and exist extraordinary related party transactions. To comply with related party transactions, according to No. 36 of the "Accounting Standards for Enterprises", RPTs are recognized if a party has the power to, directly or indirectly, control, jointly control or exercise significant influence over another

W.-J. Hung · Y.-M. Wang (✉) · Q. Luo
Minjiang University, Fuzhou, China
e-mail: yaminwang1018@163.com

Q. Luo
e-mail: 1959534859@qq.com

T.-L. Chiang
Fu Jen Catholic University, New Taipei City, Taiwan

party, or if two or more parties are subject to control, joint control or significant influence from a same party.

A lot of literature is researching about the related party transactions, focusing on transfer pricing, corporate governance, and company performance, in recent years. The empirical research on the tax burden and related party transactions is insufficient. Especially, in last period of BEPS (Base Erosion and Profit Shifting), the influence between related party transactions and tax planning, tax authorities around the world, like IRS (Internal Revenue Service), is a hot issue of concern. With regard to tax authorities of all countries, how to mining the key factors of affiliated party transactions is helpful for tax revenue and safeguarding financial stability.

Therefore, our study first develops a regression model of relation between the tax planning and related party transactions.

2 Literature Review and Hypothesis Developments

2.1 Background on Related Party Transactions

FASB ASC 850-10-50-1 indicated the Material RPTs are required to be disclosed, the nature of the relationship, a description of the transaction, the dollar amounts of the transactions for each income statement period presented and amounts due to or from related parties at each balance sheet date in the financial statements. Meanwhile, the issue about RPTs, has separate disclosure requirements, that have been gradually emphasized by the Securities and Exchange Commission (SEC).[1] Specifically, companies must disclose the information of certain relationships and related transactions in their SEC filing, such as registration statements, annual reports, and proxy statements. Disclosure (in the financial statements or other SEC filing) is required for any RPT involving more than $120,000 and in which the related persons had a direct or indirect material interest, naming such person and indicating the person's relationship to the registrant, the nature of such person's interest in the transaction, the amount of such transaction and, where practicable, the amount of such person's interest in the transaction (SEC 2004b, subsection 229.400a). However, why do most countries around the world require different forms of additional monitoring over RPTs? Caused by RPTs are constantly deemed as potential self-dealing activities between insiders and the company. According to No. 36 of the "Accounting Standards for Enterprises", RPTs are recognized if a party has the power to, directly or indirectly, control, jointly control or exercise significant influence over another party, or if two or more parties are subject to control, joint control or significant influence from the same party. Moreover, "affiliated party transaction" refers to an event whereby a transfer of resources, labor services, or obligations takes place between affiliated parties, regardless money is charged or not. In general, the general types of affiliated

---

[1] SEC 2004b, subsection 229.400a.

party transactions include purchases or sales of goods, purchases or sales of assets other than goods, rendering or receiving labor services, guarantees, providing capital (including loans or equity contributions), leasing, agency, transfer of research and development projects, license agreements, settling debts on behalf of an enterprise or by this enterprise that represents another party and the emoluments for key managerial personnel. Therefore, in this study we use various forms of RPTs to explore how they affect tax planning. In other words, the question of whether the investor could see through firm's use of RPTs and manifest in tax planning.

2.2 Background on Tax Planning

Hundal [2] argues that, tax planning is perceived to increase after-tax profit and enhance shareholders wealth, but tax planning represents a serious loss of revenue to governments. Minnicka and Noga [3] argue that, tax planning can lead to managerial opportunism, where opportunistic managers employ tax planning strategies to advance managerial wealth rather than shareholders. Apart from the agency problem, there are other costs (including costs directly related to tax planning, additional compliance costs, and non-tax cost) associated with tax planning activities [4]. Empirically, there have been studies conducted in both developed and developing countries that stress on the importance of reducing firm's costs and increasing after-tax profit through tax planning [5, 6].

We try to differentiate the related party transactions as four types by form of transactions, sales, purchase, non-operation, and contingent, trying to understand their relationship with tax planning. Based on the above discussion, we establish the following hypothesis:

Hypothesis 1: Tax planning is positive correlated with RPT_{total}.

Hypothesis 2: The negative relation between RPTs and tax planning.

Hypothesis 3: $RPT_{purchase}$ and tax planning are positively correlated.

Hypothesis 4: $RPT_{non\text{-}operating}$ and tax planning are positively correlated.

Hypothesis 5: Tax planning has a significant positive correlation with $RPT_{contingent}$.

3 Research Design and Empirical Study

Using 31,075 observations of Chinese listed companies of the Shenzhen and Shanghai Stock Exchange in the sampling period of 2007–2017. The data is jointly collected from the China Stock Market & Accounting Research database (CSMAR) and the Taiwan Economic Journal Database (TEJ). The choice of 2007 as the inception year is due to the fact that the China Securities Regulatory Commission imposed

Table 1 Descriptive statistics

	No	Mean	Median	S.D	Q1	Q2
TP	31,075	0.005	0.004	0.014	−0.000	0.010
RPT_{Totall}	31,075	0.459	0.032	18.042	0.001	0.174
RPT_{sales}	31,075	0.125	0.005	1.894	0.000	0.061
$RPT_{Purchase}$	31,075	0.085	0.004	2.197	0.000	0.042
$RPT_{Nonoperating}$	31,075	0.008	0.000	0.742	0.000	0.001
$RPT_{Contingent}$	31,075	0.241	0.000	17.627	0.000	0.000
SGR	31,075	3.331	0.131	347.263	−0.003	0.305
Size	31,075	14.794	14.675	1.432	13.813	15.627
CAPINT	31,075	0.227	0.192	0.171	0.093	0.325
EM	31,075	0.012	0.009	0.095	−0.032	0.054
LEV	31,075	0.084	0.036	0.190	0.008	0.119

a strict regulation to tackle the embezzlement problems of the large shareholders. Ever since, RPTs have been extensively used by listed firms. Our research further excludes financial industry from the sample due to the concern that financial firms are subjected to different regulations. Moreover, missing or incomplete data of RPTs, tax planning, financial numbers are excluded from the sample. Furthermore, we also winsorize 1% of the extreme samples.

The summary statistics of variables are reported in Table 1. The first variable of Tax Planning (hereinafter called TP) is following Wahab and Holland [5], this study measures TP as the difference between STR and ETR expressed by the following equations:

$$ETR = CTE/PBT$$

where ETR = Effective tax rates, CTE = Current tax expense, and PBT = Profit before tax. This measure of ETR has been used by a number of tax burden studies, for example, [7]. By subtracting the ETR from the STR, and multiplying the difference with PBT, the TP can be derived as in the following equation:

$$TP = PBT * (STR - ETR)$$

where TP = Tax planning and STR = China statutory main corporate tax rate. A positive value of the difference between China corporate tax rates and effective tax rates (STR–ETR) implies tax benefits arising from TP activities by profit-making firms. By Table 2, we know that the mean of TP 0.005, indicating that most firms didn't engage TP. RPT_{total} is the sum of related-party transactions revealed in CSMAR divided by total sales. The average value of it is 0.459. We define as the sum of related-party processing income, processing cost, accounts receivables, accounts payables, gains

Table 2 Regression analysis of tax planning on RPT

	Panel A					Panel B				
	(1)	(2)	(3)	(4)	(5)	(1)	(2)	(3)	(4)	(5)
Intercept	−0.0019 (−0.347)	−0.0015 (−0.298)	−0.0019 (−0.347)	−0.0002 (−0.038)	−0.0017 (−0.318)	0.0027 (0.26)	0.0048 (0.487)	−0.0012 (−0.113)	0.0007 (0.067)	0.0005 (0.049)
RPT_{Totall}	0.0000 (0.147)					0.0000 (0.804)				
RPT_{sales}		−0.0104*** (−4.089)					−0.0096*** (−4.399)			
$RPT_{purchase}$			0.0000 (0.164)					0.0000 (0.239)		
$RPT_{nonoperating}$				0.0617 (1.224)					0.0643 (1.389)	
$RPT_{contingent}$					0.0001** (2.049)					0.0007*** (3.514)
Size						−0.0003 (−0.557)	−0.0004 (−0.756)	−0.0003 (−0.523)	−0.0002 (−0.377)	−0.0002 (−0.329)
CAPINT						0.0187*** (2.909)	0.0178*** (2.863)	0.0187*** (2.909)	0.0217*** (3.209)	0.0171*** (2.716)
EM						0.0673*** (9.869)	0.0667*** (10.101)	0.0673*** (9.858)	0.0668*** (9.797)	0.0662*** (9.900)
LEV						0.0001 (0.055)	0.0005 (0.667)	0.0008 (0.923)	0.0008 (0.923)	−0.0155*** (−3.291)
SGR						0.0000 (0.216)	0.0000 (0.116)	0.0000 (0.219)	0.0000 (0.124)	0.0000 (0.264)
Industry	Yes	Yes	Yes	Yes	Yes	Yes	Yes	Yes	Yes	Yes

(continued)

Table 2 (continued)

	Panel A					Panel B				
	(1)	(2)	(3)	(4)	(5)	(1)	(2)	(3)	(4)	(5)
Year	Yes	Yes	Yes	Yes	Yes	Yes	Yes	Yes	Yes	Yes
Adj.R^2	0.101	0.151	0.101	0.105	0.114	0.334	0.376	0.332	0.337	0.361
Observations	31,075	31,075	31,075	31,075	31,075	31,075	31,075	31,075	31,075	31,075

***, **, and * denote the significance level of 1%, 5%, and 10%

and losses from assets sales, non-operation income and expenditure, divided by total sale. Moreover, we also calculate (RPT_{sales}) as the total related-party sales divided by total sales. The mean value of RPT_{Totall} is 0.125. $RPT_{purchase}$ being defined as the total related-party purchase divided by total sales is 0.085 on average. Furthermore, $RPT_{non\text{-}operating}$ being defined as related-party non-operating income minus non-operating expenditure, divided by total sales, is 0.008 on average. $RPT_{contingent}$ including guarantee on loans and project performance, is 0.241, gauged with respect to total sales. For control variables, we include Sales Growth Rate (SGR), Size, Capital intensity (CAPINT), Earnings Management (EM), and Leverage (LEV). SGR is defined as the current period's sales and subtracts the past period's sales. Divide the result by the past period's sales. The mean value of it is 3.331. Size denotes the natural logarithm of total assets is 14.794 on average. Capital Intensity is defined as ratio of gross machinery and equipment to total assets. The CAPINT of these sampling firms is 0.227. EM for earnings management is 0.012. LEV for leverage is 0.084.

4 Empirical Results

To investigate whether different types of related party transaction influence the extent of firm TP, we estimate the following model as our baseline regression.

In Table 2, we report the regression result using the univariate regression analysis in Panel A and the multiple regression analysis in Panel B. In Table 2—Panel A-(1) and Panel B-(1) The results show that RPTs in total are not significantly correlated with TP, indicating that firms with excessive use of RPTs are less likely to engage TP. The relation is not significant at 1% level, and sustainable for univariate regression analysis of RPTs and for multiple regression analysis RPTs. The result is not consistent with the prediction of Hypothesis 1, that is rejecting our null hypothesis of no association, indicating that TP is not positively correlated with RPTs. Consistent with our expectations, the coefficients on the control variables CAPINT and EM are positive. Insinuated that such companies tend to implement TP.

In reference to Clogg et al. [8], we further conduct test of inequality of the regression coefficients. Specifically, we are interested in the comparison of the impacts between RPT_{sales} and $RPT_{purchase}$, and between $RPT_{nonoperating}$ and $RPT_{contingent}$. These classifications enable us to explore differences in TP of RPTs types. These classifications enable us to explore differences in market valuation of RPTs types. The separation into types reduces the power of our tests, so caution should be exercised in interpreting each category as we have limited observations for certain types of RPT, and some firms have more than one type of RPT resulting in the effects being spread across types. In Table 3—Panel A-(2) and Panel B-(2) show RPT_{sales} documents a significant negative association between TP. The result shows that external monitoring effect reduces financial report window dressing by RPT_{sales}, and then reduces the risk of tax investigation by tax authorities. The result is consistent with the prediction of Hypothesis 2. Control variables CAPINT and EM are positive, as

Table 3 Regression analysis of tax planning on RPT

	Panel A					Panel B				
	(1)	(2)	(3)	(4)	(5)	(1)	(2)	(3)	(4)	(5)
Intercept	−0.0019 (−0.347)	−0.0015 (−0.298)	−0.0019 (−0.347)	−0.0002 (−0.038)	−0.0017 (−0.318)	0.0027 (0.26)	0.0048 (0.487)	−0.0012 (−0.113)	0.0007 (0.067)	0.0005 (0.049)
RPT_{Totall}	0.0000 (0.147)					0.0000 (0.804)				
RPT_{sales}		−0.0104*** (−4.089)					−0.0096*** (−4.399)			
$RPT_{purchase}$			0.0000 (0.164)					0.0000 (0.239)		
$RPT_{nonoperating}$				0.0617 (1.224)					0.0643 (1.389)	
$RPT_{contingent}$					0.0001** (2.049)					0.0007*** (3.514)
Size						−0.0003 (−0.557)	−0.0004 (−0.756)	−0.0003 (−0.523)	−0.0002 (−0.377)	−0.0002 (−0.329)
CAPINT						0.0187*** (2.909)	0.0178*** (2.863)	0.0187*** (2.909)	0.0217*** (3.209)	0.0171*** (2.716)
EM						0.0673*** (9.869)	0.0667*** (10.101)	0.0673*** (9.858)	0.0668*** (9.797)	0.0662*** (9.900)
LEV						0.0001 (0.055)	0.0005 (0.667)	0.0008 (0.923)	0.0008 (0.923)	−0.0155*** (−3.291)
SGR						0.0000 (0.216)	0.0000 (0.116)	0.0000 (0.219)	0.0000 (0.124)	0.0000 (0.264)
Industry	Yes	Yes	Yes	Yes	Yes	Yes	Yes	Yes	Yes	Yes

(continued)

Table 3 (continued)

	Panel A					Panel B				
	(1)	(2)	(3)	(4)	(5)	(1)	(2)	(3)	(4)	(5)
Year	Yes	Yes	Yes	Yes	Yes	Yes	Yes	Yes	Yes	Yes
Adj.R^2	0.101	0.151	0.101	0.105	0.114	0.334	0.376	0.332	0.337	0.361
Observations	31,075	31,075	31,075	31,075	31,075	31,075	31,075	31,075	31,075	31,075

***, **, and * denote the significance level of 1%, 5%, and 10%, respectively

same as previous paragraph. In Table 3—Panel A-(3), (4) and Panel B-(3), (4) show that there is no significant correlation between $RPT_{purchase}$ and $RPT_{nonoperating}$ by the evidences of our research. Therefore, we reject the prediction of Hypothesis 3 and Hypothesis 4. Finally, Table 3—Panel A-(5) and Panel B-(5) indicate $RPT_{contingent}$ documents a significantly positive association between TP. Since $RPT_{contingent}$ is a much salient manifest of wealth tunneling and critical factor of financial frauds than the other types of RPTs, especially, when external financing market is less developed that provide firms an important financing channel. The firms engaged energetical TP to transfer the financial assets by the manipulation of $RPT_{contingent}$. This empirical result supports the null hypothesis, i.e., accepting the prediction of Hypothesis 5. At the part of control variables, in addition to CAPINT and EM are positive, LEV is negative significant correlation between TP.

5 Additional Analyses

The final additional test for robustness in our research (Table 4), we note that our TP measures are constructed based on the firms' sales this fiscal year to the acquisition announcement date. Drawing a conclusion of our final test that is consistent with the above empirical conclusions. That is, the conclusion supports the aforementioned all hypothesis.

6 Conclusion

RPT and tax planning represents a prevalent form of business transaction for firms in most countries. In our study we use two different dimensions, RPTs' total amount and different types of RPTs, to explain the relationship with tax planning. In consistent with prior studies, e.g., [9–11] indicating RPT is primarily deemed a proxy of wealth expropriation or earnings management, the empirical results from a sample consisting of 31,075 firm-year observations from listed firms in China in the sampling period of 2007–2017.

Therefore, we can get a suggestion that relevant departments can strengthen the disclosure of RPTs, especially the disclosure of RPTs through or guaranty, so as to regulate RPTs, so that enterprises can consciously pay taxes on a fixed basis, tax evasion and taxation, and safeguard national interests.

Table 4 Robustness test-regression analysis of tax planning on RPT

	Panel A					Panel B				
	(1)	(2)	(3)	(4)	(5)	(1)	(2)	(3)	(4)	(5)
Intercept	−0.0019 (−0.347)	−0.0011 (−0.22)	−0.0019 (−0.347)	−0.0002 (−0.045)	−0.0017 (−0.325)	0.0019 (0.188)	0.0032 (0.323)	−0.0012 (−0.113)	0.001 (0.094)	−0.0011 (−0.109)
RPT_{Totall}	0.0000 (0.147)					0.0000 (−0.101)				
RPT_{sales}		−0.0121*** (−5.106)					−0.0095*** (−4.596)			
$RPT_{purchase}$			0.0000 (0.157)					0.0000 (0.231)		
$RPT_{nonoperating}$				0.1737 (1.051)					0.1687 (1.104)	
$RPT_{contingent}$					0.0016* (1.762)					0.0097** (2.029)
Size						−0.0003 (−0.498)	−0.0003 (−0.618)	−0.0003 (−0.523)	−0.0002 (−0.406)	−0.0002 (−0.304)
CAPINT						0.0187*** (2.909)	0.0173*** (2.795)	0.0187*** (2.909)	0.021*** (3.115)	0.0178*** (2.782)
EM						0.0673*** (9.849)	0.0633*** (9.543)	0.0673*** (9.858)	0.0668*** (9.78)	0.0672*** (9.911)
LEV						0.0008 (0.922)	0.0009 (1.116)	0.0008 (0.923)	0.0008 (0.942)	−0.0094* (−1.845)
SGR						0.0000 (0.221)	0.0000 (0.213)	0.0000 (0.219)	0.0000 (0.068)	0.0000 (0.239)
Industry	Yes	Yes	Yes	Yes	Yes	Yes	Yes	Yes	Yes	Yes

(continued)

Table 4 (continued)

	Panel A					Panel B				
	(1)	(2)	(3)	(4)	(5)	(1)	(2)	(3)	(4)	(5)
Year	Yes	Yes	Yes	Yes	Yes	Yes	Yes	Yes	Yes	Yes
Adj.R^2	0.101	0.177	0.101	0.104	0.110	0.332	0.380	0.332	0.335	0.342
Observations	31,075	31,075	31,075	31,075	31,075	31,075	31,075	31,075	31,075	31,075

***, **, and * denote the significance level of 1%, 5%, and 10%, respectively

References

1. Ryngaert, M., Thomas, S.: Not all related party transactions (RPTs) are the same: ex-ante versus ex-post RPTs. J. Account. Res. **50**, 845–882 (2012)
2. Hundal, S.: Why Tax Avoidance is Among the Biggest Issues of our Generation. Liberal Conspiracy (2011)
3. Minnick, K., Noga, T.: Do vorporate governance characteristics influence tax management? J. Corp. Finance **16**(5), 703–718 (2010)
4. Zemzem, A., Khaoula, F.: Moderating effects of board of directors on the relationship between tax planning and bank performance: evidence from Tunisia. Eur. J. Bus. Manag. **5**(32) (2013)
5. Wahab, A., Holland, K.: Tax planning, corporate governance and equity value. Br. Account. Rev. **44**(2), 111–124 (2013)
6. Iyoha, F., Oyerinde, D.: Accounting infrastructure and accountability in the management of public expenditure in developing countries: a focus on Nigeria. Crit. Perspect. Account. **21**(5), 361–373 (2010)
7. Dyreng, S., Hanlon, M., Maydew, E.: Long-run corporate tax avoidance. Account. Rev. **83**, 61–82 (2008)
8. Clogg, C.C., Petkova, E., Haritou, A.: Statistical methods for comparing regression coefficients between models. Am. J. Sociol. **100**(5), 1261–1293 (1995)
9. Friedman, E., Johnson, S., Mitton, T.: Propping and tunneling. J. Comp. Econ. **31**(4) (2003)
10. Jian, M., Wong, T.J.: Propping through related party transactions. Rev. Acc. Stud. **15**(1), 70–105 (2010)
11. Aharony, J., Wang, J., Yuan, H.: Tunneling as an incentive for earnings management during the IPO process in China. J. Account. Public Policy **29**(1), 1–26 (2010)

Mind-Media System: A Consumer-Grade Brain-Computer Interface System for Media Applications

Chang Liu, Yijie Zhou, and Dingguo Yu

Abstract Brain science has made many achievements in the media area. However, the brain-computer interface (BCI) is not yet available to the general public because of the cost of the equipment. This paper investigates a consumer-grade brain-computer interface system for media applications, which only cost $45. By the electroencephalogram (EEG) processing method studied in this paper, the EEG signals corresponding to different commands can be identified so that subjects could control the media by their mind without any body movements. The validation experiment on music players demonstrates the effectiveness of our mind-media system.

1 Introduction

Nowadays, with the development of brain science, it has made many achievements in image quality assessment and media content generation [1–6]. As a practical technology in brain science, brain-computer interface (BCI) is attracting the attention of experts in the field of media and has a positive impact. BCI is a communication technology that directly connects the human or animal brain and external devices to realize the information exchange between the brain and devices [7–9]. The electroencephalogram (EEG) cap is currently the most popular research device because it is economical and portable [9, 10]. However, it is still too expensive for daily use. For example, the Epoc X, consumer-grade EEG cap produced by Emotiv costs $849

C. Liu · Y. Zhou · D. Yu (✉)
Institute of Intelligent Media Technology, Communication University of Zhejiang, Hangzhou, China
e-mail: yudg@cuz.edu.cn

C. Liu
e-mail: changliu@cuz.edu.cn

Y. Zhou
e-mail: yjzhou@cuz.edu.cn

D. Yu
School of Media Engineering, Communication University of Zhejiang, Hangzhou, China

© The Author(s), under exclusive license to Springer Nature Singapore Pte Ltd. 2022
J.-F. Zhang et al. (eds.), *Advances in Intelligent Systems and Computing*, Smart Innovation, Systems and Technologies 268, https://doi.org/10.1007/978-981-16-8048-9_8

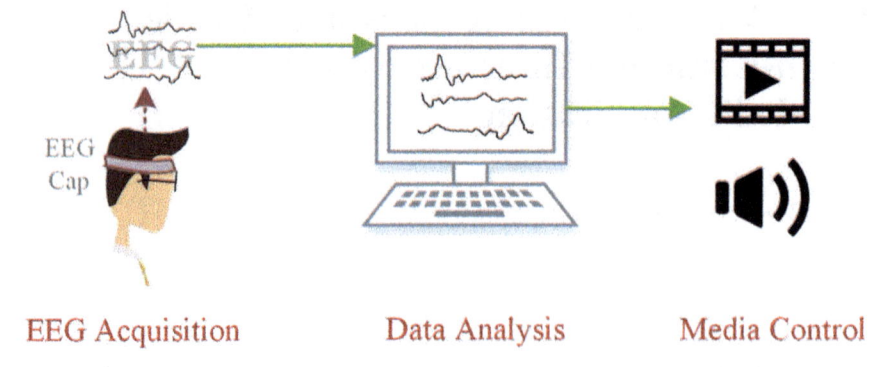

Fig. 1 Overview of mind-media system

[11]. It limits the popularity of this new media interaction among the general public. Consequently, we will explore an affordable way of brain-computer interaction and apply it to media interaction called Mind-Media System.

Traditional EEG caps are usually equipped with electrodes that cover the whole brain, which makes them expensive. In fact, brain regions do their work to a certain extent, so it is possible to reduce electrodes by designing a specific paradigm to evoke EEG signals in specific brain regions. On that basis, we adopted the TGAM (ThinkGear AM) produced by Neurosky [12, 13] to construct our Mind-Media System, and the scheme of our system design is shown in Fig. 1. In this paper, EEG data was collected by TGAM and analyzed in real-time to output command to control the media.

2 Method and Material

2.1 Hardware Design

The EEG headband is composed of TGAM, Bluetooth, a dry EEG electrode, and two reference electrodes. The EEG electrode is used to measure EEG signals in the frontal lobe, and the reference electrodes are placed at each ear. The EEG signals are transmitted via Bluetooth to a computer for real-time processing and control of the media (Fig. 2).

2.2 EEG Recognition Method

In this paper, a spectrum energy analysis method was investigated to identify the intention of the subject. As shown in Eq. (1), the feature F of each EEG data unit

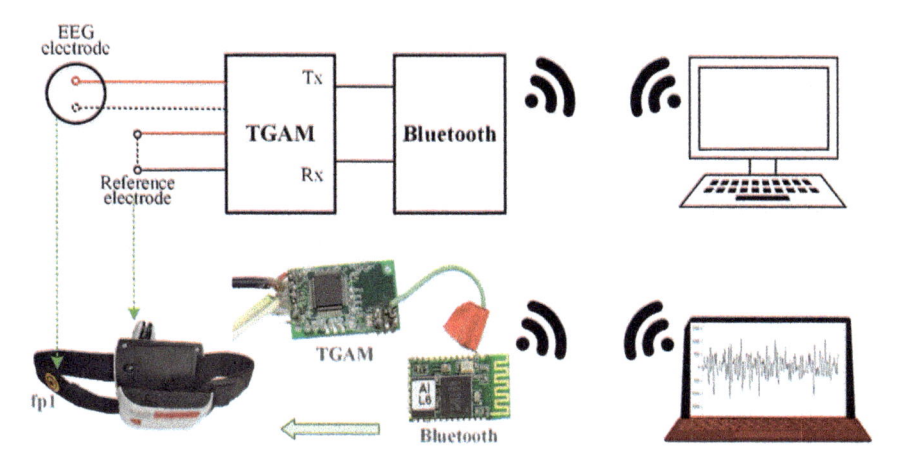

Fig. 2 Hardware implementation of mind-media system

$\{X(i), i = 1, \ldots, L\}$ recorded by EEG cap is extracted, where $L = T_s \times f_s$ is the data length, f_s is the sampling rate, T_N is usually set to 4~6, m is the independent variable in linear regression, Y is the dependent variable in linear regression, k is the slope of linear regression, and b is the intercept of linear regression.

$$\begin{cases} m = \{1, 2, \cdots, \dfrac{T_s \times f_s}{T_N}\} \\ Y(m) = \sum_{i=1}^{T_N(m-1)+T_N} |X(i)| \\ F = \arg\min_{k} \|Y - km - b\| \end{cases} \quad (1)$$

The feature F is recognized by support vector machine [14, 15] (SVM) (train: test = 7:3, linear kernel, 30 runs), so as to determine whether there is a significant change in the unique frequency EEG band.

3 Validation and Discussion

In order to validate the effectiveness of our system, we instantiated a media control task, i.e., a music player control task. In our system, we associate the energy in the delta band with the stop/play command, the energy in the beta band with the cut command, and the attention level with volume. In Fig. 3, the spectrum analysis of EEG signals is shown in the blue bar, and the difference of EEG spectrum in different control modes is significant. Furthermore, the distribution of the feature of EEG data extracted by Eq. (1) is shown in Fig. 4, and it is obvious that the features of EEG

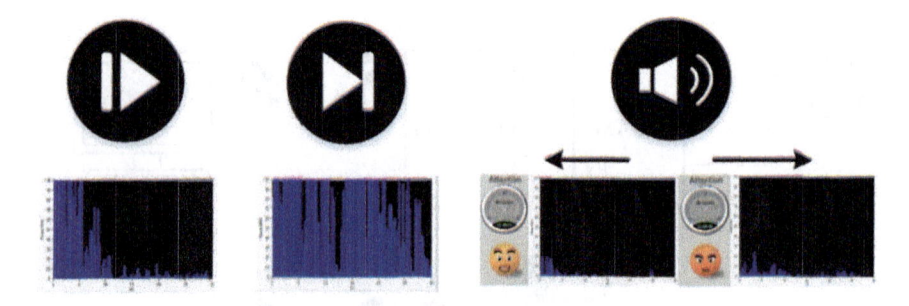

Fig. 3 Music player controlled by EEG signals

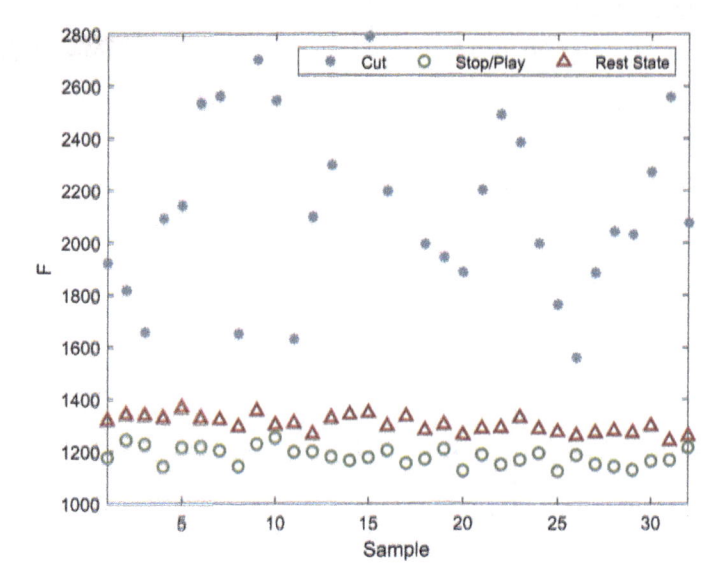

Fig. 4 The feature distribution of different commands

signals vary a great deal from command to command. Consequently, the intention of the subject can be easily distinguished by EEG signals.

The implementation of the Mind-Media system is shown in Fig. 5, and the subjects could control the music player by their mind. They play or stop the music by clicking their eyes and cut the music by gritting. Moreover, if they want to tune up the volume, they need to concentrate their attention while relaxing to cut down the volume. The validation experiment proves that the Mind-Media system proposed in this paper can complete the task of controlling the music player.

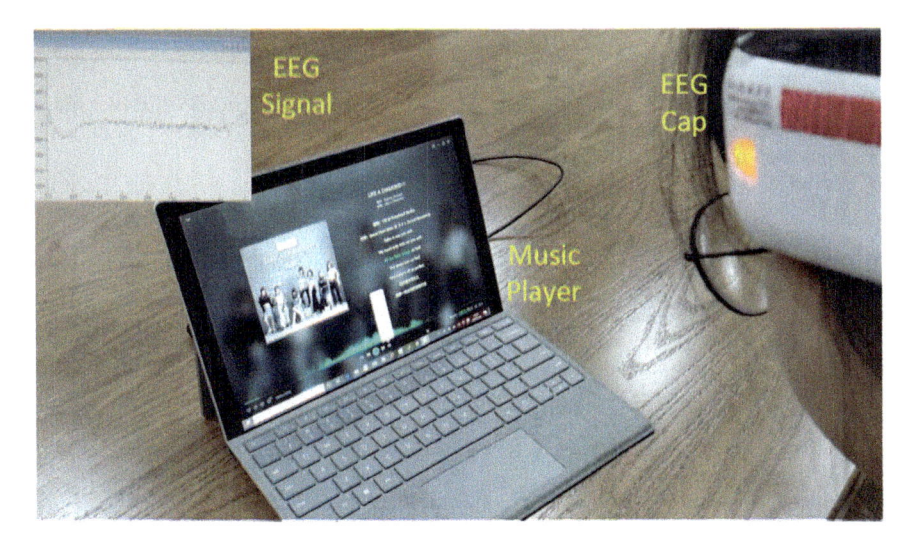

Fig. 5 The implementation of the mind-media system

4 Conclusion

In this paper, we investigated a consumer-grade brain-computer interface system for media applications, which only cost $45. To some extent, this work solves the problem that BCI technology is difficult to be applied in daily life. By the feature extracted method proposed in this paper, the EEG signals corresponding to different commands can be identified so that subjects could control the media by their mind without any body movements. The validation experiment on music players demonstrates the effectiveness of our mind-media system, which is a beneficial attempt of brain-computer interface technology in the field of intelligent media.

References

1. Scholler, S., Bosse, S., Treder, M.S., Blankertz, B., Curio, G., Muller, K., Wiegand, T.: Toward a direct measure of video quality perception using EEG. IEEE Trans. Image Process. **21**, 2619 (2012)
2. Blankertz, B., Acqualagna, L., Dahne, S., Haufe, S., Schultzekraft, M., Sturm, I., Uscumlic, M., Wenzel, M., Curio, G., Muller, K.: The Berlin brain-computer interface: progress beyond communication and control. Front. Neurosci. **10**, 530 (2016)
3. Kamitani, Y., Tong, F.: Decoding the visual and subjective contents of the human brain. Nat. Neurosci. **8**, 679 (2005)
4. Rashkov, G., Bobe, A., Fastovets, D., Komarova, M.: Natural image reconstruction from brain waves: a novel visual BCI system with native feedback. bioRxiv 787101 (2019)
5. Ponce, C.R., Xiao, W., Schade, P.F., Hartmann, T.S., Kreiman, G., Livingstone, M.S.: Evolving images for visual neurons using a deep generative network reveals coding principles and neuronal preferences. Cell **177** (2019)

6. Shen, G., Horikawa, T., Majima, K., Kamitani, Y.: Deep image reconstruction from human brain activity. PLoS Comput. Biol. **15** (2019)
7. Liu, C., Xie, S., Xie, X., Duan, X., Meng, Y.: Design of a video feedback SSVEP-BCI system for car control based on improved MUSIC method. In: International Conference on Brain and Computer Interface. GangWon, South Korea, p. 1 (2018)
8. Edelman, B.J., Meng, J., Suma, D., Zurn, C., Nagarajan, E., Baxter, B.S., Cline, C.C., He, B.: Noninvasive neuroimaging enhances continuous neural tracking for robotic device control. Sci. Robot. **4**, w6844 (2019)
9. Zhang, M., Tang, Z., Liu, X., Van der Spiegel, J.: Electronic neural interfaces. Nat. Electron. **3**, 191 (2020)
10. Klimesch W.: EEG alpha and theta oscillations reflect cognitive and memory performance: a review and analysis. Brain Res Brain Res Rev. 29. 169 (1999).
11. EPOC X. https://www.emotiv.com/product/emotiv-epoc-x-14-channel-mobile-brainwear/
12. EEG hardware platforms. http://neurosky.com/biosensors/eeg-sensor/biosensors/
13. Liu, C., Yu, D., Zhang, J., Xie, S.: A utility human machine interface using low cost EEG cap and eye tracker. In: 2021 9th International Winter Conference on Brain-Computer Interface (BCI), vol. 1. IEEE, (2021)
14. Cortes, C., Vapnik, V.: Support-vector networks. Mach. Learn. **20**, 273 (1995)
15. Chen, X., Peng, X., Li, J., Peng, Y.: Overview of deep kernel learning based techniques and applications. J. Netw. Intell. **1**, 83 (2016)

Artificial Intelligence

A Cooperative Evolution Framework Based on Fish Migration Optimization

Wenqi Li, Shu-Chuan Chu, and Jeng-Shyang Pan

Abstract The Fish Migration Optimization (FMO) method was inspired by the fish swim and migration behaviors and has been proofed to be a brilliant algorithm for solving numerical optimization problems. However, FMO still has the problems of premature convergence, search stagnation and easy to fall into local optimum. In this paper, a cooperative evolution framework based on fish migration optimization (CEFMO) is proposed. By dividing the whole swarm into several subsets and introducing an evaluation function, at the end of each iteration, all the individuals are evaluated and when the evaluation result meets the conditions, the cooperative evolution is triggered. In order to verify the performance of the proposed algorithm, CEFMO was compared with serval iconic swarm intelligence algorithms, such as Particle Swarm Optimization (PSO), Whale Optimization Algorithm (WOA) and Black Hole (BH) algorithm under CEC-2013 benchmark function and the result shows that the CEFMO is slightly better than the compared algorithms.

1 Introduction

Intelligence computing, as a significant branch of artificial intelligence, has been developed rapidly in the past decades. It is inspired by the swarm intelligence movement that existed in nature [1]. As a meta-heuristic optimization, intelligence computing has been successfully applied in various fields of the engineering community [2, 3]. The scholars in the field have proposed numerous inspired by the behavior of animals and social events and physical phenomena [4]. In various existing algorithms, each of them initializes a random set of initial solutions throughout the search space, then these solutions use their unique moving, combination and evolution

---

W. Li · J.-S. Pan (✉)
Dalian Maritime University, Dalian 116026, China

W. Li
e-mail: lwq921030@dlmu.edu.cn

S.-C. Chu · J.-S. Pan
Shandong University of Science and Technology, Qingdao 266000, China

schemes of each algorithm to iterate until the end of the iteration and obtain the optimal solution. And finally, the newly proposed algorithm will be tested under different test suites to prove the performance of the algorithm.

The common techniques of soft computing include fuzzy logic, neural network, probability reasoning and meta-heuristic [5]. Meta-heuristic strategy, first proposed in 1960s, is usually a general heuristic strategy, they usually do not rely on the specific conditions of a problem, so they can be applied to a wider range of aspects. Meta-heuristic strategy usually puts forward some requirements for the search process, and then the heuristic algorithm is implemented according to these requirements. Meta-heuristic algorithms can obtain optimal solutions without having any specific requirements [6]. In the field of intelligence computing, swarm intelligence (SI) and evolutionary algorithms (EAs) [7] are two kinds of meta-heuristic algorithms. It's worth mentioning that SI comes from the study of swarm behavior of social insects such as birds, ants and bees. For example, the basic concept of Particle Swarm Optimization (PSO) [8–10], one of the most classic SI algorithms, comes from the study of foraging behavior of birds. Ant Colony Optimization (ACO) was proposed in 2006 and inspired by the behavior of ants finding paths in the process of searching for food [11, 12]. In reference [13], an Artificial Bee Colony algorithm (ABC) [14] is proposed by simulating the honey collection process of bees. The whale Optimization Algorithm [15] was proposed by simulating the hunting behavior of humpback whales. Cat Swarm Optimization (CSO) [16, 17] is a global optimization proposed by observing the behavior of cats. The Gray Wolf Optimization (GWO) [18] was proposed in 2014 by summarizing the social behavior and hunting behavior of wolves. Black Hole (BH) [25, 26] algorithm is a new heuristic algorithm that is inspired by the black hole phenomenon.

Besides SI algorithms, the inspiration of EAs comes from the phenomena of nature and the evolution of organisms, such as Genetic Algorithm (GA) [19, 20], Differential Evolution (DE) [21], Quasi-Affine Transformation Evolutionary (QUATRE) [22, 23], Multi-Universe Optimization (MVO) [24] and so on. The GA is designed and put forward according to the evolution law of organisms in nature. It is a computational model simulating the natural selection and genetic mechanism of Darwin's theory of biological evolution. It is a method to search the optimal solution by simulating the natural evolution. DE is a stochastic model simulating biological evolution. On the basis of GA, DE retains the global search strategy based on population and reduces the complexity of genetic operation. As an improvement of DE algorithm, QUATRE applies the crossover operation of DE to matrix. MVO is a global optimization inspired by nature phenomena, using white hole, black hole, and wormhole to perform exploration, exploitation, and local search.

In particular, Fish Migration Optimization (FMO) [27], as one of the SI algorithms, was first proposed in 2010, by observing the behaviors of graylings, a kind of migratory fish. It simulated the swim process, growth process and the migration process of graylings. As an excellent and efficient optimization algorithm, the formulas of FMO are based on bioscience. Compared with other meta-heuristic algorithms, FMO has a better performance. However, like most metaheuristics, FMO still has the problems of premature convergence, search stagnation, and the tendency to fall into local optimum.

In order to optimize the inadequacies of FMO algorithm, this manuscript proposed a Cooperative Evolution Framework Based on Fish Migration Optimization (CEFMO). The CEFMO draws on the experience of cooperative evolution. By splitting the initial individuals set into four subsets without intersection, and according to a certain rule, the individuals in the current solution space are evaluated after each iteration, and the individuals in the corresponding subset are selected according to the evaluation results to guide the evolution of the whole population. The rest parts of the manuscript are organized as follows. Section 2 describes the related works of the manuscript, including the original FMO algorithm and the concept of cooperative evolution. Section 3 shows the detail of the proposed CEFMO algorithm. The test results on CEC-2013 benchmark functions are shown in Sect. 4. And finally, Sect. 5 is the conclusion of the whole manuscript.

2 Related Works

2.1 Fish Migration Optimization Algorithm

Fish migration Optimization (FMO) algorithm [27] was first proposed by Pan et al. in 2010 by simulating the swim and migration behaviors of the grayling. Based on the observation of the graylings, biologists have come to the conclusion that due to food and natural enemies, some graylings will migrate to spawn, while others will continue to grow. Thus, in FMO algorithm, assume the life cycle of graylings are 4 years and the whole lifetime of the graylings are divided into 5 stages and constitutes the life cycle graph is shown in Fig. 1 and introduced as follows.

The symbols S_1, S_2, S_3, S_4 represent the survival rates of each stage and F_2, F_3, F_4 denote the fecundity rates in each stage, respectively. The graylings in each stage have different survival rates and fecundity rates where the survival rates are set to be $S_1 > S_2 > S_3 > S_4$. And on the contrary, the fecundity rates are set to be $F_2 < F_3 < F_4$ since the fecundity rate increases with the increase in age. The value of F_4 is set to be 1.0 because the FMO algorithm assumed that all the graylings must return to their

Fig. 1 Life cycle graph of the graylings

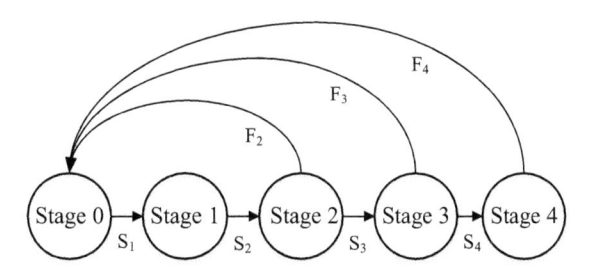

birthplace at stage 4 and F_1 is set to be 0 due to the fishes in stage 1 are not natural and have no capacity for reproducing the offspring according to research [27].

The FMO algorithm simulates the swim and migrate process while swim process is for finding best position and migrate process is for generating new individuals. The distance of the swim process follows Eq. 1, and the whole population update their position according to Eq. 2.

$$d_{offset} = \frac{E_r \cdot U_s}{a + b \cdot U_s^x} \tag{1}$$

$$P_{new} = P_{old} + d_{offset} \tag{2}$$

where E_r represents the energy on each dimension, U_s denotes the swimming speed, d_{offset} is the swim distance, a, b, x are three constants, respectively. And based on the findings in [28], the constants are set to be $a = 2.25$, $b = 36.2$ and $x = 2.23$.

As the fishes grow, some of them will migrate. The fecundity rates are set to be $F_2 = 0.05$, $F_3 = 0.10$ and $F_4 = 1$, respectively. The position of newly born fishes would be updated according to Eq. 3 and the velocity will be updated according to Eq. 4.

$$P = (d_{max} - d_{min}) \cdot rand + d_{min} \tag{3}$$

$$U = \begin{cases} \pi \cdot U_s, & F_P < F_{P_{best}} \\ U_s, & otherwise \end{cases} \tag{4}$$

where d_{max} and d_{min} denote the maximum and maximum values of all dimensions of the fish, $rand$ is a random number between 0 and 1, F_P is the fitness value of the current individual and $F_{P_{best}}$ is the best fitness values observed so far, respectively.

2.2 Cooperative Evolution

Cooperative evolution is a method for solving large-scale optimization problems, which adopts the strategy of "divide and rule". For an optimization problem, it is decomposed into several groups of problems according to variables and optimized by groups, and each group cooperates to complete the optimization of the whole problem. The complex problem is decomposed into subproblems, which are solved in evolutionary subpopulations. The evaluation of individuals depends on the cooperation among subpopulations, and a complete solution is obtained by the combination of representative individuals of each subpopulation. The fitness of individuals in subpopulations is evaluated by their participation in the complete solution.

3 Proposed CEFMO Algorithm

In this section, a Cooperative Evolution framework based on FMO (CEFMO) algorithm is introduced in detail. The cooperative evolution framework can be divided into two stages, the population evaluation stage and the coevolution stage. Unlike other parallel algorithms [29], CEFMO can evaluate the whole population to decide whether cooperative evolution is needed among the subpopulations hence it can get a better performance.

3.1 Population Evaluation Stage

Like FMO, CEFMO first randomly initializes the population set P. Then the initial population is divided into four subsets without intersection P_1, P_2, P_3, P_4, shown in Eqs. 5 and 6.

$$P = P_1 \cup P_2 \cup P_3 \cup P_4 \tag{5}$$

$$\forall i, j \in [1, 4], i \neq j, P_i \cap P_j = \varnothing, \tag{6}$$

After each iteration, the evaluation function, shown in Eq. 7, is used to evaluate the whole population.

$$EF^t\left(P_t^{'}, P_t^*\right) = \frac{\sum_{x \in P_t^*} min_{y \in P_t^{'}} dis(x, y)}{|P_t^*|} \tag{7}$$

where P_t^* is the set of individuals with the best fitness value in each subset in the *t-th* iteration, $P_t^{'}$ represents the set of the rest individuals in each subset in the *t-th* iterator. And the function *dis (x, y)* represents the Euclidean distance between *x* and *y*.

The value of EF^t represents the convergence and diversity of the current population. Therefore, after each iterator, EF^t will be compared with the evaluation function value of the previous generation EF^{t-1}. If $EF^t > EF^{t-1}$, the next iteration will trigger coevolution, otherwise, the four subsets go straight to the next iteration.

3.2 Coevolution Stage

In the stage of coevolution, we will have to choose some individuals to guide the whole population to evolve because the subset is supposed to jump out of local optimum by introducing the best members from other subsets [30]. Therefore, we

Fig. 2 Subset 1 has the best solution of all population, subset 2, 3, 4 use their best solution to replace the worst individuals in subset 1

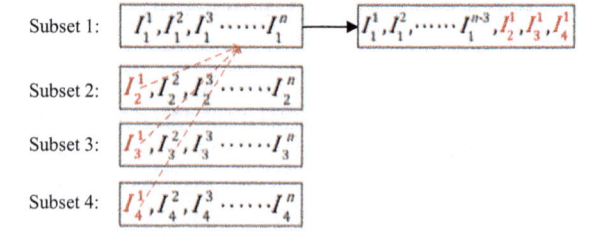

Subset 1: $I_1^1, I_1^2, I_1^3 \cdots I_1^n$ → $I_1^1, I_1^2, \cdots I_1^{n-3}, I_2^1, I_3^1, I_4^1$

Subset 2: $I_2^1, I_2^2, I_2^3 \cdots I_2^n$

Subset 3: $I_3^1, I_3^2, I_3^3 \cdots I_3^n$

Subset 4: $I_4^1, I_4^2, I_4^3 \cdots I_4^n$

Fig. 3 Subset 1 doesn't have the best solution of all population, subset 1 use its best solution to replace the worst individuals in subset 2, 3, 4

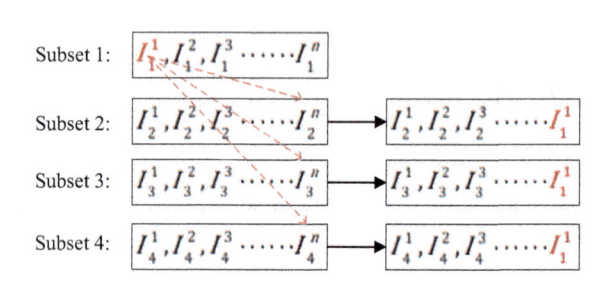

Subset 1: $I_1^1, I_1^2, I_1^3 \cdots I_1^n$

Subset 2: $I_2^1, I_2^2, I_2^3 \cdots I_2^n$ → $I_2^1, I_2^2, I_2^3 \cdots I_1^1$

Subset 3: $I_3^1, I_3^2, I_3^3 \cdots I_3^n$ → $I_3^1, I_3^2, I_3^3 \cdots I_1^1$

Subset 4: $I_4^1, I_4^2, I_4^3 \cdots I_4^n$ → $I_4^1, I_4^2, I_4^3 \cdots I_1^1$

introduce the Mean Square Error (MSE) as the measurement standard. $fitness_S^{(i,t)}$ is one of a fitness values, where S is the S-th subset, i represents the i-th solution in S-th subset, t denotes current iteration. MSE_S^t is the mean square error in S-th subset of t-th iteration and defined as Eq. 8.

$$MSE_S^t = \frac{1}{n}\sum_{i=1}^{n}(\hat{y}_i - y_i)^2 \qquad (8)$$

Equation 8 means that the smaller MSE_S^t is, the smaller difference in S-th subset is, vice versa. Therefore the coevolution stage is defined as Eqs. 9 and 10.

$$S_a^{best,t} \rightarrow S_b^{worst,t}, a \neq b, a \in [1, 4] \qquad (9)$$

$$S_b^{best,t} \rightarrow S_a^{worst,t}, a \neq b, a \in [1, 4] \qquad (10)$$

where $S_a^{best,t}$ is the solution with the best fitness value in a-th subset in current iteration and $S_b^{worst,t}$ is the solution with the worst fitness value in b-th subset in current iteration. Equation 9 means that if the b-th subset has the best solution of all population and the lowest *MSE*, then other subsets use their best solution to replace the worst individuals in subset b, and Eq. 10 means that if the b-th subset doesn't have the best solution of all population but the lowest MSE, the best solution in b-th subset is used to replace the worst individual in other subsets. Assume that the subset 1 has the lowest *MSE*, the coevolution stage graph is shown in Figs. 2 and 3.

4 Simulation Result and Discussion

In order to test the performance of the proposed algorithm, CEC-2013 benchmark function is used to compare the performance of CEFMO, PSO, WOA and BH. FMO doesn't participate in the comparison because of the poor performance under CEC-2013. The f_1 to f_5 in CEC-2013 are unimodal functions, using to test the convergence rate. The f_6 to f_{20} are multimodal functions for testing the ability of avoiding local optimal. The f_{21} to f_{28} are composition functions for testing the comprehensive performance. The simulation results are shown in Fig. 4.

Figure 4 shows that the search ability of CEFMO is better than the others and has a better convergence rate and most of the optimal values are observed before 300 iterations on unimodal functions, and the proposed algorithm observed six optimal values, on multimodal functions due to the use of cooperation evolution framework and the partition of population set, the algorithm can continuously jump out of the local optimum while ensuring the convergence speed. The CEFOM observed 5 optimal values on composition functions, the most of the four algorithms, due to the use of evaluation function guiding the whole population.

5 Conclusion

In this manuscript, a cooperation evolution framework is proposed based on FMO algorithm, called CEFMO. The initial population set is divided into several subsets, and the coevolution strategy is introduced. The evaluation function EF is used to evaluate the current population, and the cooperation evolution strategy is selected according to the evaluation results and MSE of each subset, so that CEFMO can avoid local optimization while ensuring the convergence speed and the performance of the algorithm. Finally the performance of the proposed CEFMO is tested under CEC-2013 benchmark functions and compared with three famous SI algorithms. The simulation results show that the proposed CEFMO has a better performance in comparison.

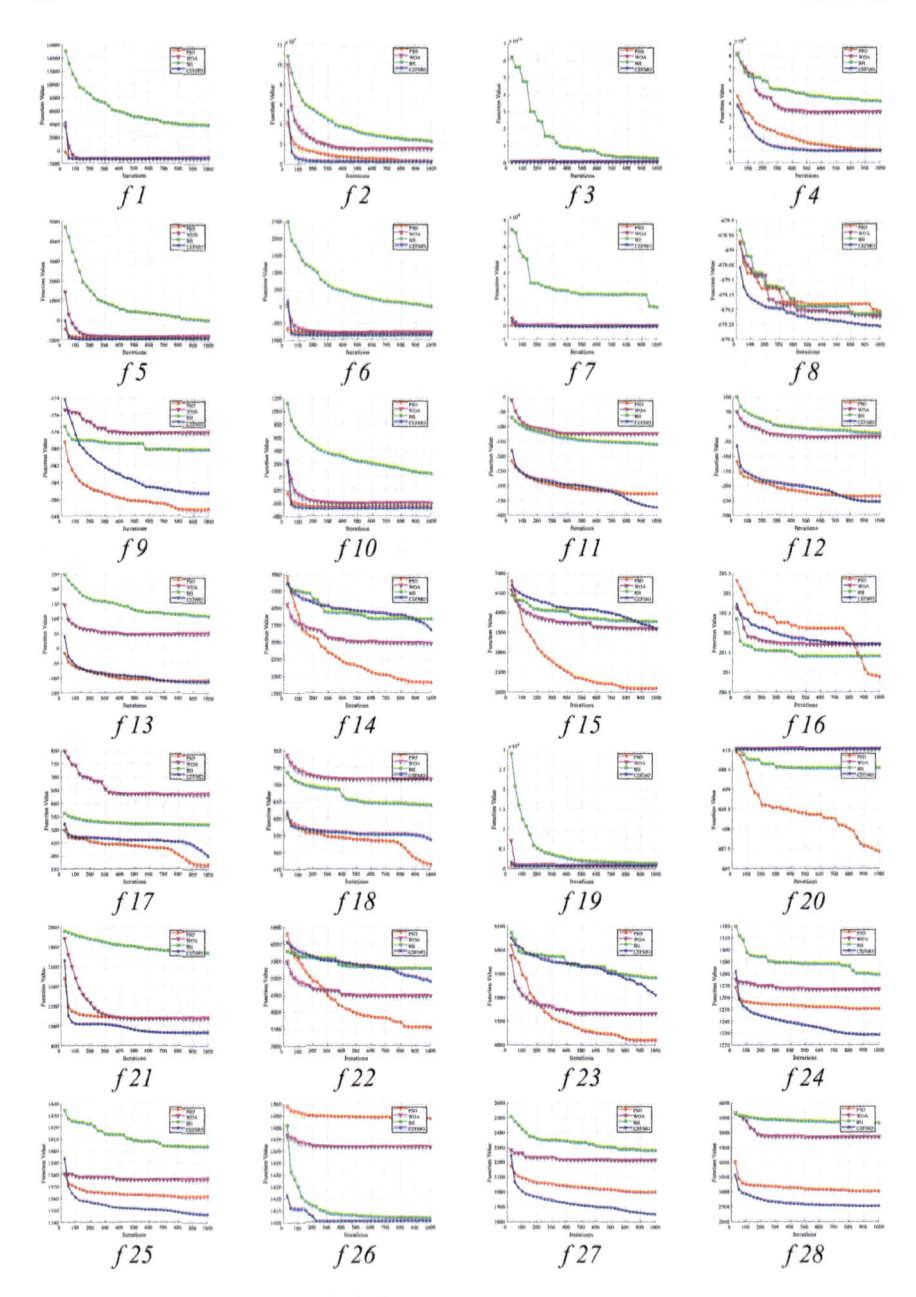

Fig. 4 Simulation Results of CEC-2013

Acknowledgment This work is supported by the National Nature Science Foundation of China (No.61772102).

References

1. Chai, Q.W., Chu, S.C., Pan, J.S., Zheng, W.M.: Applying adaptive and self assessment fish migration optimization on localization of wireless sensor network on 3-D terrain. J. Inform. Hiding Multimedia Signal Process. **11**(2), 90–102 (2020)
2. Das, S., Suganthan, P.N.: Differential Evolution: A Survey of the State-of-the-Art. IEEE Trans. Evol. Comput. **15**(1), 4–31 (Feb. 2011)
3. Mahdavi, S., Shiri, M.-E., Rahnamayan, S.: Metaheuristics in large-scale global continues optimization: a survey. Inform. Sci. **295**, 407–428 (2015)
4. Wang, X., Pan, J., Chu, S.: A parallel multi-verse optimizer for application in multilevel image segmentation. IEEE Access **8**, 32018–32030 (2020)
5. Guo, B., Zhuang, Z., Pan, J.-S., Chu, S.-C.: Optimal design and simulation for pid controller using fractional-order fish migration optimization algorithm. IEEE Access **9**, 8808–8819
6. Xue, X., Chen, J., Pan, J.-S.: Evolutionary Algorithm based Ontology Matching Technique. Science Press (2018)
7. Sallam, K.M., Elsayed, S.M., Chakrabortty, R.K., Ryan, M.J.: Evolutionary framework with reinforcement learning-based mutation adaptation. IEEE Access **8**, 194045–194071 (2020)
8. Poli, R., Kennedy, J., Blackwell, T.: Particle swarm optimization. Swarm Intell. **1**(1), 33–57 (2007)
9. Chang, K.-C., Zhou, Y.-W., Wang, H.-C., Lin, Y.-C., Chu, K.-C., Hsu, T.-L., Pan, J.-S.: Study of pso optimized bp neural network and smith predictor for mocvd temperature control in 7 nm 5g chip process. International Conference on Advanced Intelligent Systems and Informatics, pp. 568–576. Springer (2020)
10. Qin, S., Sun, C., Zhang, G., He, X., Tan, Y.: A modified particle swarm optimization based on decomposition with different ideal points for many-objective optimization problems. Complex Intell. Syst., 1–12 (2020)
11. Uthayakumar, J., Metawa, N., Shankar, K., Lakshmanaprabu, S.: Financial crisis prediction model using ant colony optimization. Int. J. Inf. Manage. **50**, 538–556 (2020)
12. Dorigo, M., Birattari, M., Stutzle, T.: Ant colony optimization. IEEE Comput. Intell. Mag. **1**(4), 28–39 (2006)
13. Karaboga, D., Basturk, B.: A powerful and efficient algorithm for numerical function optimization: artificial bee colony (ABC) algorithm. J. Global Optim. **39**(3), 459–471 (2007)
14. Tang, L., Li, Z., Pan, J., Wang, Z., Ma, K., Zhao, H.: Novel artificial bee colony algorithm based load balance method in cloud computing. J. Inf. Hiding Multimed. Sig. Process **8**(2), 460–467 (2017)
15. Mirjalili, S., Lewis, A.: The whale optimization algorithm. Adv. Eng. Softw. **95**, 51–67 (2016)
16. Chu, S.-C., Tsai, P.-W., Pan, J.-S.: Cat swarm optimization. In Pacific Rim international conference on artificial intelligence, pp. 854–858. Springer (2006)
17. Ji, X.-F., Pan, J.-S., Chu, S.-C., Hu, P., Chai, Q.-W., Zhang, P.: Adaptive cat swarm optimization algorithm and its applications in vehicle routing problems. Math. Problems Eng. (2020)
18. Mirjalili, S., Mirjalili, S.M., Lewis, A.: Grey wolf optimizer. Adv. Eng. Softw. **69**, 46–61 (2014)
19. Whitley, D.: Ageneticalgorithmtutorial. Statisticsandcomputing **4**(2), 65–85 (1994)
20. Weng, C.-J., Liu, S.-J., Pan, J.-S., Liao, L., Zeng, W.-D., Zhang, P., Huang, L. et al.: Enhanced secret hiding mechanism based on genetic algorithm. In: Advances in Intelligent Information Hiding and Multimedia Signal Processing, pp. 79–86. Springer (2020)
21. Price, K.V.: Differential evolution. In: Handbook of Optimization, pp. 187–214. Springer (2013)

22. Du, Z.-G., Pan, J.-S., Chu, S.-C., Luo, H.-J., Hu, P.: Quasi-affine transformation evolutionary algorithm with communication schemes for application of rssi in wireless sensor networks. IEEE Access **8**, 8583–8594 (2020)
23. Meng, Z., Pan, J.-S., Xu, H.: QUasi-Affine TRansformation Evolutionary (QUATRE) algorithm: a cooperative swarm based algorithm for global optimization. Knowl.-Based Syst. **109**, 104–121 (2016)
24. Mirjalili, S., Mirjalili, S.-M., Hatamlou, A.: Multi-verse optimizer: a nature-inspired algorithm for global optimization. Neural Comput. Appl. **27**(2), 495–513 (2016)
25. Hatamlou, A.: Black hole: A new heuristic optimization approach for data clustering. Inf. Sci. **222**, 175–184 (2013)
26. Pan, J.-S., Chai, Q.-W., Chu, S.-C., Wu., Ning: 3-d terrain node coverage of wireless sensor network using enhanced black hole algorithm. Sensors **20**(8), 2411 (2020)
27. Pan, J.-S., Tsai, P.-W., Liao, Y.-B.: Fish migration optimization based on the fishy biology. In: 2010 Fourth International Conference on Genetic and Evolutionary Computing, pp. 783–786. IEEE (2010)
28. Brodersen, J., Nilsson, P.A., Ammitzbøll, J., Hansson, L.A., Skov, C., Bronmark, C.: Optimal swimming speed in head currents and effects on distance movement of winter-migrating fish. PloS ONE **3**(5), e2156 (2008)
29. Chang, J.-F., Chu, S.-C., Roddick, J.-F., Pan, J.-S.: A parallel particle swarm optimization algorithm with communication strategies. J. Inf. Sci. Eng. **21**, 809–818 (2005)
30. Jiang, T.-B., Chu, S.-C., Pan, J.-S.: Parallel charged system search algorithm for energy management in wireless sensor network. 2020 2nd International Conference on Industrial Artificial Intelligence (IAI), Shenyang, China, pp. 1–6 (2020)

Improving K-Means with Harris Hawks Optimization Algorithm

Li-Gang Zhang, Xingsi Xue, and Shu-Chuan Chu

Abstract Data clustering aims at partitioning the data set into several disjoint segments, where the data inside the same segment are closely related, and the ones between two different segments are not. K-means is a popular data clustering algorithm, but it is easy to fall into the local optimum. To improve K-means' performance, in this work, a Harris Hawks Optimization algorithm (HHO), which is a newly emerging Evolutionary Algorithm with strong global search capability, is utilized to optimize the central data set. In particular, we encode all the central data in one individual, and optimize them with HHO. In the experiment, multiple data sets are utilized to test our proposal's performance. The experimental results show that our approach significantly outperforms the traditional K-means.

1 Introduction

With the rapid development of computer networks and the widespread application of database systems, the amount of information data has exploded, and these massive amounts of data are stored in the database. Although these massive amounts of data are a huge asset, our existing technology is not enough to mine its useful information, so data mining technology came into being.

Data mining is based on statistics, and cluster analysis in statistics is the core technology in data mining. The main idea of cluster analysis is "Birds of a feather flock together", which divides similar data in a database into a category for targeted analysis [7, 21]. Cluster analysis uses different methods and often leads to different conclusions. Clustering technology has been extensively developed in dif-

---

L.-G. Zhang · S.-C. Chu (✉)
College of Computer Science and Engineering, Shandong University of Science and Technology, Qingdao 266590, China
e-mail: lgangzhang@126.com

X. Xue
School of Computer Science and Mathematics, Fujian University of Technology, Fuzhou, Fujian 350118, China

© The Author(s), under exclusive license to Springer Nature Singapore Pte Ltd. 2022
J.-F. Zhang et al. (eds.), *Advances in Intelligent Systems and Computing*, Smart Innovation, Systems and Technologies 268, https://doi.org/10.1007/978-981-16-8048-9_10

ferent application fields [27]. There are many types of clustering methods, including partition-based methods [1], hierarchical-based methods [20], density-based methods [28], grid-based methods [25], and so on. The typical method based on partition is the k-means clustering algorithm.

Heuristic algorithm is a method to find the optimal solution, but it cannot guarantee that the desired result is optimal. Metaheuristic algorithms are the improvement and development of heuristic algorithms. Most of them are proposed based on physical or biological phenomena and used to solve various complex problems [4, 11, 24, 26]. However, according to the no free lunch theorem, there is no meta-heuristic algorithm that can be widely applied to various practical problems. Therefore, metaheuristic algorithms suitable for different types of problems have attracted the attention of modern researchers, and more and more excellent algorithms have been proposed, such as Genetic Algorithm (GA) [17, 23], Particle Swarm Optimization (PSO) [14], Whale Optimization Algorithm (WOA) [6], Grey Wolf Optimizer (GWO) [18], Sine Cosine Algorithm (SCA) [16], Fish Migration Optimization (FMO) [2, 22], Cat Swarm Optimization (CSO) [5], and Harris Hawks Optimizer (HHO) [3, 9, 10, 12], etc. Harris Hawks Optimizer algorithm (HHO) is a new type of swarm intelligence optimization algorithm. Harris hawks uses a variety of chasing modes to cooperate in predation. Because of its strong global search capability, it can be used to perform cluster analysis on data sets. Therefore, this paper proposes a combination of HHO algorithm and k-means algorithm to perform clustering analysis and reduce clustering errors.

The rest of this article is organized as follows. Related work introduced the principles of HHO and K-means. In the third section, the combination process of HHO algorithm and K-means algorithm is introduced. In the fourth section, the performance of HK-means is tested and compared with HHO and K-means algorithms. Finally, conclusions are given in Sect. 5.

2 Related Work

2.1 HHO

The idea of HHO is derived from the cooperative predation behavior of Harris hawks. According to the dynamic characteristics of prey escape, Harris hawks includes a variety of chasing modes. HHO mathematically realizes this dynamic chasing and predation process. HHO includes exploration phase, transition phase, and development phase.

Exploration Phase. In the exploration phase, the position of the i-th individual in the Harris hawks population at time t is $X_i(t)$, when the i-th individual in its position observes the prey on the ground through its keen eyes, according to the two strategies of equation (1) in order to make random position adjustments. The position of the i-th individual at time t+1 is

$$X_i(t+1) = \begin{cases} X_{rand}(t) - r_1 |X_{rand}(t) - 2r_2 X_i(t)| & q \geq 0.5 \\ (X_{rabbit}(t) - X_m(t)) - r_3 (LB + r_4(UB - LB)) & q < 0.5 \end{cases} \quad (1)$$

where $X_{rand}(t)$ is a random position in the hawks population at time t, r_1 r_2 r_3 and r_4 are all random numbers between 0 and 1, $X_{rabbit}(t)$ is the location of the prey at time t, LB and UB are the predatory range of motion, and the average position of the hawks population at time t is $X_m(t)$, which is calculated as follows:

$$X_m(t) = \frac{1}{N} \sum_{i=1}^{N} X_i(t) \quad (2)$$

where N is the number of Harris hawks in the population.

Transition from Exploration to Exploitation. In HHO, we control the transformation of the algorithm from global search to local search by the amount of energy contained in the rabbit's body. So in order to simulate the dynamic change process of the rabbit energy, we model the rabbit energy as

$$E = 2E_0 \left(1 - \frac{t}{T}\right) \quad (3)$$

Among them, E is the energy of the rabbit, which is the physical energy to escape from hunting. In the iterative process, E gradually decreases with the increase of time. T is the maximum number of iterations, t is the current number of iterations, and E_0 is the initial value of the rabbit energy, which is obtained as a random value by the following equation:

$$E_0 = 2rand() - 1 \quad (4)$$

Exploitation Phase. Here, we use the energy E of the rabbit proposed in the transition phase and a random number r to control the strategy selected in the development phase. There are four encirclement strategies in the development stage. *Soft besiege.* During the process of escaping, the rabbit will make random jumps to avoid the fatal blow. The jumping strength of the random jump is a random value, which is determined by the Eq. (5).

$$J = 2(1 - r_5) \quad (5)$$

In the soft besiege phase, at this time $|E| \geq 0.5$ and $r \geq 0.5$, the prey has enough energy to avoid the pursuit of natural enemies by jumping randomly, and the Harris hawks adjusts its position to approach the prey according to the Eqs. (6) and (7).

$$X(t+1) = \Delta X(t) - E |J X_{rabbit}(t) - X(t)| \quad (6)$$

$$\Delta X(t) = X_{rabbit}(t) - X(t) \quad (7)$$

where $\Delta X(t)$ is the distance between the Harris hawks and the rabbit, and r_5 is a random number in (0, 1).

Hard besiege. In the hard besiege stage, at this time $|E| < 0.5$ and $r \geq 0.5$, the prey does not have enough energy to avoid the pursuit of natural enemies. The Harris hawks uses a hard besiege to catch the prey, and uses the Eq. (8) to update the position

$$X(t+1) = X_{rabbit}(t) - E|\Delta X(t)| \tag{8}$$

Soft besiege with progressive rapid dives. At this stage, when $|E| \geq 0.5$ and $r < 0.5$, the rabbit still has enough energy to escape. We use the following equation for location update:

$$Y = X_{rabbit}(t) - E|JX_{rabbit}(t) - X(t)| \tag{9}$$

$$Z = Y + S \times LF(D) \tag{10}$$

$$X(t+1) = \begin{cases} Y, if\, F(Y) < F(X(t)) \\ Z, if\, F(Z) < F(X(t)) \end{cases} \tag{11}$$

where S is a random vector, D is the dimension of the problem, and LF is the Levy flight function.

Hard besiege with progressive rapid dives. At this stage, when $|E| < 0.5$ and $r < 0.5$, the rabbit does not have enough energy to escape The Harris hawks constructs a hard besiege to round up its prey before making progressive rapid dives. The following rules are enforced under hard besiege conditions:

$$X(t+1) = \begin{cases} Y, if\, F(Y) < F(X(t)) \\ Z, if\, F(Z) < F(X(t)) \end{cases} \tag{12}$$

The definitions of F and Z are the same in the Soft besiege with progressive rapid dives, where F is the fitness function. Y is obtained by using the new rule in Eq. (13).

$$Y = X_{rabbit}(t) - E|JX_{rabbit}(t) - X_m(t)| \tag{13}$$

2.2 K-Means

Clustering is an unsupervised learning algorithm that uses distance as an index of similarity to perform clustering [13, 15, 19].

The Concept of K-Means. The process of the K-means clustering algorithm is as follows: First, manually set the K value, which is the number of categories of the problem to be solved. Then specify an initial centroid for each class, that is, each class randomly selects an element as the cluster center. Calculate the distance between each element and all cluster centers, and classify the element closest to a certain cluster center into this cluster center. Here, we calculate the distance between the element and its cluster center using the Euclidean distance, the following equation:

$$D = \sqrt[2]{\sum_{i=1}^{n}(x_i - c_i)^2} \tag{14}$$

where D is the Euclidean distance between the element and the classification center, X_1, X_2, X_n indicate that an element has n-dimensional feature items.

Then iteratively optimize the divided classes, that is, iteratively update the centroid of each class to the mean value of all samples in that class. Finally, for simple problems, the optimal result is obtained when the clustering center does not change anymore; however, for data with a large amount of data and a large number of dimensions, we consider the final classification result to be optimal when the number of iterations is repeated a fixed number of times or the sum of squared distance errors is less than a certain threshold when the similarity between the elements in the class is the greatest.

3 HHO Combined with K-Means

The HHO algorithm can perform cluster analysis on data sets. The method is to form a vector of the center points of all classifications, use this vector as an individual in the population, and use the HHO evolution algorithm to optimize each individual according to the fitness value, and then find the best individual. This individual is the collection of the best classification center points. The fitness value here is the sum of squared distance errors, that is, to find a set of center points with the smallest fitness value, and the equation for solving the fitness value is as follows:

$$Fitness = \sqrt[2]{\sum_{j=1}^{m} x_j - c_j}(k = 1, 2, \ldots m) \tag{15}$$

where m is the total number of elements, x_j is the j-th data, and c_j is the cluster center corresponding to the j-th data. By calculating the fitness from the element to the cluster center, after multiple iterations, an individual with the smallest fitness is selected, and then this individual is the best cluster center point.

However, the efficiency of the HHO algorithm is low, and the K-means algorithm can easily fall into a local optimal solution. The two algorithms, HHO and K-means, cannot quickly and accurately obtain the correct results when clustering separately. Therefore, in order to speed up the efficiency of the HHO algorithm in calculating the center point and prevent the local optimal situation in the calculation of the K-means algorithm, we will propose HK-means, that is, combining HHO with K-means. The idea of HK-means is to use K-means to adjust the position of the center point, and use the population collaboration and information sharing capabilities of the HHO algorithm to select and optimize the center point of classification, thereby, the problem of local optimal solutions in K-means and the slow running speed and low efficiency of the HHO algorithm are reduced.

The steps of the HK-means algorithm: First initialize the classification number K, generate the initial K center points through the K-means algorithm, and combine all the center points into a vector as an individual in the Harris hawks population. If the data dimension is two-dimensional and the number of classifications k = 4, there will be four two-dimensional center points, and the Harris hawks population individual is a 1*8 vector. The HHO evolutionary algorithm is used to optimize each individual according to the fitness function, and the individual with the smallest fitness value is found after multiple iterations. The flow of the HK-means algorithm is shown in Fig. 1.

4 HHO Combined with K-Means

In this section, we use five different dimensional data sets to test the effectiveness of the proposed hybrid algorithm.

4.1 Parameter Configuration

To verify the results, we compared HK-means with K-means and HHO. The test parameters of the function are given in Table 1.

First, experiment on the yeast dataset of the UCI database. The total amount of data is N = 1484, the data dimension is D = 8, and the number of clusters in the experiment is set to K = 10, and the HK-means, K-means, and HHO algorithms are run 20 times. The average results are 6.9169, 7.0160, and 13.9049, respectively, and the variances are 0.2038, 0.2289, and 1.6069, respectively.

It can be seen that the proposed HK-means algorithm has better average results than the original K-means algorithm on the 8dim data set, and the results obtained are more stable.

Experimentation on the image dataset of the bridge in paper [8]. The total amount of data is N = 4096, the data dimension is D = 16, and the number of clusters in the experiment is set to K = 10. The HK-means, K-means, and HHO algorithms were run 20 times, and the average results were 5422.65, 5423.37, and 10206.41, respectively, and the variances were 17.9497, 18.4505, and 474.0712.

From the above data, it can be seen that the average distance error obtained by the HK-means algorithm is smaller, proving that the HK-means algorithm is more competitive than the original K-means algorithm and the HHO algorithm in terms of accuracy and stability on a 16dim dataset.

Next, we consider the HK-means algorithm for clustering high-dimensional data. We validate the performance of the proposed new algorithm with three high-

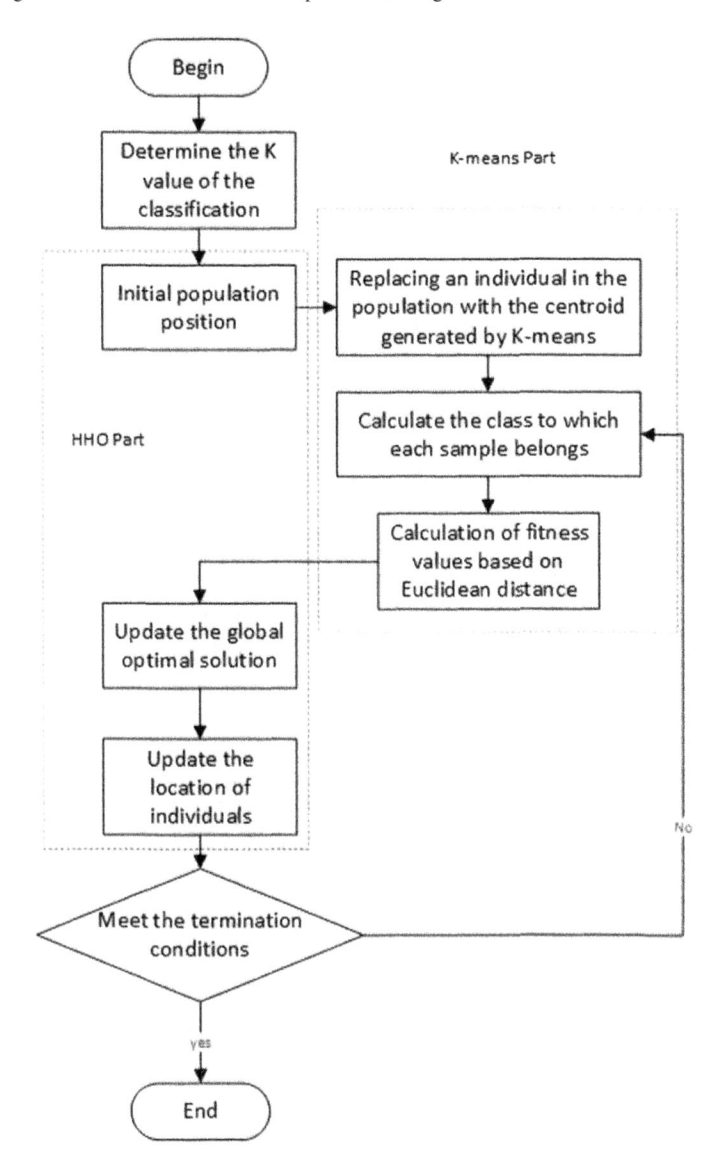

Fig. 1 HHO combined with K-means flow chart

Table 1 Parameter settings for each related algorithm

Name	Parameter
HK-means	Np = 30, Iter = 200, Runtimes = 20
K-means	Runtimes = 20
HHO	Np = 30, Iter = 200, Runtimes = 20

Table 2 Experimental results on the Dim032, Dim064 and Dim128 datasets

	HK-means		K-means		HHO	
	Mean	Std	Mean	Std	Mean	Std
Dim032	5805.24	359.40	5870.25	357.41	9206.09	88.29
Dim064	8252.27	499.08	8291.61	545.69	13303.08	82.11
Dim128	12284.06	652.56	12446.60	955.44	18970.45	56.56

dimensional datasets Dim032, Dim064, and Dim128, all of which are from the paper [8].

On the datasets Dim032, Dim064, and Dim128, all of which have a total data volume of $N = 1024$, the experiments were set to a classification number of $K = 10$ and run 20 times on the HK-means, K-means, and HHO algorithms, with the average results and variance in the Table 2.

From the data in the above table, it can be analyzed that the HK-means algorithm is significantly better than the original K-means and HHO algorithms, and it is more competitive in the clustering of large-dimensional data. The HHO algorithm has the best stability but the worst performance. The stability of HK-means is better than the K-means algorithm.

5 Conclusion

By analyzing the advantages and disadvantages of K-means algorithm and HHO algorithm, we found that K-means algorithm is faster in solving the problem of clustering center, but it is very easy to fall into the local optimal solution; HHO algorithm has the advantage of strong global searchability, but the solution efficiency is low. Therefore, in this paper, we propose a hybrid algorithm HK-means, which combines the solution efficiency and searchability of K-means algorithm and HHO algorithm, based on the advantages of each of the two algorithms. The K-means algorithm is used to initialize the centroids of the data set, and then all the centroids of the data set are treated as an individual in the Harris hawks population, and the distance from the data elements to their respective centroids in the data set is used as a criterion to optimize each individual with the HHO algorithm, and finally the optimal centroid is obtained after several iterations. The proposed algorithm is analyzed and compared with the original K-means and the original HHO algorithm on 8dim, 16dim, 32dim, 64dim, and 128dim datasets, respectively, and it is found that HK-means has better performance on each dataset.

References

1. Bose, S., Das, C., Chakraborty, A., Chattopadhyay, S.: Effectiveness of different partition based clustering algorithms for estimation of missing values in microarray gene expression data. In: Meghanathan, N., Nagamalai, D., Chaki, N. (eds.) Advances in Computing and Information Technology, pp. 37–47. Springer, Berlin (2013)
2. Chai, Q.W., Chu, S.C., Pan, J.S., Zheng, W.M.: Applying adaptive and self assessment fish migration optimization on localization of wireless sensor network on 3-d terrain. J. Inf. Hiding Multimedia Signal Process. **11**(2), 90–102 (2020)
3. Chen, H., Heidari, A.A., Chen, H., Wang, M., Pan, Z., Gandomi, A.H.: Multi-population differential evolution-assisted harris hawks optimization: Framework and case studies. Future Gener. Comput. Syst. **111**, 175–198 (2020). https://doi.org/10.1016/j.future.2020.04.008. https://www.sciencedirect.com/science/article/pii/S0167739X19313263
4. Chu, S.C., Huang, H.C., Roddick, J.F., Pan, J.S.: Overview of algorithms for swarm intelligence. In: Jędrzejowicz, P., Nguyen, N.T., Hoang, K., (eds.) Computational Collective Intelligence. Technologies and Applications, pp. 28–41. Springer, Berlin (2011)
5. Chu, S.C., Tsai, P.W., Pan, J.S.: Cat swarm optimization. In: Yang, Q., Webb, G. (eds.) PRICAI 2006: Trends in Artificial Intelligence, pp. 854–858. Springer, Berlin (2006)
6. Dao, T., Pan, T., Pan, J.: A multi-objective optimal mobile robot path planning based on whale optimization algorithm. In: 2016 IEEE 13th International Conference on Signal Processing (ICSP), pp. 337–342 (2016). https://doi.org/10.1109/ICSP.2016.7877851
7. Edwards, A.W.F., Cavalli-Sforza, L.L.: A method for cluster analysis. Biometrics **21**(2), 362–375 (1965). http://www.jstor.org/stable/2528096
8. Franti, P., Virmajoki, O., Hautamaki, V.: Fast agglomerative clustering using a k-nearest neighbor graph. IEEE Trans. Pattern Anal. Mach. Intell. **28**(11), 1875–1881 (2006). https://doi.org/10.1109/TPAMI.2006.227
9. Heidari, A.A., Mirjalili, S., Faris, H., Aljarah, I., Mafarja, M., Chen, H.: Harris hawks optimization: algorithm and applications. Future Gener. Comput. Syst. **97**, 849–872 (2019). https://doi.org/10.1016/j.future.2019.02.028. https://www.sciencedirect.com/science/article/pii/S0167739X18313530
10. Houssein, E.H., Hosney, M.E., Oliva, D., Mohamed, W.M., Hassaballah, M.: A novel hybrid harris hawks optimization and support vector machines for drug design and discovery. Comput. Chem. Eng. **133**, 106,656 (2020). https://doi.org/10.1016/j.compchemeng.2019.106656. https://www.sciencedirect.com/science/article/pii/S0098135419309330
11. Huang, H.C., Chu, S.C., Pan, J.S., Huang, C.Y., Liao, B.Y.: Tabu search based multi-watermarks embedding algorithm with multiple description coding. Inf. Sci. **181**(16), 3379–3396 (2011). https://doi.org/10.1016/j.ins.2011.04.007. https://www.sciencedirect.com/science/article/pii/S0020025511001757
12. Jia, H., Lang, C., Oliva, D., Song, W., Peng, X.: Dynamic harris hawks optimization with mutation mechanism for satellite image segmentation. Remote Sens. **11**(12) (2019). https://doi.org/10.3390/rs11121421. https://www.mdpi.com/2072-4292/11/12/1421
13. Kanungo, T., Mount, D.M., Netanyahu, N.S., Piatko, C.D., Silverman, R., Wu, A.Y.: An efficient k-means clustering algorithm: analysis and implementation. IEEE Trans. Pattern Anal. Mach. Intell. **24**(7), 881–892 (2002). https://doi.org/10.1109/TPAMI.2002.1017616
14. Kennedy, J., Eberhart, R.: Particle swarm optimization. In: Proceedings of ICNN'95 - International Conference on Neural Networks, vol. 4, pp. 1942–1948 (1995). https://doi.org/10.1109/ICNN.1995.488968
15. Likas, A., Vlassis, N., J. Verbeek, J.: The global k-means clustering algorithm. Pattern Recogn. **36**(2), 451–461 (2003). https://doi.org/10.1016/S0031-3203(02)00060-2. https://www.sciencedirect.com/science/article/pii/S0031320302000602. Biometrics
16. Mirjalili, S.: Sca: A sine cosine algorithm for solving optimization problems. Knowl. Based Syst. **96**, 120–133 (2016). https://doi.org/10.1016/j.knosys.2015.12.022. https://www.sciencedirect.com/science/article/pii/S0950705115005043

17. Mirjalili, S.: Genetic Algorithm, pp. 43–55. Springer International Publishing, Cham (2019). https://doi.org/10.1007/978-3-319-93025-1_4
18. Mirjalili, S., Mirjalili, S.M., Lewis, A.: Grey wolf optimizer. Adv. Eng. Softw. **69**, 46–61 (2014). https://doi.org/10.1016/j.advengsoft.2013.12.007. https://www.sciencedirect.com/science/article/pii/S0965997813001853
19. Na, S., Xumin, L., Yong, G.: Research on k-means clustering algorithm: an improved k-means clustering algorithm. In: 2010 Third International Symposium on Intelligent Information Technology and Security Informatics, pp. 63–67 (2010). https://doi.org/10.1109/IITSI.2010.74
20. Nazari, Z., Kang, D., Asharif, M.R., Sung, Y., Ogawa, S.: A new hierarchical clustering algorithm. In: 2015 International Conference on Intelligent Informatics and Biomedical Sciences (ICIIBMS), pp. 148–152 (2015). https://doi.org/10.1109/ICIIBMS.2015.7439517
21. Pan, J., McInnes, F., Jack, M.: Fast clustering algorithms for vector quantization. Pattern Recogn. **29**(3), 511–518 (1996). https://doi.org/10.1016/0031-3203(94)00091-3. https://www.sciencedirect.com/science/article/pii/0031320394000913
22. Pan, J., Tsai, P., Liao, Y.: Fish migration optimization based on the fishy biology. In: 2010 Fourth International Conference on Genetic and Evolutionary Computing, pp. 783–786 (2010). https://doi.org/10.1109/ICGEC.2010.198
23. Pan, J.S., Kong, L., Sung, T.W., Tsai, P.W., Snfor Wireless Sensor Networks Basediel, V.: A clustering scheme for wireless sensor networks based on genetic algorithm and dominating set. J. Internet Technol. **19**(4), 1111–1118 (2018). https://jit.ndhu.edu.tw/article/view/1729
24. Pan, J.S., Nguyen, T.T., Chu, S.C., Dao, T.K., Giang, N.: Diversity enhanced ion motion optimization for localization in wireless sensor network. J. Inf. Hiding Multimedia Signal Process. **10**, 221–229 (2019)
25. Pilevar, A., Sukumar, M.: Gchl: A grid-clustering algorithm for high-dimensional very large spatial data bases. Pattern Recogn. Lett. **26**(7), 999–1010 (2005). https://doi.org/10.1016/j.patrec.2004.09.052. https://www.sciencedirect.com/science/article/pii/S0167865504002946
26. Song, P.C., Chu, S.C., Pan, J.S., Yang, H.: Phasmatodea population evolution algorithm and its application in length-changeable incremental extreme learning machine. In: 2020 2nd International Conference on Industrial Artificial Intelligence (IAI), pp. 1–5 (2020). https://doi.org/10.1109/IAI50351.2020.9262236
27. Thuy, Q.D.T., Huu, Q.N., Lan, P.N.T., Quoc, T.N., Ngo, M.H.: Improve the efficiency of content-based image retrieval through incremental clustering. J. Inf. Hiding Multim. Signal Process. **11**, 103–115 (2020)
28. Wang, Y., Wang, D., Pang, W., Miao, C., Tan, A.H., Zhou, Y.: A systematic density-based clustering method using anchor points. Neurocomputing **400**, 352–370 (2020). https://doi.org/10.1016/j.neucom.2020.02.119. https://www.sciencedirect.com/science/article/pii/S0925231220303702

Weighted Multi-task Sparse Representation Classifier for 3D Face Recognition

Linlin Tang, Zhangyan Li, Tao Qian, Shuhan Qi, Yang Liu, Jiajia Zhang, Shuaijie Shi, Churan Liu, and Jingyong Su

Abstract Rapid development of 3D face recognition can help people overcome some bottlenecks in 2D recognition. But still susceptible to changes in facial expressions. At the same time, due to the large number of 3D point clouds, the calculation speed is also greatly affected. This paper mainly proposes a method to classify 3D human faces according to the characteristics of their semi-rigid and non-rigid regions to enhance the robustness of recognition of 3D facial expression changes. At the same time, improve the expression of the 3D point cloud face, reduce the number of points involved in the calculation, and increase the speed of the algorithm. Experimental results show that the algorithm not only has a higher recognition rate but also has stronger robustness to changes in facial expression.

L. Tang (✉) · Z. Li · T. Qian · S. Qi · Y. Liu · J. Zhang · S. Shi · C. Liu · J. Su
Harbin Institute of Technology, Shenzhen, China
e-mail: hittang@126.com

Z. Li
e-mail: 1223288931@qq.com

T. Qian
e-mail: qt41@qq.com

S. Qi
e-mail: shuhanqi@cs.hitsz.edu.cn

Y. Liu
e-mail: liu.yang@hit.edu.cn

J. Zhang
e-mail: zhangjiajia@hit.edu.cn

S. Shi
e-mail: 823766945@qq.com

C. Liu
e-mail: 710859740@qq.com

J. Su
e-mail: sujingyong@hit.edu.cn

1 Introduction

Geometric information contained in 3D face point cloud data or 3D mesh data is more substantial and reliable than 2D. It also has strong robustness to zoom, rotation, and illumination changes, and moreover, result of pose estimation in a 3D face is more accurate than pose estimation in a 2D face. There has been a big change in 3D face recognition technology since it had been introduced, and we can divide 3D face recognition into two major stages: in early days, method of 3D face recognition is mainly focused on entire 3D face, which use geometric information as feature, without any subdivision. And the feature similarity of two 3D face is usually used to calculate the similarity. The algorithms at this stage are known collectively as the 3D face recognition algorithm based on overall features, common methods in this categories include Principal Component Analysis (PCA) [1], Deformation Model [2], Signed Shape Differential Graph [3], Spherical Harmonic Function [4], etc. The second stage is the algorithm frequently researched today. This type of algorithm extracts feature vectors from local areas, such as nose and eyes. It is called 3D face recognition algorithm based on local features. 3D key points are mainly selected from those more prominent positions, and 3D key points are detected according to some geometric information on surface. It mainly involves two steps: key point detection and feature description. Algorithm based on key points is robust to occluded and missing data sets, but its calculation cost is high. This is due to large amount of key point calculations, so selection of key points is very important.

2 Related Work

2.1 Feature Point Extraction

We represent 3D facial surface as $S = \{p_i\}_i^N$. Let $N_{bhd(p)}$ be neighborhood of p. For Local Reference Frame (LRF) has good robustness to clutter and occlusion, local neighborhood $N_{bhd(p)}$ is used to extract LRF. At the same time, Hotelling transform or principal component analysis (PCA) is used to transform each point p_i in $N_{bhd(p)}$ into a locally aligned neighborhood, represented as $A_{nbhd(p_i)}$. Let $X = \{x_1, x_2, ..., x_l\}$ and $Y = \{y_1, y_2, ..., y_l\}$ be X and Y components of $A_{nbhd(p_i)}$, where l is the length of $A_{nbhd(p_i)}$.

Face surface change index ϑ is a ratio between axis X and axis Y of local alignment neighborhood with the center of key point p, which can be formulated as:

$$\vartheta = \frac{max(X) - min(X)}{max(Y) - min(Y)} \qquad (1)$$

ϑ is the geometric change of local neighborhood of point p. For any symmetric local point set, such as a plane or a sphere, its ϑ is equal to 1; for any asymmetric local point set, its ϑ would be greater than 1. Therefore, point p can be viewed as a candidate key point, when its ϑ is greater than an empirical threshold ε_ϑ.

C_p denote the covariance matrix of local neighborhood of each point p. By performing eigenvalue decomposition on C_p, eigenvalue $\lambda_1(p) > \lambda_2(p) > \lambda_3(p)$ can be obtained. So eigenvalue change rate ρ can be obtained by formula. Then, point p can be viewed as candidate key point if its ρ is greater than threshold $\varepsilon_{\_\rho}$.

$$\rho = \frac{\lambda_1(p)}{\lambda_2(p)} \quad (2)$$

So far, a point p is selected as a key point only if its ϑ and ρ satisfies the condition.

$$\vartheta > \varepsilon_\vartheta \vee \rho > \varepsilon_\rho \quad (3)$$

2.2 Covariance Descriptor

By selecting key points by method in Sect. 2.1, covariance matrix of points in neighborhood with radius r and center key point p. First of all, face surface is divided into several regions of neighborhood of p_i, using $P = \{p_i, i = 1, 2, \ldots, m\}$ to denote the 3D face surface, where $p_i = (x_i, y_i, z_i)^T$ is key point. For each key point $p_j \in N_{bhd(p_i)}$, feature vector can be represented as follows.

$$f_j = [x_j, y_j, z_j, k_j^1, k_j^1, D_j] \quad (4)$$

where x_j, y_j, z_j is coordinate location of key point p_j; k_j^1, k_j^2 is the maximum and the minimum curvature of p_j, D_j is distance from p_i to p_j. With feature vector f, covariance descriptor can be calculated by formula (5).

$$X_i = \frac{1}{n}\sum_{j}^{n}(f_j - \mu_i)(f_j - \mu_i)^T \quad (5)$$

Here μ_i is means of feature vector of $p_j \in N_{bhd(p_i)}$. Diagonal of X_i is variation of each feature, remain non-diagonal elements of a X_i is the covariation between features.

2.3 Sparse Representation

Sparse representation has been used for signal processing. It uses fewer but important coefficients to replace origin signal. It's firstly introduced in face recognition by Wright [5]. using transposed vector of pixel vector of face image to calculate its sparse representation, and then to classify.

In this paper, the idea of sparse representation is introduced into 3D face recognition task [9–11]. Assume there are k-class samples, to be classified, there are n_i 3D face samples in ith-class. Each training sample is converted to an m-dimensional column vector through the feature extraction stage in previous section, then i^{th}-class training set's learning dictionary can be represented as formula (6).

$$A_i = [v_{i,1}, v_{i,2}, \ldots, v_{i,n_i}] \in IR^{m \cdot n_i} \quad (6)$$

For test sample $y \in IR^m$ and $y \in i$th-class, it can be linearly represented as:

$$y = \alpha_{i,1} v_{i,1} + \alpha_{i,2} v_{i,2} + \ldots + \alpha_{i,n_i} v_{i,n_i} \quad (7)$$

where $\alpha_{i,j} \in IR (j = 1, 2, \ldots, n_i)$ is a constant. However, it can't be known the true label of y, so it's necessary to define a new matrix that represents the entire training set which is concatenated with all train sample.

$$A = [A_1, A_2, \ldots, A_k] = [v_{1,1}, v_{1,2}, \ldots, v_{1,n_1}, \ldots, v_{k,n_k}] \quad (8)$$

Then the test sample y can be linearly represented by all training sets.

$$y = A x_i \in IR^m \quad (9)$$

Here $x_i = [0, \ldots, 0, \alpha_{i,1}, \alpha_{i,2}, \ldots, \alpha_{i,n_i}, 0 \ldots, 0]^T \in IR^n$ is sparse coefficients of test sample y, which values are equal to zeros except for the corresponding position of class i. So that x_i can be the class which y belong to. This problem is a solution of $y = Ax$. For $m > n$, so the entire dictionary A of the train set is an over-complete dictionary, which ensure solution of x is unique. So the problem is to solve this minimized ℓ^1-paradigm:

$$\hat{x}_1 = \arg \min \|x\|_1 \quad s.t. \ Ax = y \quad (10)$$

In formula (10), a sparse representation $\hat{x}_1$ coefficient is obtained, which approximately represents the sample y to be $\hat{y}_i = A \delta_i (\hat{x}_1)$. Classification result is based on the nearest neighbor of the train set as shown in the formula.

$$\text{identity}(y) = \arg \min_i \| y - A \delta_i (\hat{x}_1) \|_2 \quad (11)$$

3 Proposed Method

3.1 Face Partition

Bosphorus 3D face database is used here, each person has up to six expression changes: neutral, happy, angry, surprised, scared, and disgusted. When expression changes, geometric information will also be deformed, which causes difference in calculation results of the same area to become larger. At the same time, deformation of different areas is also affected by expression. Therefore, expression and expressionless areas on the 3D face have a great influence on the algorithm recognition.

This section mainly studies how to partition faces. Because the rigid area is not deformed, there is no rigid area on the face strictly speaking. Therefore, it can be divided into semi-rigid area and non-rigid area according to the degree of influence by expression. Through observation and experimentation, we can classify the areas that are less affected by expressions such as the forehead, nose, cheeks, and chin as semi-rigid regions, and regions that are easily affected by changes in expressions are classified as non-rigid regions, as shown in Table 1.

Divided 3D face is shown in Fig. 1b, 3D face is divided into 11 regions according to definition of semi-rigid area and non-rigid area, and corresponding names and properties of these 11 regions are shown in Table 2.

3.2 Multi-task Sparse Representation Classifier

If face sparse representation classifier proposed in Sect. 2.1 is directly used for face recognition, it will have great limitations, because not only the face must be of a uniform size, but the built dictionary also needs to have consistent dimensions. But for 3D faces, these are uncertain. Therefore, a multi-task sparse representation classifier to improve the limitations of the face sparse representation classifier is proposed. The specific algorithm steps are as follows:

(1) Construct the learning dictionary for the training set. Suppose there are k classes in training set, and there are n_i key points in jth -class, then the learning dictionary of jth-class can be described as:

$$A_i = \left[d_{i,1}, d_{i,2}, \ldots, d_{i,n_j}\right] \in IR^{m \cdot n_i} = [A_1, A_2, \ldots, A_k] \in IR^{m \cdot K} \quad (12)$$

In formula (12), d_{i,n_i} is covariance matrix corresponding to key point n_i, which belongs to jth training set sample and d_{i,n_i} is a column vector with m dimension. We can calculate a learning dictionary for each category, and this learning dictionary can be concatenated into a learning dictionary for entire training set, where K represents the number of key points. Like what we introduced in Sect. 3.1, K is a very big number,

Table 1 The result of face partition

ID	R1	R2	R3	R4	R5	R6	R7	R8	R9	R10	R11
Name	Fore head	Right eye	Nose bridge	Left eye	Top of the right face	Nose	Top of the left face	Bottom of the right face	Mouth	Bottom of the left face	Under chin

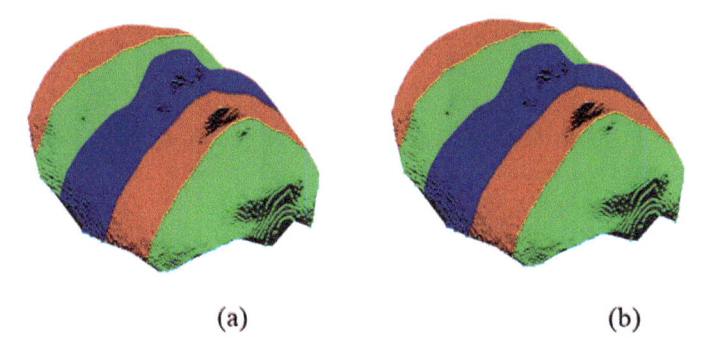

Fig. 1. 3D face partition results

Table 2 Comparison results

Method	Accuracy
Li [6]	0.954
Zhang [7]	0.926
Kim [8]	0.938
WMSP-N	0.973

and this shows that learning dictionary A of the training set fits the over-complete condition. In other words, any sample to be tested can be linearly represented by learning dictionary A, and we can get a spatial solution from the above formula.

(2) Multi-task spatial representation. According to key points extract algorithm introduced before, every 3D face will extract many key points, that is, we will get many covariance matrices. For any 3D face sample to be tested, we denote it by $Y = [y_1, y_2, ..., y_n] \in IR^{m \cdot n}$ where m represents the number of dimensions of column vector which is converted from covariance matrix, n represents the number of key points which is extracted from sample to be tested. Then, we convert ℓ^1-paradigm minimization problem into n ℓ^1-paradigms minimization problem.

$$X = \arg\min_{X} \sum_{i=1}^{n} \|x_i\|_1, \ s.t. \ AX = Y \tag{13}$$

In the above formula (13), let $X = (x_1, x_2, ..., x_n) \in IR^{K \cdot n}$ be sparse coefficient matrix of sample Y which to be tested. Solving formula is equivalent to solve problem of minimizing n ℓ^1-paradigms, and to every key point.

$$\hat{x}_i = \arg\min_{x_i} \|x_i\|_1, \ s.t. \ Ax_i = y_i \tag{14}$$

We change the formula to make it available for multi-task sparse representation classifier as follows:

$$\min_k r_k(Y) = \frac{1}{n}\sum_{i=1}^{n}\left\|y_i - A_k\delta_k(\hat{x}_i)\right\|_2^2 \quad (15)$$

The following classification idea is similar to original sparse representation classifier, it uses average residual calculated according to formula (15) for classification, and this idea can be expressed as formula (16).

$$identity(Y) = \min_k r_k(Y) \quad (16)$$

As we know, human face can be divided into half-rigid area and non-rigid area, we can divide these areas into two categories: $C1$ and $C2$, because their contributions to 3D face recognition are not the same. We assign different weights to $C1$ and $C2$, respectively, named as w_1 and w_2. Weighted Multi-task Sparse Representation Classifier (WMSP) is introduced. For each 3D face, total key points number is n, number of key points in $C1$ and $C2$ is n_1 and n_2. And we build the extracted feature F by putting feature vector extracted from $C1$ into the former part of F by column and putting feature vector extracted from $C2$ into the second half of $C2$ by column. This process can be shown as in formula (17).

$$F = [F_1, F_2] \in IR^{m \cdot n} \quad (17)$$

In formula (17), $F_1 \in R^{m \times n_1}$, $F_2 \in R^{m \times n_2}$, $n = n_1 + n_2$, m is dimensions number of feature. After different weights are assigned, feature F_w extracted from this 3D face can be represented as follows:

$$F_w = [w_1 F_1, w_2 F_2] \in IR^{m \cdot n} \quad (18)$$

Learning dictionary of i-th training set, which get from the i-th feature of sample in training set after weighted, can be represented as below.

$$A_{iw} = [F_{1w}, F_{2w}, ..., F_{n_i w}] \in IR^{m \cdot n_i} \quad (19)$$

The learning dictionary of training set (20) has k categories.

$$A_w = [A_{1w}, A_{2w}, ..., A_{kw}] \in IR^{m \cdot K} \quad (20)$$

The classification method is not changed, formulas (21) and (22) are used as base of the weighted muti-task sparse representation classifier.

$$\hat{x}_{iw} = \arg\min\|x_{iw}\|_1, \ s.t. \ A_w x_{iw} = y_i \quad (21)$$

$$identity\,(Y) = \arg\min \frac{1}{n}\sum_{i=1}^{n}\|y_i - A_{kw}\delta_{kw}(x_{iw})\|_2^2 \quad (22)$$

There is an important metric in sparse representation classifier called sparse standard, and the sparse representation coefficient $\hat{x}_{iw}$ (21) is used to check sparseness. Most data in $\hat{x}_{iw}$ should be close to 0, except the element corresponding to the sparse representation coefficient. Formula (24), is used to calculate the sparseness of $\hat{x}_{iw}$.

$$sparisity(\hat{x}_{iw}) = \frac{K \cdot Max(\delta_{kw}(\hat{x}_{iw}))/\|\hat{x}_{iw}\|_1}{K-1} \in [0, 1] \qquad (24)$$

When $sparisity(\hat{x}_{iw}) = 1$, it means the sample to be tested from a category with only one training sample. And when $sparisity(\hat{x}_{iw}) = 0$, sparse representation coefficient will evenly be distributed in each category of training set. Based on this analysis, this paper select a threshold $\tau \in (0, 1)$, and when $sparisity(\hat{x}_{iw}) \geq \tau$(this is an empirical value, in this experiment we choose $\tau = 0.8$), the sparse representation vector will sparse enough for the next classification.

4 Experimental Result

The dataset used in this experiment is Borphorus 3D face dataset. Borphorus Dataset includes 105 subjects with 9 kinds of different pose and expression. Experiments focus on verifying the effect of different face areas on 3D face recognition algorithm. We divide 3D face into 11 regions, which can be roughly divided into two types: semi-rigid regions and non-rigid regions. Experiments are mainly to verify impact of these two parts on the recognition effect.

Face images of each person in Borphorus's dataset can be divided into neutral pose and expressions and others express. We used N to denote set of neutral pose and expressions, and E to denote set of all pose and expression. All pose classes E are added into test set to verify performance. Followed by three comparative experiments:

1. Only compute the semi-rigid regions' feature vectors w_0 noted as **N**.
2. Only compute the non-rigid regions' feature vectors w_1 w_1, noted as **S**.
3. Simultaneously consider the semi-rigid regions and non-rigid regions' feature vector, noted as **NS**.

In feature extraction stage, neighbor of key point p required a radius r, which is an empirical value. Therefore, we must firstly determine an optimal radius r, which is obtained by getting the best results on experiment on semi-rigid regions. Value r is the ratio of the selected radius r and the radius r of the entire 3D face. Experiment results are as shown in Fig. 2. It's easier to find when neighborhood radius r is 20% of the entire 3D face, the best recognition result can be obtained, here **N–N** represents both training and testing images belong to category **N**, **N-E** represents that training image belongs to category N, the testing image belongs to category **E**.

After deciding radius of neighbor to r = 0.2, we can perform comparative experiments: **N**, **S**, **NS**. Experiment results are shown in Fig. 3. In this experiment, train set is a set of neutral pose and expression with 105 persons, which sums up with 264

Fig. 2 Change of Neighborhood radius

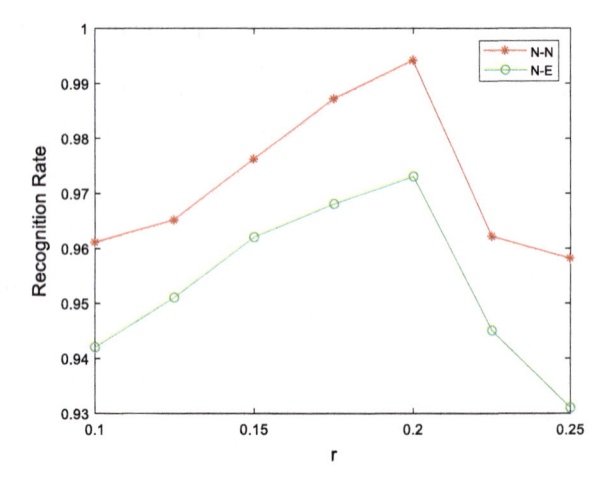

Fig. 3 Classification effect of semi-rigid and non-rigid regions

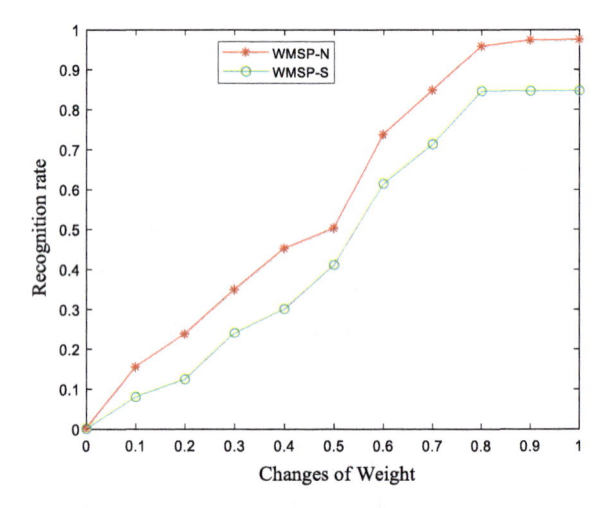

3D face images. Test set is a set of all expression with 105 persons, which sums up with 453 3D face images. First of all, a learning dictionary is computed for all train sets, and then the 3D face data's label in test set is predicted by solving the equation. Experiment results in Fig. 3 show that the recognition precision based on semi-rigid region is much higher than that based on non-rigid region features, while they are both affected by weight changes.

The algorithm firstly computes a learning dictionary with all training samples, and then predicts each sample in the test set. Experimental results in Fig. 3 show that recognition accuracy based on semi-rigid region is much higher than that of based on the non-rigid region features. They are both affected by weight changes. In Fig. 4, x-axis denotes value of w_0, y-axis is recognition accuracy. Each curve in Fig. 4 represents different values of w_1. Results show that non-rigid region has a negative

Fig. 4 Effect of non-rigid regioin on recognition rate

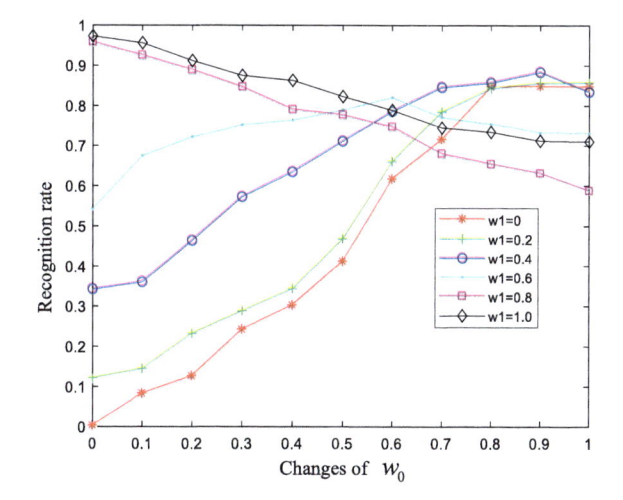

impact on the recognition accuracy. Table 2 gives the comparison between ours and other state-of-the-art algorithms based on Bosphorus dataset.

5 Conclusions

A Weighted Multi-task Sparse Representation classification algorithm based on sparse representation classification algorithm has been proposed in this paper, which solves problems of inconsistent number of data and inconsistent number of the feature point. Results show different influence of semi-rigid and non-rigid regions. When using feature points for 3D face classification, the extraction effect of semi-rigid area is better than that of non-rigid area or whole face regions, and it also has certain robustness to facial expression change at the same time.

Acknowledgments This work was supported by Shenzhen Science and Technology Plan Fundamental Research Funding JCYJ20180306171938767 and Shenzhen Foundational Research Funding JCYJ20180507183527919.

References

1. Russ, T., Boehnen, C., Peters, T.: 3D face recognition using 3D alignment for PCA[C]. Computer Vision and Pattern Recognition, 2006 IEEE Computer Society Conference on. IEEE, vol. 2, pp. 1391–1398 (2006)
2. Lu, H., Forin, A.: Automatic processor customization for zero-overhead online software verification[J]. IEEE Trans. Very Large Scale Integration (VLSI) Syst. **16**(10), 1346–1357 (2008)

3. Wang, Y., Liu, J., Tang, X.: Robust 3D face recognition by local shape difference boosting[J]. IEEE Trans. Pattern Anal. Mach. Intell. **32**(10), 1858–1870 (2010)
4. Liu, P., Wang, Y., Huang, D., et al.: Learning the spherical harmonic features for 3-D face recognition[J]. IEEE Trans. Image Process. **22**(3), 914–925 (2013)
5. Joshi, S.H., Srivastava, A., Klassen, E., et al.: A novel representation for computing geodesics between n-dimensional elastic curves[C]. IEEE Conference on Computer Vision and Pattern Recognition (CVPR), vol. 4, pp. 1–7 (2007)
6. Li, H., Huang, D., Morvan, J.M., et al.: Expression-robust 3D face recognition via weighted sparse representation of multi-scale and multi-component local normal patterns[J]. Neurocomputing **133**, 179–193 (2014)
7. Li, Y., Wang, Y.H., Liu, J., et al.: Expression-insensitive 3D face recognition by the fusion of multiple subject-specific curves[J]. Neurocomputing **275**, 1295–1307 (2018)
8. Kim, J., Han, D., Hwang, W., et al.: 3D face recognition via discriminative keypoint selection[C]. Ubiquitous Robots and Ambient Intelligence (URAI), 2017 14th International Conference on, pp. 477–480. IEEE (2017)
9. Wang, E.K., Chen, C.M., Hassan, M.M., Almogren, A.: A deep learning based medical image segmentation technique in internet-of-medical-things domain[J]. Futur. Gener. Comput. Syst. **108**, 135–144 (2020)
10. Wang, K., Chen, C.M., Hossain, M.S., Muhammad, G., Kumar, S., Kumari, S.: Transfer reinforcement learning-based road object detection in next generation IoT domain [J]. Comput. Netw. 108078 (2021)
11. Tseng, K.K., Zhang, R., Chen, C.M.: MM Hassan DNetUnet: a semi-supervised CNN of medical image segmentation for super-computing AI service [J]. J. Supercomput. **77**(4), 3594–3615 (2021)

Multiple Kernel Clustering with Direct Consensus Graph Learning

Yanlong Wang and Zhenwen Ren

Abstract Multiple kernel graph-based clustering (MKGC) has achieved impressive experimental results, primarily due to the superiority of multiple kernel learning (MKL) and the outstanding performance of graph-based clustering. However, many present MKGC methods face the following two disadvantages that pose challenges for further improving clustering performance: (1) these methods always rely on MKL to learn a consensus kernel from multiple base kernels, which may lose some important graph information since graph learning is the key to graph-based clustering, not kernel learning; (2) these methods perform affinity graph learning and subsequent graph-based clustering in two separate steps, which may not be optimal for clustering tasks. To tackle these problems, this paper proposes a new MKGC method for multiple kernel clustering. By directly learning a consensus affinity graph rather than a consensus kernel from multiple base kernels, the important graph information can be preserved. Moreover, by utilizing rank constraint, the cluster indicators are obtained directly without performing the k-means clustering and any graph cut technique. Extensive experiments on benchmark datasets demonstrate the superiority of the proposed method.

1 Introduction

Clustering has long been a fundamental topic in computer vision and machine learning domains. The goal of clustering is to partition data points into their own clusters. Recently, there has been a surge of graph-based clustering methods [3, 4, 7, 8, 10, 12, 15], which consist of first constructing an affinity graph based on graphical rep-

---

Y. Wang
College of Media Engineering, Communication University of Zhejiang, Zhejiang 310018, China
e-mail: wangyl@cuz.edu.cn

Z. Ren (✉)
Department of National Defence Science and Technology, Southwest University of Science and Technology, Mianyang 621010, China
e-mail: rzw@njust.edu.cn

resentations of the relationships among data points, and then applying spectral or graph-theoretic algorithms to accomplish clustering.

It is well known that the performance of such graph-based methods is heavily dependent on the quality of the affinity graph [8]. However, in practice, the data points may not fit exactly to linear patterns, the affinity graph constructed on the raw nonlinear data is unable to reveal the similarity structure of the data. To efficiently handle nonlinear data, some methods have consequently extended linear hypothesis to nonlinear counterparts by kernel trick, namely: single kernel learning (SKL) [11, 12]. Although SKL is a powerful technology to efficiently handle nonlinear data, its performance is crucially determined by the choice of kernel function and parameters, this is not user-friendly since the most suitable kernel and the associated parameters for a specific dataset are usually challenging to decide [3, 16]. But multiple kernel learning (MKL) [14] is an efficient way for automatic kernel selection and integrating complementary information. When integrating MKL and graph-based clustering, this is known as multiple kernel graph-based clustering (MKGC).

Recently, MKGC has produced several state-of-the-art clustering methods [1–3, 5, 16, 17]. These methods typically work as follows: (1) constructing multiple base kernel Gram matrices relied on the given multiple base kernels, (2) learning a consensus kernel and an affinity graph, and (3) producing the clustering results on this affinity graph. However, these existing methods still suffer from the following drawbacks: (1) these methods always pay more attention to consensus kernel rather than affinity graph, this violates the fact that the affinity graph is the key of graph-based clustering problems; significantly, some important graph information from each candidate kernel may be lost, thus impairing the final clustering performance greatly and (2) these methods require an additional clustering step to produce the final clusters after learning an affinity graph.

To tackle these problems, in this paper, a new MKC method is proposed, termed Direct Consensus Graph Learning (DCGL). In summary, our main contributions are threefold:

- Unlike the existing MKGC methods, which learn a consensus kernel and an affinity graph dispersedly, DCGL concentrates fully on graph learning directly. Especially, we propose to learn multiple candidate kernel graphs and a consensus kernel affinity graph simultaneously. The former is derived from kernel self-expressiveness, and the latter is learned by an auto-weighted graph fusion strategy. By doing so, the resulting consensus affinity graph can integrate complementary information of different candidate kernel graphs, meanwhile avoid losing important graph information.
- DCGL does not need to run an additional clustering algorithm on the learned consensus affinity graph to produce the final clusters. Especially, a rank constraint on the graph Laplacian matrix of the consensus affinity matrix is introduced, which partitions the data points naturally into the required number of clusters.
- To the best of our knowledge, the highest clustering performance is obtained to date reported for multiple kernel clustering.

2 Related Works

To avoid selecting kernel function and tuning kernel parameters when handling nonlinear data clustering, a number of MKGC methods have been proposed in recent years [1–3, 5, 9, 16, 17]. For instance, affinity aggregation for spectral clustering (AASC) [2] replaces the single affinity matrix in spectral clustering with multiple kernel matrices, which can be considered as a MKL version of spectral clustering. Self-weighted multiple kernel learning (SMKL) [3] assumes that the consensus kernel is a neighborhood of multiple base kernels, then proposes a new MKL model to fuse the given multiple kernels. Spectral clustering with multiple kernels (SCMK) [4] uses the convex combination of multiple base kernels to learn a better consensus kernel by using the fact that the optimal consensus kernel is a linear combination of base kernels. By considering the neighborhood structure among base kernels, neighbor-kernel-based MKL has been proposed [17]. By considering the low-rank property of the samples, low-rank kernel learning graph-based clustering (LKGr) [5] has been proposed to impose a low-rank constraint on kernel matrix. Robust multiple kernel subspace clustering (JMKSC) [16] uses the block diagonal regularizer (BDR) and the self-expressiveness property to learn an affinity matrix with optimal block diagonal property. Local structural graph and low-rank consensus multiple kernel learning (LLMKL) [10] integrates the global and local structure preserving into a unified MKL optimization model, and has produced promising results.

3 Proposed Methodology

3.1 Notations

Throughout this paper, matrices and vectors are denoted as boldface capital letters and boldface lowercase letters, respectively. For an arbitrary matrix A, a_{ij} denotes the (i, j)-th entry and a_i denotes the i-th column or the transpose of the i-th row. For an arbitrary vector a, a_i denotes the i-th entry.

3.2 Direct Consensus Graph Learning

Based on the fact that one sample from one subspace can be represented as a linear combination of other samples in the same subspace, the well-know self-expressiveness [6] is

$$\min_{G} \|X - XG\|_F^2 + \alpha \mathcal{R}(G), \tag{1}$$

where $X \in \mathbb{R}^{d \times n}$ is the data matrix that contains n samples and d features from c clusters, $G \in \mathbb{R}^{n \times n}$ is the coefficient matrix and $\mathcal{R}(G)$ is a regularization term. Various choices of $\mathcal{R}(G)$ have been used to regularize subspace clustering [13]. After obtaining graph G, the affinity matrix (also known as affinity graph) $Z \in \mathbb{R}^{n \times n}$ can be constructed by $Z = (|G|^T + |G|)/2$ to eliminate unbalance.

However, (1) cannot handle nonlinear data efficiently. To address this issue, linear subspace clustering can be extended to nonlinear counterparts by kernel tricks [12]. Let $\phi : \mathbb{R}^d \to \mathcal{H}$ be a mapping from the input space to the reproducing kernel Hilbert space $\mathcal{H}$, and $(K \succeq 0) \in \mathbb{R}^{n \times n}$ be a kernel Gram matrix whose elements are computed as

$$K_{ij} = \phi(x_i)^T \phi(x_j) = ker(x_i, x_j), \qquad (2)$$

where $ker : \mathbb{R}^d \times \mathbb{R}^d \to \mathbb{R}$ is a kernel function. Consequently, a kernel graph can be learned by upgrading (1) to

$$\min_G \|\phi(X) - \phi(X)G\|_F^2 + \alpha \mathcal{R}(G) = \min_G \text{Tr}(K - 2KG + G^T KG) + \alpha \mathcal{R}(G). \qquad (3)$$

Obviously, (3) requires the user to select and tune a pre-defined kernel function, but the most suitable kernel for a specific task is usually challenging to decide. By contrast, MKL [14] can automatically select or combine a kernel from a set of base kernels. Based on MKL, in this paper, we fuse m candidate kernel graphs $\{G_r\}_{r=1}^m$ derived from (3) and automatically assign a suitable weight w_r to the graph G_r; meanwhile, a consensus kernel affinity graph Z in $\mathcal{H}$ is directly learned. Thus, this thought can be formulated as

$$\min_{G_r, Z, w} \sum_{r=1}^m \text{Tr}(G_r^T K_r G_r - 2K_r G_r) + \alpha \sum_{r=1}^m \|G_r\|_F^2 + \beta \sum_{r=1}^m w_r \|Z - G_r\|_F^2 \qquad (4)$$
$$\text{s.t. } G_r \geq 0, z_{ij} \geq 0, z_i^T \mathbf{1} = 1, w_r \geq 0, w^T \mathbf{1} = 1.$$

In (4), the first two terms are used to learn m candidate kernel graphs, and the third term is used to fuse these graphs. These terms can collaboratively promote each of them to reach the optimum condition. In addition, $G_r \geq 0$ can boost the representation power of homogeneous samples while limiting the representation power of heterogeneous samples; $z_{ij} \geq 0, z_i^T \mathbf{1} = 1$ is to guarantee the probability property of Z; $w_r \geq 0, w^T \mathbf{1} = 1$ is to obtain more meaningful weight and to avoid scale change. Note that we let $\mathcal{R}(G) = \|G\|_F^2$ since $\|G\|_F^2$ can encourage G_r to preserve grouping effect and ensure connectedness property [13].

For accurate clustering purpose, the affinity graph Z with accurate c connected components is strongly desired. Based on Theorem 1, the connectivity of a graph can be replaced with a rank constraint, i.e., $\text{rank}(L_Z) = n - c$.

Theorem 1 *[6] For any affinity matrix Z, the multiplicity c of the eigenvalue 0 of the corresponding Laplacian matrix ($L_Z = \mathrm{Diag}(Z1) - Z$) equals the number of connected components in Z.*

Motivated by such a theorem, denote σ_i is the i-th smallest eigenvalue of L_Z, there is $\sigma_i \geq 0$ due to $L_Z \succeq 0$. Accordingly, if $\mathrm{rank}(L_Z) = n - c$, i.e., $\sum_{i=1}^{c} \sigma_i = 0$, graph Z will contain exact c connected components. According to rank theory and Fan's theorem [8], we have

$$\mathrm{rank}(L_Z) = n - c \Rightarrow \min_{F \in \mathbb{R}^{n \times c}, F^T F = I} \mathrm{Tr}(F^T L_Z F). \tag{5}$$

Essentially, the matrix F is also the class indicator matrix that is exactly what spectral clustering [2] tends to learn. Thus, (5) can be integrated into a unified objective function to avoid the need for additional clustering steps.

Regarding the problems mentioned above, DCGL is formulated as

$$\min_{\substack{G_r, Z, \\ F, w}} \sum_{r=1}^{m} \mathrm{Tr}(G_r^T K_r G_r - 2 K_r G_r) + \alpha \sum_{r=1}^{m} \|G_r\|_F^2 + \beta \sum_{r=1}^{m} w_r \|Z - G_r\|_F^2 + \lambda \mathrm{Tr}(F^T L_Z F)$$

$$\text{s.t.} \ \forall i, z_{ij} \geq 0, z_i^T 1 = 1, w_r \geq 0, w^T 1 = 1, G_r \geq 0, F \in \mathbb{R}^{n \times c}, F^T F = I. \tag{6}$$

3.3 Optimization

(1) Update F as given w, G_r, and Z: The sub-problem w.r.t. F is

$$F^* = \arg\min_{F} \lambda \mathrm{Tr}(F^T L_Z F) \quad \text{s.t.} \ F^T F = I. \tag{7}$$

The optimal solution of (7) is the eigenvectors corresponding to the smallest c eigenvalues of L_Z.

For obtaining an optimal affinity graph with exact c diagonal blocks, in each iteration, we automatically double λ (i.e., $\lambda = 2\lambda$) or halve λ (i.e., $\lambda = \lambda/2$) when the blocks of the learned graph are smaller or greater than the number of clusters c, respectively, until to achieve $\mathrm{rank}(L_Z) = n - c$.

(2) Update G_r as given w, Z, and F: For G_r, we need to solve

$$\min_{G_i \geq 0} \sum_{r=1}^{m} \mathrm{Tr}(G_r^T K_r G_r - 2 K_r G_r) + \alpha \sum_{r=1}^{m} \|G_r\|_F^2 + \beta \sum_{r=1}^{m} w_r \|Z - G_r\|_F^2. \tag{8}$$

Taking derivative of (8) w.r.t. G_r and setting to zero gives

$$G_r^* = \lfloor (K_r + \alpha I + \beta w_r I)^{-1} (K_r + \beta w_r Z) \rfloor_+. \tag{9}$$

(3) Update Z as given w, G_r, and F: We need to solve

$$\min_Z \beta \sum_{r=1}^{m} w_r \|Z - G_r\|_F^2 + \lambda \operatorname{Tr}(F^T L_Z F) \; s.t. \; z_{ij} \geq 0, z_i^T \mathbf{1} = 1. \tag{10}$$

Due to $\operatorname{Tr}(F^T L_Z F) = \sum_{i,j=1}^{n} \|f_i - f_j\|_2^2 z_{ij}$, (10) can also be rewritten as

$$\min_{\substack{z_{ij} \geq 0 \\ z_i^T \mathbf{1} = 1}} \beta \sum_{i,j=1}^{n} \sum_{r=1}^{m} w_r (z_{ij} - p_{ij}^r)^2 + \lambda \sum_{i,j=1}^{n} \|f_i - f_j\|_2^2 z_{ij}, \tag{11}$$

where $p_{ij}^r = (G_r)_{ij}$. Then, we have

$$\min_{z_{ij} \geq 0, z_i^T \mathbf{1}=1} \sum_{r=1}^{m} \|z_i - p_i^r + \frac{\lambda}{2 m \beta w_r} q_i\|_2^2. \tag{12}$$

where $q_{ij} = \|f_i - f_j\|_2^2$. For ease of exploration, we denote p_i^r and q_i as the vectors with the j-th element as p_{ij}^r and q_{ij}, respectively. Analogously to [8], such a problem can be efficiently solved by an ALM scheme within $\mathcal{O}(n^2)$.

After obtaining Z, to avoid manually tuning the weight $w = \{w_1, w_2, \cdots, w_m\}$, a self-weighted strategy is introduced via Proposition 1.

Proposition 1 *The weight of the r-th candidate graph is determined by $w_r = \frac{1}{2\sqrt{\|Z - G_r\|_F^2}}$.*

Proof Motivated by iteratively re-weighted technique [8], we define an auxiliary problem without w as follows:

$$\min_Z \sum_{r=1}^{m} \sqrt{\|Z - G_r\|_F^2} + \lambda \operatorname{Tr}(F^T L_Z F) \tag{13}$$

$$s.t. \; z_{ij} \geq 0, z_i^T \mathbf{1} = 1, F \in \mathbb{R}^{n \times c}, F^T F = I.$$

The Lagrange function of (13) is $\sum_{r=1}^{m} \sqrt{\|Z - G_r\|_F^2} + \lambda \operatorname{Tr}(F^T L_Z F) + \Phi(\Lambda, Z)$, where Λ is Lagrange multiplier, and $\Phi(\Lambda, Z)$ indicates the indicator function of Z from the constraints. Then, we have

$$\sum_{r=1}^{m} \widehat{w}_r \frac{\partial \|Z - G_r\|_F^2}{\partial Z} + \frac{\partial \Omega(Z)}{\partial Z} = 0, \tag{14}$$

where $\Omega(Z) = \lambda \operatorname{Tr}(F^T L_Z F) + \Phi(\Lambda, Z)$ and $\widehat{w}_r = 1/(2\|Z - G_r\|_F)$. Obviously, (14) is the same as the derivation of the Lagrange function of (10). Thus, $\widehat{w}_r$ can be considered as the w_r in (8). To avoid dividing by zero in theory, $\widehat{w}_r$ can be transformed into

$$w_r = \frac{1}{2\|Z - G_r\|_F + \zeta}, \tag{15}$$

where ζ is infinitely close to zero. □

Algorithm 1 Algorithm to the proposed DCGL method.

Require: Data matrix X, and parameters α and β.
Output: The clustering results: ACC, NMI, purity.
    Initialize: $w = \frac{1}{r}, \lambda = 10^{-4}, k = \frac{n}{c}$, and $mt = 10^2$.
1: Construct r base kernel matrices $\{K_r\}_{r=1}^m$ via (2).
2: **while** (rank(L_Z)! $= n - c$) and ($++t < mt$) **do**
3:     Update $F^{(t+1)}, G_r^{(t+1)}, Z^{(t+1)}$, and w via (7), (9), (12), and Proposition 1, respectively.
4: **end while**
5: Use graphconncomp function to find the connected components of graph Z.
---

4 Experiments and Analysis

4.1 Datasets and Settings

Nine real benchmark datasets are used for evaluation [12, 14], including 4 face image sets (i.e., Yale, Jaffe, AR, and ORL), 2 object/character image sets (i.e., binaryalphadigs (BA) and COIL20 (COIL)), and 3 text corporas (i.e., TR11, TR41, and TR45). Following [1, 14], we construct 12 base kernels that contain 7 radial basis function (RBF) ones, 4 polynomial ones, and 1 cosine one.

4.2 Comparison Methods

The proposed DCGL method is compared with 6 state-of-the-art MKGC methods, including SCMK [4], LKGr [5], SMKL [3], JMKSC [16], and LLMKL [10]. For fair comparison, the parameters of these methods are carefully tuned by following the recommended experimental settings provided by their respective authors. Three widely used metrics are applied for clustering evaluation, i.e., accuracy (ACC), normalized mutual information (NMI), and purity [1, 16].

Table 1 Clustering results of all the compared MKGC methods

Dataset	Metric	LKGr [5]	SCMK [4]	SMKL [4]	JMKSC [16]	LLMKL [10]	DCGL
Yale	ACC	0.540(0.030)	0.582(0.025)	0.585(0.017)	0.630(0.006)	0.655(0.009)	**0.667(0.000)**
	NMI	0.566(0.025)	0.576(0.012)	0.614(0.015)	0.631(0.006)	0.646(0.007)	**0.655(0.000)**
	Purity	0.554(0.029)	0.610(0.014)	0.667(0.014)	0.673(0.007)	0.683(0.009)	**0.693(0.000)**
Jaffe	ACC	0.861(0.052)	0.869(0.022)	0.967(0.000)	0.967(0.007)	**1.000(0.000)**	**1.000(0.000)**
	NMI	0.869(0.031)	0.868(0.021)	0.951(0.000)	0.952(0.010)	**1.000(0.000)**	**1.000(0.000)**
	Purity	0.859(0.038)	0.882(0.023)	0.967(0.000)	0.967(0.007)	**1.000(0.000)**	**1.000(0.000)**
AR	ACC	0.314(0.015)	0.544(0.024)	0.263(0.009)	0.609(0.007)	0.853(0.013)	**0.906(0.000)**
	NMI	0.648(0.007)	0.775(0.009)	0.568(0.014)	0.820(0.002)	0.935(0.003)	**0.959(0.000)**
	Purity	0.330(0.014)	0.642(0.014)	0.530(0.014)	0.656(0.010)	0.897(0.003)	**0.910(0.000)**
ORL	ACC	0.616(0.016)	0.656(0.015)	0.573(0.032)	0.725(0.014)	0.800(0.003)	**0.885(0.000)**
	NMI	0.794(0.008)	0.808(0.008)	0.733(0.027)	0.852(0.012)	0.891(0.003)	**0.937(0.000)**
	Purity	0.658(0.017)	0.699(0.015)	0.648(0.017)	0.753(0.012)	0.839(0.009)	**0.898(0.000)**
COIL	ACC	0.618(0.051)	0.591(0.028)	0.487(0.031)	0.696(0.016)	0.636(0.010)	**0.866(0.000)**
	NMI	0.766(0.023)	0.726(0.011)	0.628(0.018)	0.818(0.007)	0.806(0.004)	**0.938(0.000)**
	Purity	0.650(0.039)	0.635(0.013)	0.683(0.004)	0.806(0.010)	0.714(0.010)	**0.888(0.000)**
BA	ACC	0.444(0.018)	0.384(0.014)	0.246(0.012)	0.484(0.015)	0.482(0.010)	**0.541(0.000)**
	NMI	0.604(0.009)	0.544(0.012)	0.486(0.011)	0.621(0.007)	0.619(0.007)	**0.654(0.000)**
	Purity	0.479(0.017)	0.606(0.009)	0.623(0.011)	0.563(0.018)	0.593(0.009)	**0.625(0.000)**
TR11	ACC	0.607(0.043)	0.549(0.015)	0.708(0.033)	0.737(0.002)	0.718(0.001)	**0.771(0.000)**
	NMI	0.597(0.031)	0.371(0.018)	0.557(0.068)	0.673(0.002)	0.633(0.002)	**0.714(0.000)**
	Purity	0.776(0.030)	0.783(0.011)	0.835(0.048)	0.819(0.001)	0.801(0.002)	**0.855(0.000)**
TR41	ACC	0.595(0.020)	0.650(0.068)	0.671(0.002)	0.689(0.004)	0.689(0.004)	**0.812(0.000)**
	NMI	0.604(0.023)	0.492(0.017)	0.625(0.004)	0.660(0.003)	0.666(0.003)	**0.756(0.000)**
	Purity	0.759(0.031)	0.758(0.034)	0.761(0.003)	0.799(0.003)	0.817(0.003)	**0.860(0.000)**
TR45	ACC	0.663(0.042)	0.634(0.058)	0.671(0.004)	0.687(0.036)	0.745(0.000)	**0.752(0.000)**
	NMI	0.671(0.020)	0.584(0.051)	0.622(0.007)	0.690(0.022)	0.726(0.000)	**0.755(0.000)**
	Purity	0.800(0.026)	0.728(0.048)	0.816(0.004)	0.822(0.031)	0.797(0.000)	**0.859(0.000)**

4.3 Experimental Results

The average clustering results with standard deviations are presented in Table 1. It can be seen that our DCGL consistently obtains the best performance. More precisely, DCGL improves by 6.91%, 4.96%, and 4.58%, respectively, compared to LLMKL (the best comparison method) in terms of ACC, NMI, and purity.

In addition, to assess the quality of the learned graph Z, we illustrate Z on the Jaffe dataset in Fig. 1, for example. Obviously, the matrix Z of our DCGL has better block diagonal property and inter-cluster separability than the compared methods.

(a) LKGr (b) SMKL (c) SCMK (d) JMKSC (e) LLMKL (f) DCGL

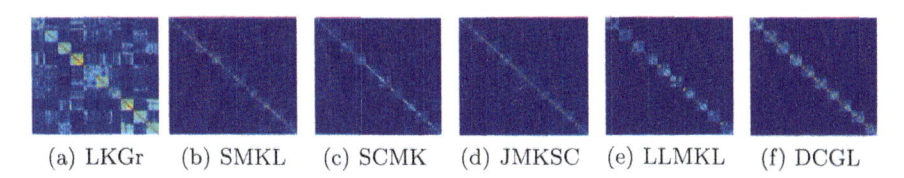

Fig. 1 Visualization of the learned affinity graph Z on the Jaffe dataset

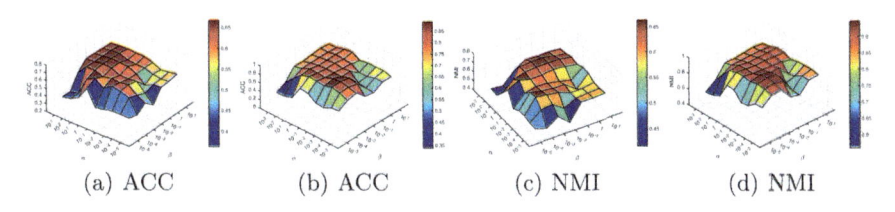

(a) ACC (b) ACC (c) NMI (d) NMI

Fig. 2 ACC and NMI of DCGL w.r.t. α and β on the Yale and ORL datasets

4.4 Parameter Sensitivity

Two parameters (i.e., α and β) are needed to be tuned in (6), where α balances the effect of the term $\|G_r\|_F^2$ to encourage the data grouping effect in $\mathcal{H}$, and β controls the term $\sum_{r=1}^{m} w_r \|Z - G_r\|_F^2$ that fuses m candidate kernel graphs to learn an optimal consensus kernel affinity graph Z for clustering purpose. By employing a grid search technique, α and β are selected from $\{10^{-5}, \cdots, 10^3\}$ and $\{10^{-5}, \cdots, 10\}$, respectively. Additionally, the initial λ is set to 10^{-4}, and k is fixed as n/c, where n and c are the number of samples and clusters, respectively.

Taking the ORL dataset for example, as shown in Fig. 2, the proposed DCGL method can achieve the best performance when setting $\alpha, \beta \in [10^{-4}, 10^{-2}]$.

4.5 Convergence Study

The convergence curves and the clustering ACC curves of the proposed DCGL method evaluated on the Yale, ORL, COIL20, and TR41 datasets are shown in Fig. 3. As seen, Algorithm 1 can converge quickly and obtain the desired c block diagonal components within about 30 iterations. Especially, it spends 16, 9, 28, 17 iterators for the four datasets, respectively. This validates that DCGL has very strong and stable convergence behavior.

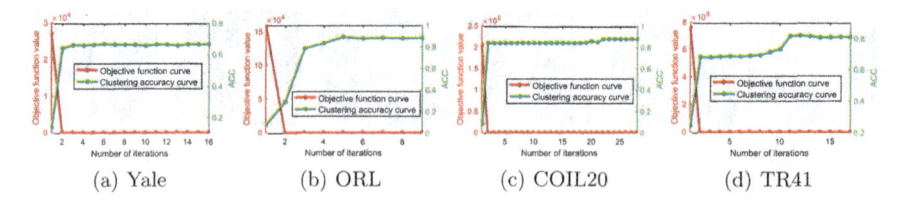

Fig. 3 Convergence curve and clustering ACC of DCGL

5 Conclusion

A new graph-based method is proposed for multiple kernel clustering tasks. Different from existing methods, our learning goal is shifted from multiple kernel learning to multiple graph learning, which can address some known problems, such as graph information loss and additional clustering burden. The experiments on nine benchmark datasets demonstrate the effectiveness and efficiency of the proposed method.

Acknowledgements This research was supported by the Key Lab of Film and TV Media Technology of Zhejiang Province (Grant no. 2020E10015), and the Zhejiang Province Public Welfare Technology Application Research Project (Grant no. LGF21F020003).

References

1. Du, L., Zhou, P., Shi, L., Wang, H., Fan, M., Wang, W., Shen, Y.D.: Robust multiple kernel k-means using l21-norm, pp. 3476–3482 (2015)
2. Huang, H.C., Chuang, Y.Y., Chen, C.S.: Affinity aggregation for spectral clustering. In: 2012 IEEE Conference on Computer Vision and Pattern Recognition (CVPR), pp. 773–780. IEEE (2012)
3. Kang, Z., Lu, X., Yi, J., Xu, Z.: Self-weighted multiple kernel learning for graph-based clustering and semi-supervised classification. IJCAI, pp. 2312–2318 (2018)
4. Kang, Z., Peng, C., Cheng, Q., Xu, Z.: Unified spectral clustering with optimal graph. In: Proceedings of the AAAI Conference on Artificial Intelligence, vol. 32 (2018)
5. Kang, Z., Wen, L., Chen, W., Xu, Z.: Low-rank kernel learning for graph-based clustering. Knowl.-Based Syst. **163**, 510–517 (2019)
6. Lu, C., Feng, J., Lin, Z., Mei, T., Yan, S.: Subspace clustering by block diagonal representation. IEEE Trans. Pattern Anal. Mach. Intell. **41**(2), 487–501 (2018)
7. Lu, J., Ni, J., Li, L., Luo, T., Chang, C.: A coverless information hiding method based on constructing a complete grouped basis with unsupervised learning. J. Netw. Intell. **6**(1), 29–39 (2021)
8. Nie, F., Wang, X., Deng, C., Huang, H.: Learning a structured optimal bipartite graph for co-clustering. In: Advances in Neural Information Processing Systems, pp. 4129–4138 (2017)
9. Ren, Z., Lei, H., Sun, Q., Yang, C.: Simultaneous learning coefficient matrix and affinity graph for multiple kernel clustering. Inf. Sci. **547**, 289–306 (2021)
10. Ren, Z., Li, H., Yang, C., Sun, Q.: Multiple kernel subspace clustering with local structural graph and low-rank consensus kernel learning. Knowl.-Based Syst. 105040 (2019)

11. Ren, Z., Mukherjee, M., Bennis, M., Lloret, J.: Multi-kernel clustering via non-negative matrix factorization tailored graph tensor over distributed networks. IEEE J. Select. Areas Commun. (2020)
12. Ren, Z., Sun, Q.: Simultaneous global and local graph structure preserving for multiple kernel clustering. IEEE Trans. Neural Netw. Learn. Syst. (2020)
13. Ren, Z., Sun, Q., Wu, B., Zhang, X., Yan, W.: Learning latent low-rank and sparse embedding for robust image feature extraction. IEEE Trans. Image Process. **29**, 2094–2107 (2019)
14. Ren, Z., Yang, S.X., Sun, Q., Wang, T.: Consensus affinity graph learning for multiple kernel clustering. IEEE Trans. Cybern. (2020)
15. Wang, R., An, Z., Wang, W., Yin, S., Xu, L.: A multi-stage data augmentation approach for imbalanced samples in image recognition. J. Netw. Intell. **6**(1), 94–106 (2021)
16. Yang, C., Ren, Z., Sun, Q., Wu, M., Yin, M., Sun, Y.: Joint correntropy metric weighting and block diagonal regularizer for robust multiple kernel subspace clustering. Inf. Sci. **500**, 48–66 (2019)
17. Zhou, S., Liu, X., Li, M., Zhu, E., Liu, L., Zhang, C., Yin, J.: Multiple kernel clustering with neighbor-kernel subspace segmentation. IEEE Trans. Neural Netw. Learn. Syst. **31**(4), 1351–1362 (2019)

Density Peaks Clustering Algorithm Based on K Nearest Neighbors

Shihao Yin, Runxiu Wu, Peiwu Li, Baohong Liu, and Xuefeng Fu

Abstract Density peaks clustering algorithms calculate the local density based on the cutoff distance and the global distribution of the sample. They cannot capture the local characteristics of the sample well, and are prone to appear errors in the selection of density peaks; additionally, the allocation strategy has poor fault tolerance. Once a sample is allocated incorrectly, subsequent allocations will magnify the error. Hence, we proposed a density peaks clustering algorithm based on k-nearest neighbors (DPC-KNN). First, the k-nearest neighbors information of the sample is used to define the local density of the sample in order to find the cluster centers accordingly; the sample with the distance between cluster centers and k-nearest neighbors sample less than the set threshold is defined as the core sample, and the core sample is classified into the corresponding cluster to construct the core area of the cluster; after the degree of attribution of the remaining samples and various clusters are calculated, they are allocated to clusters with high degree of attribution. In order to verify the effectiveness of the proposed algorithm, eight synthetic datasets and ten UCI datasets are selected for experiments, and the proposed algorithm is compared with FKNN-DPC, DPCSA, FNDPC, DPC and DBSCAN. The experimental results indicated that the proposed algorithm had better clustering performance.

S. Yin · R. Wu (✉) · P. Li · B. Liu · X. Fu
School of Information Engineering, Nanchang Institute of Technology, Nanchang 330099, China
e-mail: wrx@nit.edu.cn

S. Yin
e-mail: yinshihao97@163.com

P. Li
e-mail: lpw@nit.edu.cn

X. Fu
e-mail: fxfcn@126.com

1 Introduction

As a key technology in data mining, clustering divides original data into multiple clusters based on sample similarity so that samples in the same cluster have high similarities, while samples in different clusters have low similarities [1–3]. Aiming to recognize the internal hiding mode of data, clustering has been widely applied in image processing, network security, pattern recognition, bioinformatics and document retrieval [4]. According to the clustering principles, clustering methods can be divided as division-based, hierarchy-based, density-based, network-based and model-based ones. Among them, density-based clustering algorithms have attracted great attention since they can effectively identify clusters with random shapes and reduce noises.

Density peaks clustering (DPC) [5] algorithm is facile yet effective than density-based clustering methods. DPC algorithms are based on two assumptions: (1) cluster centers are surrounded by samples with low density; (2) cluster centers are relatively far away from each other.

First, local density and relative distance of samples are calculated and cluster centers are identified using decision graph. Then, samples are allocated to clusters with identical nearest neighbors but higher sample density. The main advantages of DPC algorithm include no need to specify the number of clusters, only one parameter, and high computational efficiency (compared with other density-based clustering methods). However, DPC algorithm has the following disadvantages: (1) The local density does not reflect the sample distribution well, and can easily lead to the wrong selection of density peaks; (2) The allocation strategy has poor fault tolerance. Once a sample is allocated incorrectly, the rest of the samples allocated to the sample will continue to magnify the error.

Aiming at the defect of definition of local density, Duet et al. [6] estimated local density using k-nearest neighbors and proposed density peaks clustering algorithm based on k-nearest neighbors (DPC-KNN). Xie et al. [7] used a exponential kernel function with width $\delta = 1$ to define the local density of samples. The above improved methods are based on the local distribution of the samples, which reflect the true distribution information of the samples better than the conventional definition of local density, and the parameter K in the algorithm is better determined than d_c. However, the degree of differentiation between cluster centers and other samples is not high.

For allocation strategy, Liu et al. [8] introduced the concepts of inevitable subordination and possible subordination and assigned the labels of non-cluster center samples. Lu et al. [9] divided the samples into core and non-core samples and used different allocation strategy to allocate samples. Seyedi et al. [10] use graph-based label propagation to assign labels to non-cluster centers. The fuzzy weighted allocation strategy [11] can reduce the propagation of clustering label errors. The key step is to calculate the probability p_i^c that the sample i belongs to the cluster c, and then allocate the sample to the cluster with the largest p_i^c. Xue et al. [12] proposed a method combining k-nearest neighbors with iterative ideas to realize the allocation of samples and avoided dividing a cluster into multiple clusters by mistake through the cluster merging strategy.

In this study, we propose a DPC-KNN algorithm. First, the k-nearest neighbors and Gaussian kernel are combined to calculate the local density of samples, and the decision graph is used to find the cluster centers; then the sample with the distance between cluster centers and k-nearest neighbors sample less than the set threshold is defined as the core sample, and the core samples are classified into the corresponding cluster to construct the backbone area of the cluster; after the degree of attribution of the remaining samples and various clusters are calculated, they are allocated to the cluster with a high degree of attribution. The proposed algorithm designed a new density peaks searching method and the two-step allocation strategy, which solved the problem of inaccurate selection of cluster centers and poor fault tolerance of the allocation strategy in DPC algorithm. The experimental results indicated that the proposed algorithm had good adaptability and high accuracy.

2 DPC Algorithm

Each sample of DPC algorithm includes two attributes: local density ρ_i and relative distance δ_i. DPC algorithms are based on two assumptions. First, the local density of cluster centers is higher than other samples in the same cluster, and cluster centers have a relatively large distance from any sample with higher local density. For Sample i, local density ρ_i has two definition methods. Large-scale datasets use truncated kernel, that is, Eq. (1) calculates local density, and small-scale datasets use Gaussian kernel, that is, Eq. (2) calculates local density.

$$\rho_i = \sum_j \chi(d_{ij} - d_c) \quad \chi(x) = \begin{cases} 0 & x \geq 0 \\ 1 & x < 0 \end{cases} \quad (1)$$

$$\rho_i = \sum_j \exp\left(-\left(\frac{d_{ij}}{d_c}\right)^2\right) \quad (2)$$

where d_{ij} represents the Euclidean distance between Samples i and j; d_c is the cutoff distance and the only input parameter.

For Sample i, the distance δ_i is defined by Eq. (3):

$$\delta_i = \min_{j:\rho_j > \rho_i}(d_{ij}) \quad (3)$$

If Sample i has maximum local density, $\delta_i = \max_{i \neq j}(\delta_j)$.

DPC believes that when ρ_i and δ_i are both large, the sample can be used as candidate cluster centers. The sample with a larger decision value γ is selected as the cluster centers in the decision graph. The definition of the decision value γ is as follows:

$$\gamma_i = \rho_i \times \delta_i \tag{4}$$

After selecting cluster centers, the remaining samples are allocated to the cluster of the sample with a local density larger than it and the closest distance to it.

3 DPC-KNN Algorithm

The definition of local density in DPC algorithm does not reflect the sample distribution well, and DPC algorithm is prone to appear errors in the selection of density peaks, the allocation strategy is too simple and the fault tolerance is poor. DPC algorithm is improved in two aspects: one is to combine k-nearest neighbors with Gaussian kernel to define the local density of samples; the second is to change the allocation strategy of the remaining samples after finding the cluster centers. Based on the local distribution of the samples, the local density is calculated through the Gaussian kernel, the decision graph is used to find the cluster centers; the sample with the distance between cluster centers and k-nearest neighbors sample less than the set threshold is defined as the core sample, and the core sample is classified into the corresponding cluster, and the core area of the cluster is constructed based on this; after the degree of attribution of the remaining samples and various clusters is calculated, the remaining samples are allocated to the cluster with a high degree of attribution.

3.1 Definition of Local Density

DPC algorithm calculates the local density based on the cutoff distance dc and samples in global range; however, the density peaks of each cluster are the samples with higher density in the cluster, that is, the samples with higher density in the local range. DPC algorithm only considers the global structure of samples, cannot capture the local structural characteristics of the sample well, and cannot reflect the distribution of the sample well. DPC algorithm is prone to appear incorrect selection of density peaks. Hence, DPC-KNN combines k-nearest neighbors with Gaussian kernel to redefine the local density of samples. The local density of Sample i is defined by:

$$\rho_i = \sum_{j \in KNN_i} \exp\left(-\left(\frac{d_{ij}}{\delta_i^K}\right)^2\right) \qquad (5)$$

where d_{ij} is the Euclidean distance between samples i and j, KNN_i represents the set of k-nearest neighbors of Sample i, and δ_i^K is the KNN distance of the sample i:

$$\delta_i^K = \max_{j \in KNN_i} d_{ij} \qquad (6)$$

DPC-KNN starts from the local distribution of the samples, introduces δ_i^K to adjust the density contribution of each sample to the current sample, and calculates the local density of samples in the form of Gaussian kernel. This definition of local density makes full use of the distribution information of its k-nearest neighbor samples, which can reflect the actual distribution of the sample objectively. Meanwhile, it can identify cluster centers with a higher degree of discrimination to achieve better clustering performance.

Figure 1a shows the synthetic datasets consisting of five clusters, Fig. 1b–d shows the γ decision graph corresponding to different definitions of local density, the parameters are dc = 2 and K = 13. As observed, Fig. 1b, c select four density peaks, Fig. 1d selects five density peaks; Fig. 1d has a greater discrimination between density peaks and non-density peaks. Hence, the proposed definition of local density can identify cluster centers with a high degree of discrimination.

3.2 Design of Allocation Strategy

The allocation strategy of DPC is simple and efficient, but has poor fault tolerance. For this reason, Xie et al. [11] proposed robust clustering by detecting density peaks and assigning points based on fuzzy weighted K-nearest neighbors (FKNN-DPC). FKNN-DPC proposed strategy 1 based on KNN and strategy 2 based on fuzzy weighted KNN, and defined thresholds to divide the samples into outliers and non-outliers samples, and performed the allocation of samples in two steps. Strategy 1 marked density peaks as allocated samples, classified the non-outlier samples with distance from density peaks not greater than the average distance from its k-nearest neighbor samples into this cluster. Strategy 2 introduces fuzzy weighted k-nearest neighbor technology to define the degree of attribution between samples in order to allocate samples.

Strategy 1 of FKNN-DPC searches for the k-nearest neighbor samples of density peaks and classifies non-outlier samples with distance from the density peaks not greater than the average distance from its k-nearest neighbor samples into this cluster, which may result in too few samples in some core areas. In this paper, strategy 1 is changed. After the threshold is reset to classify the samples, the core samples of k-nearest neighbor points of the cluster centers are classified into the corresponding

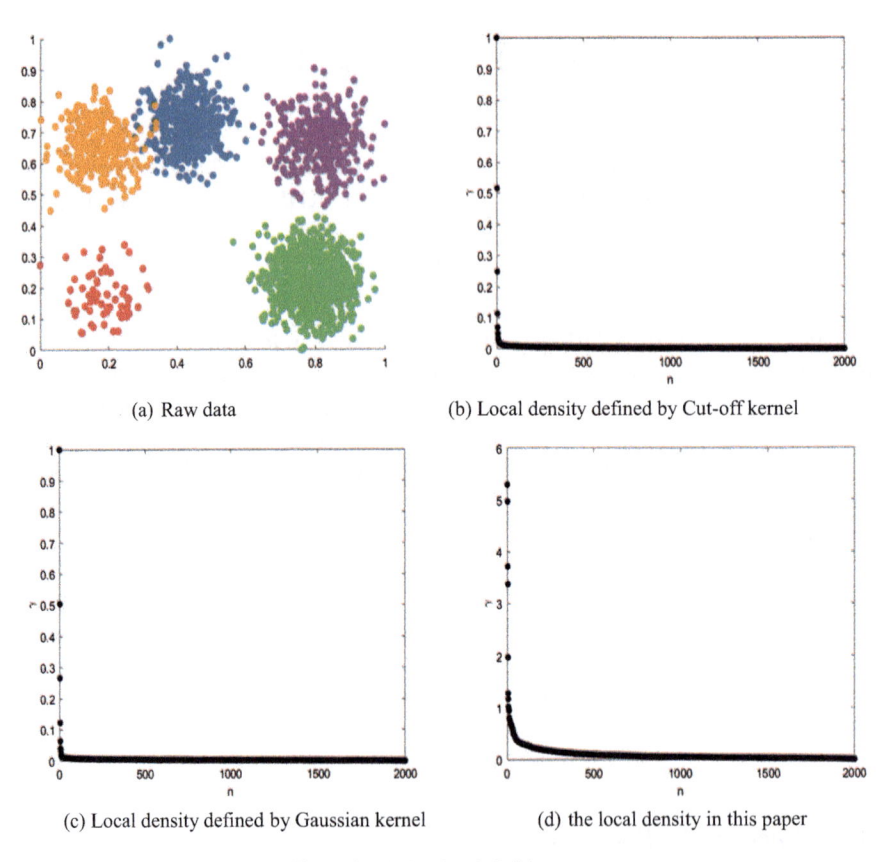

Fig. 1 Decision graphs with different local density definitions

(a) Raw data
(b) Local density defined by Cut-off kernel
(c) Local density defined by Gaussian kernel
(d) the local density in this paper

clusters, which ensured that the cluster core area obtained after the first step allocation has a certain scale of samples, and the samples are compact. The allocation process of the second step is the same as strategy 2 of FKNN-DPC. The threshold τ defined in this paper is shown in (9), δ_i^K is the KNN distance of Sample i, m is the average of the distance of all samples, N is the number of samples in the dataset, and σ is the mean square error of all samples δ_i^K and m. This threshold makes the selection of core area points of cluster more compact, which is conducive to the subsequent allocation of samples.

$$m = \frac{1}{N}\sum_{i=1}^{N} \delta_i^K \tag{7}$$

$$\sigma = \sqrt{\frac{1}{N}\sum_{i=1}^{N} (\delta_i^K - m)^2} \tag{8}$$

$$\tau = \frac{m}{1+\sigma} \tag{9}$$

4 Experimental Results and Analysis

To verify the performance of DPC-KNN, we used eight classical synthetic datasets and ten UCI datasets to test DPC-KNN, and compared DPC-KNN with FKNN-DPC [11], DPCSA [13], FNDPC [14], DPC [5] and DBSCAN [15]. The basic properties of all datasets involved in this study are given in Tables 1 and 2. The experimental environment is Win 10 64bit operating system, MATLAB R2020a software, 12.0 GB memory, and Intel(R)Core (TM) i5-10210U CPU @ 1.60 GHz processor.

In order to test the clustering performance of various algorithm more objectively, we performed parameter tuning for each algorithm. DPC algorithm determines the cutoff distance by selecting the percentage in the distance matrix, and the ratio was selected from 0.01%-5%. The parameter ε of FNDPC was selected between 0.01 and 1. The K of FKNN-DPC was selected between 2 and 50. DBSCAN needs to set two parameters including the radius ε of the field and the minimum number of samples Minpts. The radius ε of the field, with 0.01 as the step size, was selected

Table 1 Synthetic datasets

Dataset	Source	No. of records	No. of attributes	No. of clusters
Jain	[18]	373	2	2
Pathbased	[19]	300	2	3
Aggregation	[20]	788	2	7
Flame	[21]	240	2	2
R15	[22]	600	2	15
Spiral	[19]	312	2	3
D31	[22]	3100	2	31
S2	[23]	5000	2	15

Table 2 UCI datasets

Dataset	Source	No. of records	No. of attributes	No. of clusters
Iris	[24]	150	4	3
Wine	[24]	178	13	3
Wdbc	[25]	569	30	2
Seeds	[26]	210	7	3
Balance	[24]	625	4	3
Sonar	[24]	208	60	2
Ionosphere	[27]	351	34	2
pima	[24]	768	8	2
Ecoli	[24]	336	8	8
Dermatology	[24]	366	33	6

between 0.01 and 1; the minimum number of samples Minpts included in the field, was selected between 1 and 100. The K of DPC-KNN was selected between 2 and 100.

This paper selected three evaluation indicators independent of the absolute value of the label to evaluate clustering performance. Clustering evaluation indicators include Adjusted Rand Index (ARI) [16], Adjusted Mutual Information (AMI) [16] and Fowlkes-Mallows Index (FMI) [17]. The best results of the three indicators are all 1. The closer the value is to 1, the better the clustering performance is.

4.1 Data Pre-processing

The synthetic datasets and practical datasets selected in this paper are widely used to test the performance of various clustering algorithm. Datasets involved in this study have different sample scales, number of attributes and number of clusters. Generally, synthetic datasets have a higher number of clusters, and practical datasets have a high number of features. Tables 1 and 2 show the datasets involved in this study in detail. In order to eliminate the influence of the numerical range of the sample attribute values on the experimental results, Eq. (10) is used to normalize each sample attribute.

$$x'_{ij} = \frac{x_{ij} - \min(x_{ij})}{\max(x_j) - \min(x_j)} \qquad (10)$$

where x'_{ij} is the normalized data of the i-th row and the j-th column, x_{ij} is the original data of the i-th row and the j-th column, and x_j is the collection of all the data in the j-th column.

4.2 Experimental Results of Synthetic Datasets

Synthetic datasets are all 2D datasets, with different sample distributions, including manifold, multi-scale, cross-winding and other characteristics. Its inherent characteristics are very suitable for testing and comparing the performances of various clustering algorithm. The clustering performance figure can reflect clustering performance more intuitively. Table 3 shows the clustering results of 6 clustering algorithm on 8 synthetic datasets. In the table, "Arg-" is the parameter value when each algorithm obtains the optimal clustering result, and "-" means that no input parameters are required. The blacked values in the table indicate the optimal clustering results. The dots in different colors in the figure represent different clusters. Except for DBSCAN, the cluster centers of other algorithm are represented by "hexagonal stars", and the "cross" in DBSCAN means the noise points determined by this algorithm. Due to

Table 3 The clustering results of six clustering algorithms on eight synthetic datasets

Algorithm	AMI	ARI	FMI	Arg-	AMI	ARI	FMI	Arg-
	Aggregation				Spiral			
DPC-KNN	**0.9950**	**0.9971**	**0.9977**	30	1	1	1	2
FKNN-DPC	0.9905	0.9949	0.9960	20	1	1	1	6
DPCSA	0.9537	0.9581	0.9673	–	1	1	1	–
FNDPC	0.9864	0.9913	0.9932	0.02	1	1	1	0.07
DPC	0.9922	0.9956	0.9966	4.00	1	1	1	1.8
DBSCAN	0.9681	0.9779	0.9827	0.04/6	1	1	1	0.04/2
	Pathbased				Flame			
DPC-KNN	0.6465	0.5902	0.7325	3	0.9267	0.9666	0.9845	8
FKNN-DPC	**0.9305**	**0.9499**	**0.9665**	9	0.9267	0.9666	0.9845	5
DPCSA	0.7073	0.6133	0.7511	–	1	1	1	–
FNDPC	0.5751	0.5067	0.7065	0.01	1	1	1	0.13
DPC	0.5212	0.4717	0.6664	3.8	1	1	1	2.8
DBSCAN	0.8721	0.9011	0.9340	0.08/10	0.8665	0.9388	0.9712	0.09/8
	Jain				R15			
DPC-KNN	1	1	1	2	**0.9938**	**0.9929**	**0.9933**	40
FKNN-DPC	0.7092	0.8224	0.9359	43	**0.9938**	0.9928	**0.9933**	25
DPCSA	0.2167	0.0442	0.5924	–	0.9885	0.9857	0.9866	–
FNDPC	0.5961	0.7257	0.9051	0.47	**0.9938**	0.9928	**0.9933**	0.03
DPC	0.6183	0.7146	0.8819	0.9	**0.9938**	0.9928	**0.9933**	0.7
DBSCAN	0.9281	0.9758	0.9906	0.08/2	0.9832	0.9758	0.9799	0.04/12
	D31				S2			
DPC-KNN	**0.9656**	**0.9523**	**0.9538**	24	**0.9480**	**0.9406**	**0.9445**	23
FKNN-DPC	0.9654	**0.9523**	**0.9538**	28	0.9180	0.8889	0.8963	22
DPCSA	0.9552	0.9353	0.9374	–	0.9333	0.9152	0.9209	-
FNDPC	0.9555	0.9364	0.9385	0.04	0.9431	0.9351	0.9395	0.03
DPC	0.9554	0.9365	0.9385	0.6	0.9437	0.9352	0.9395	1.5
DBSCAN	0.9034	0.8101	0.8168	0.04/47	0.8781	0.7510	0.7767	0.04/30

the length of the paper, only the clustering performances of 6 clustering algorithms on Aggregation and Jain datasets are listed.

For the eight synthetic datasets, DPC-KNN achieved the best clustering performance on 6 datasets, only the clustering results of Path-based and Flame datasets were not ideal, and the clustering performances of other datasets were better than those of other algorithm. DBSCAN had the same clustering results on the Spiral dataset as DPC-KNN, and better clustering results on Path-based dataset than DPC-KNN. DPCSA was better than DPC-KNN on Flame and Path-based datasets, and had the same clustering results on Spiral dataset as DPC-KNN. DPC and FNDPC had better clustering results on Flame dataset than DPC-KNN, and had the same clustering results on Spiral dataset as DPC-KNN. Compared with FKNN-DPC, DPC-KNN was superior in five datasets, comparable on two datasets, and inferior on one dataset.

Figure 2 shows the clustering results of six clustering algorithms on Jain dataset. Jain dataset consists of two crescent clusters, one is sparse and the other is compact. For Jain dataset, DPC-KNN can obtain accurate clustering performance; although FKNN- DPC and FNDPC found the correct cluster centers, however, during the sample allocation process, some samples in the sparse cluster were incorrectly allocated to the compact cluster; DPCSA and DPC did not find the correct cluster centers, resulting in poor clustering performance of the algorithm. DBSCAN successfully performed clustering on compact clusters, meanwhile, it marked some samples of sparse clusters as noise, resulting in slightly worse clustering results (Fig. 3).

Figure 3 shows the clustering results of six algorithms on the Aggregation dataset. The Aggregation dataset consists of seven clusters of different sizes and shapes, and the distances between the clusters are not equal. As observed, except for DPCSA and DBSCAN, other algorithm can obtain good clustering performances on datasets. However, the results in Table 3 show that clustering performance of DPC-KNN was the best. Other algorithm had more edge sample allocation errors in the processing of edge points. DPCSA incorrectly allocated some core samples of the lower right cluster to upper right cluster. DBSCAN incorrectly marked part of the edge points of the upper left cluster and the lower right cluster as noise, resulting in slightly worse results.

In order to reflect the comprehensive performance of each algorithm on synthetic datasets, the Friedman test was used to perform the mean rank test on AMI, ARI and FMI, respectively. The mean rank is equivalent to the comprehensive score of the algorithm. The larger the value is, the better the performance of the corresponding algorithm is. The Friedman test values of the evaluation indicators of 6 algorithms on synthetic datasets are shown in Table 4. As observed, the mean rank of DPC-KNN ranks first among the three clustering evaluation indicators. Therefore, among the six comparative clustering algorithms, DPC-KNN had the best overall performance.

4.3 Analysis of Experimental Results on Practical Datasets

In order to better verify the clustering performance and applicability of 6 clustering algorithms, 10 practical datasets with different scales, different features and different numbers of clusters were selected for experiments. The basic features of all datasets involved in this study are shown in Table 2. In order to further verify the effectiveness of the proposed algorithm, the proposed algorithm is compared with other 5 algorithms on 10 UCI datasets. Table 5 shows the clustering results of 6 clustering algorithm on UCI datasets. As can be seen from Table 5, each clustering evaluation index of Wine, E. coli, Dermatology and Wdbc datasets, DPC-KNN is better than that

(a) DPC-KNN

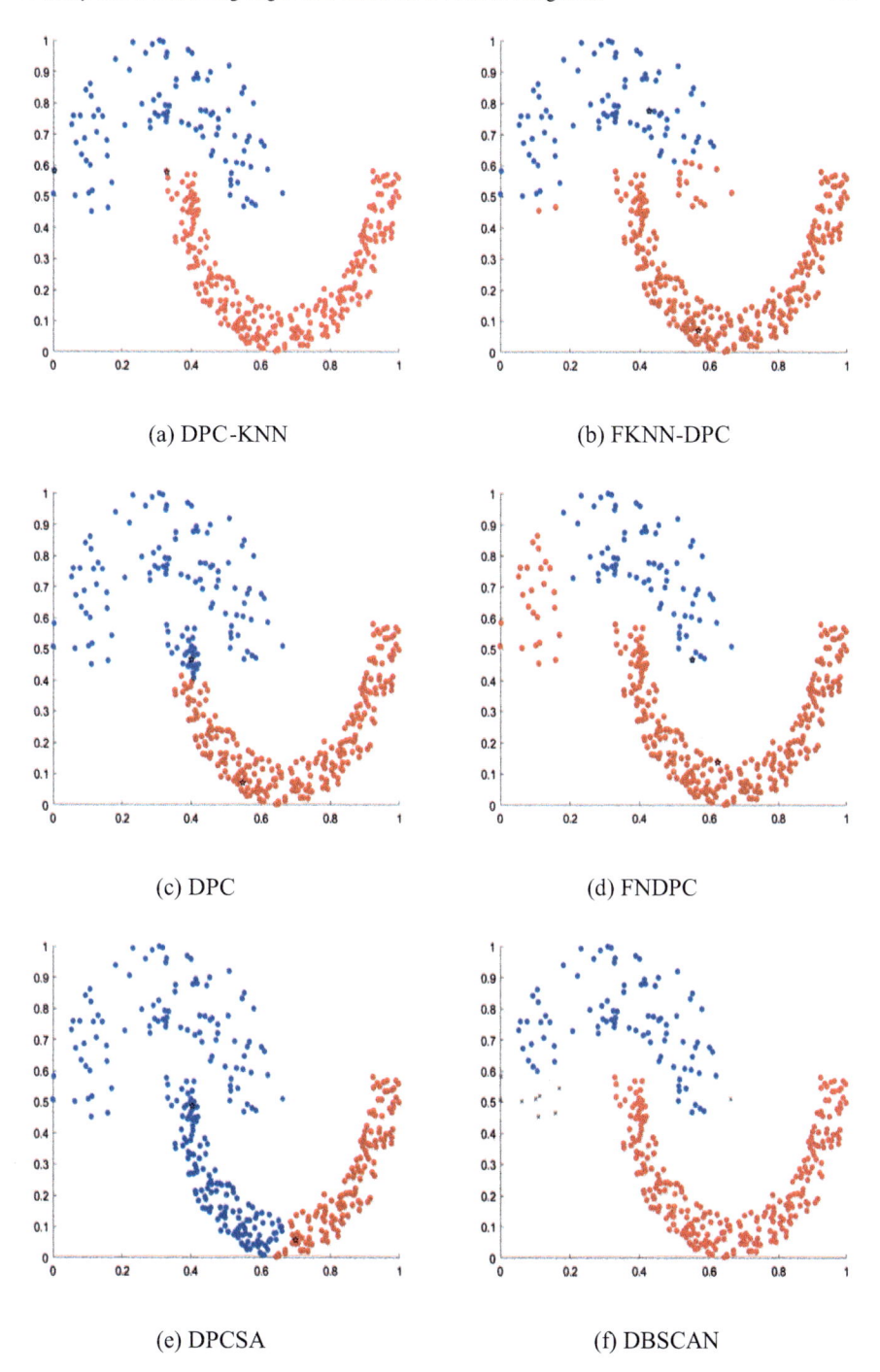

(b) FKNN-DPC

(c) DPC

(d) FNDPC

(e) DPCSA

(f) DBSCAN

Fig. 2 Clustering results of six clustering algorithms on Jain dataset

(a) DPC-KNN

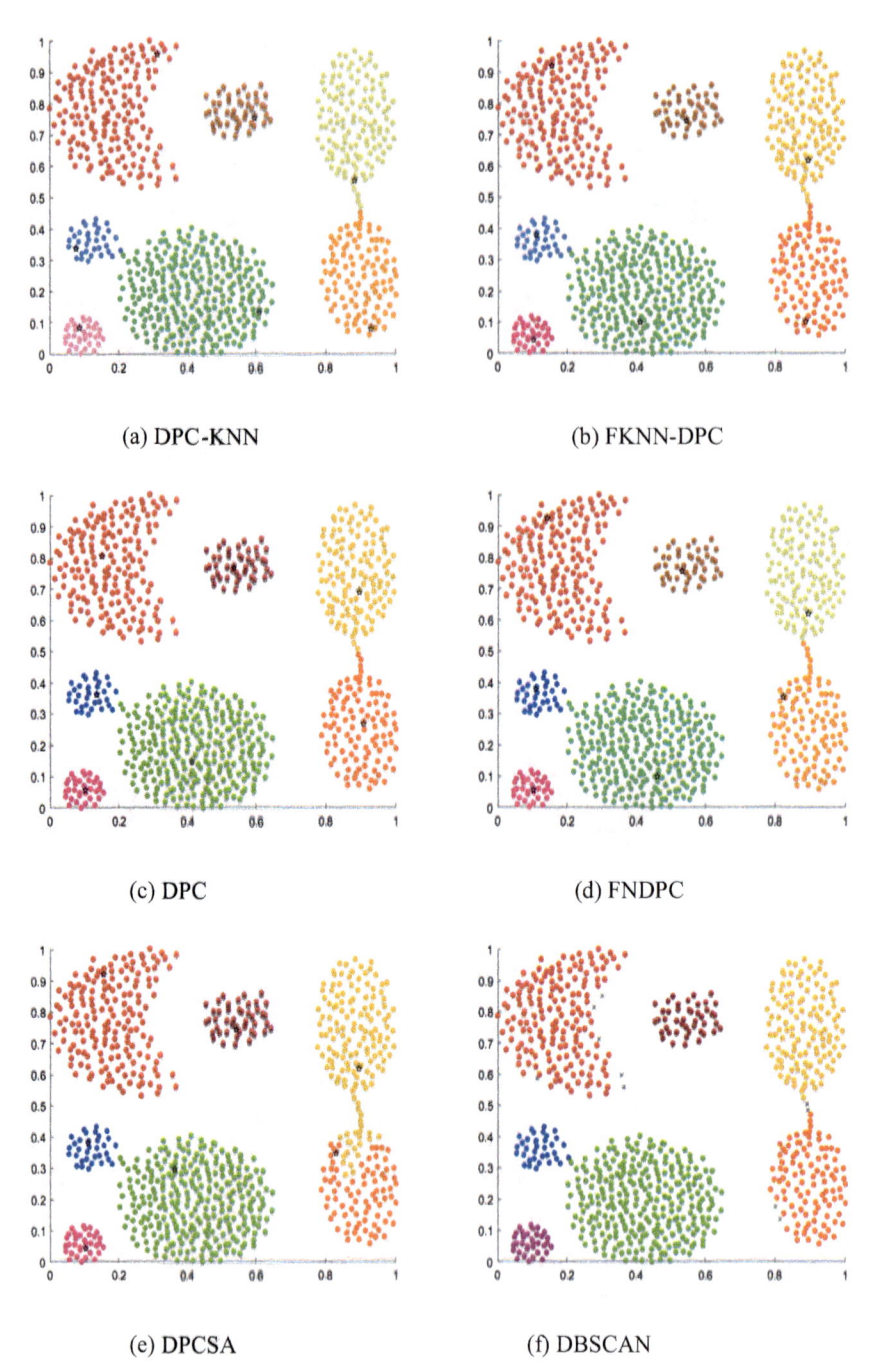

(b) FKNN-DPC

(c) DPC

(d) FNDPC

(e) DPCSA

(f) DBSCAN

Fig. 3 Clustering results of six clustering algorithms on Aggregation dataset

Table 4 Friedman test values of the evaluation indicators of six algorithms on synthetic datasets

AMI		ARI		FMI	
Algorithms	Mean ranks	Algorithms	Mean ranks	Algorithms	Mean ranks
DPC-KNN	**4.69**	DPC-KNN	**4.81**	DPC-KNN	**4.69**
FKNN-DPC	3.94	FKNN-DPC	3.94	FKNN-DPC	4.06
FNDPC	3.50	FNDPC	3.44	FNDPC	3.69
DPC	3.75	DPC	3.69	DPC	3.44
DPCSA	2.69	DPCSA	2.69	DPCSA	2.69
DBSCAN	2.44	DBSCAN	2.44	DBSCAN	2.44

of the comparison algorithm. FKNN-DPC and DBSCAN had the best performances on Seeds and Ionosphere dataset, respectively. The clustering performances of the 6 clustering algorithms on other datasets are very poor, however, DPC-KNN had the best performance on the sonar and pima datasets, and the clustering performance of FNDPC on the Balance dataset was the best.

The Friedman test values of the evaluation indicators of the six algorithms on UCI datasets are shown in Table 6. As observed, on the three cluster evaluation indicators, the rank average of DPC-KNN, FKNN-DPC and FNDPC ranks first, second and third, respectively. Overall, DPC-KNN has the best overall performance.

5 Conclusions

Aiming at problems in the clustering process of DPC algorithm, we propose a DPC-KNN algorithm. This algorithm introduces k-nearest neighbors, calculates local density of samples and finds density peaks; after finding density peaks, the threshold is set to divide the samples into core samples and non-core samples, the core samples in the k-nearest neighbors samples of cluster centers are merged into corresponding clusters; after the degree of attribution of the remaining samples and various clusters are calculated, they are allocated to the clusters with high degrees of attribution. DPC-KNN effectively solves the problem of allocation errors in the sample allocation process of DPC algorithm and difficulty in determining cluster centers. Through experiments on multiple synthetic datasets and UCI datasets, it is found that DPC-KNN can quickly search and find cluster centers of datasets of arbitrary shape, and the algorithm has good clustering performance. The proposed algorithm has a parameter K. The value of K, which is an integer, is easier to be determined compared with that of dc. However, K still needs to be determined manually, and the adaptive determination of the parameter K is the next research direction.

Table 5 The clustering results of six clustering algorithms on UCI datasets

Algorithm	AMI	ARI	FMI	Arg-	AMI	ARI	FMI	Arg-
	Iris				Wine			
DPC-KNN	**0.8831**	**0.9038**	**0.9355**	10	**0.9055**	**0.9309**	**0.9541**	26
FKNN-DPC	**0.8831**	**0.9038**	**0.9355**	22	0.8481	0.8839	0.9229	8
DPCSA	**0.8831**	**0.9038**	**0.9355**	–	0.748	0.7414	0.8283	–
FNDPC	**0.8831**	**0.9038**	**0.9355**	0.11	0.7898	0.8025	0.8686	0.26
DPC	0.7247	0.7037	0.8032	0.2	0.7065	0.6724	0.7835	2
DBSCAN	0.6401	0.6120	0.7291	0.12/5	0.5905	0.5292	0.7121	0.50/21
	Seeds				Ecoli			
DPC-KNN	0.7142	0.7340	0.8234	74	**0.6734**	**0.7591**	**0.8311**	9
FKNN-DPC	**0.7757**	**0.8024**	**0.8682**	9	0.5878	0.5894	0.7027	2
DPCSA	0.6609	0.6873	0.7918	–	0.4406	0.4593	0.6467	–
FNDPC	0.7136	0.7545	0.8361	0.07	0.4833	0.5618	0.7178	0.35
DPC	0.7298	0.7670	0.8444	0.7	0.4978	0.4465	0.5775	0.4
DBSCAN	0.5912	0.5291	0.6711	0.24/16	0.5169	0.5367	0.6692	0.20/22
	Dermatology				Wdbc			
DPC-KNN	**0.9034**	**0.8826**	**0.9068**	18	**0.6749**	**0.7860**	**0.9018**	52
FKNN-DPC	0.8066	0.8361	0.8709	35	0.6423	0.7613	0.8894	2
DPCSA	0.7451	0.6062	0.6896	–	0.3361	0.3771	0.7595	–
FNDPC	0.7898	0.7995	0.8418	0.17	0.6076	0.7305	0.8758	0.05
DPC	0.6086	0.6110	0.7056	1.5	0.0007	0.0028	0.7257	1.2
DBSCAN	0.6205	0.4152	0.5385	0.99/3	0.3602	0.4786	0.7570	0.46/38
	Inonsphere				Balance			
DPC-KNN	0.1739	0.2650	0.6620	3	0.0777	0.1099	0.5036	2
FKNN-DPC	0.3485	0.4790	0.7716	8	0.0351	0.0236	0.5548	9
DPCSA	0.1335	0.2135	0.6390	–	0.119	0.172	0.4924	–
FNDPC	0.1630	0.2483	0.6513	0.06	**0.1545**	**0.2142**	0.5318	0.96
DPC	0.1355	0.2183	0.6432	0.5	0.0613	0.0893	0.452	1.1
DBSCAN	**0.5627**	**0.6835**	**0.8575**	0.78/9	0	0	**0.6557**	0.8/1
	Sonar				Pima			
DPC-KNN	0.0149	**0.0163**	0.6547	6	**0.171**	**0.058**	0.6635	13
FKNN-DPC	**0.0767**	0.0135	0.6386	11	0.0012	0.0131	0.7106	7
DPCSA	−0.0019	−0.0031	0.6693	–	0.0017	0.0143	0.7119	–
FNDPC	0.0018	0.0066	0.6389	0.02	0.0056	0.0144	0.7323	0.03
DPC	0.0005	0.0014	0.7041	4	0.0132	0.0452	0.5917	2.8
DBSCAN	0	0	**0.707**	0.1/8	0.0042	0.0023	**0.738**	0.5/2

Table 6 Friedman test values of the evaluation indicators of six algorithms on UCI datasets

AMI		ARI		FMI	
Algorithms	Mean ranks	Algorithms	Mean ranks	Algorithms	Mean ranks
DPC-KNN	**5.15**	DPC-KNN	**5.15**	DPC-KNN	**4.45**
FKNN-DPC	4.45	FKNN-DPC	4.45	FKNN-DPC	4.35
FNDPC	3.85	FNDPC	4.05	FNDPC	3.75
DPC	2.70	DPC	2.70	DPC	2.30
DPCSA	2.45	DPCSA	2.65	DPCSA	2.85
DBSCAN	2.40	DBSCAN	2.00	DBSCAN	3.30

Acknowledgments This research was supported by the Science and Technology Project of Jiangxi Province Department of Education (No. GJJ180940),the National Natural Science Foundation of China (61762063),the Scientific research project of the Department of Education (No. GJJ170991).

References

1. Berkhin, P.: A survey of clustering data mining techniques. Grouping Multidimensional Data **43**(1), 25–71 (2006)
2. Xu, R., Wunsch, D.: Survey of clustering algorithm. IEEE Trans. Neural Netw. **16**(3), 645–678 (2005)
3. Xu, D., Tian, Y.: A comprehensive survey of clustering algorithm. Annals Data Sci. **2**(2), 165–193 (2015)
4. Jain, A.K., Murty, M.N., Flynn, P.J.: Data clustering: a review. ACM Comput. Surv. **31**(3), 264–323(1999)
5. Rodriguez, A., Laio, A.: Clustering by fast search and find of density peaks. Science **344**(6191), 1492–1496 (2014)
6. Du, M., Ding, S., Jia, H.: Study on density peaks clustering based on K-nearest neighbors and principal component analysis. Knowl. Based Syst. **99**, 135–145 (2016)
7. Xie, J., Gao, H., Xie, W.: K-nearest neighbors optimized clustering algorithm by fast search and finding the density peaks of a dataset. Scientia Sinica Informationis **46**(2), 258–280 (2016)
8. Liu, R., Wang, H., Yu, X.: Shared-nearest-neighbor-based clustering by fast search and find of density peaks. Inf. Sci. **450**, 200–226 (2018)
9. Lu, J., Zhu, Q.: An effective algorithm based on density clustering framework. IEEE Access **5**, 4991–5000 (2017)
10. Seyedi, S.A., Lotfi, A., Moradi, P., Qader, N.N.: Dynamic graph-based label propagation for density peaks clustering. Expert Syst. Appl. **115**, 314–328 (2019)
11. Xie, J., Gao, H., Xie, W., Liu, X., Grant, P.W.: Robust clustering by detecting density peaks and assigning points based on fuzzy weighted K-nearest neighbors. Inf. Sci. **354**, 19–40 (2016)
12. Xue, X., Gao, S., Peng, H.: Density peaks clustering algorithm based on K-nearest neighbors and classes-merging. J. jilin Univ. (Science Edition) **57**(1), 111–120 (2019)
13. Yu, D., Liu, G., Guo, M., Liu, X., Yao, S.: Density peaks clustering based on weighted local density sequence and nearest neighbor assignment. IEEE Access **7**, 34301–34317 (2019)
14. Du, M., Ding, S., Xue, Y.: A robust density peaks clustering algorithm using fuzzy neighborhood. Int. J. Mach. Learn. Cybern. **9**(7), 1131–1140 (2018)
15. Ester, M.: A density-based algorithm for discovering clusters in large spatial databases with noise. In: Proceedings of the Second International Conference on Knowledge Discovery and Data Mining, pp. 226–231. Palo Alto, AAAI Press(1996)

16. Vinh, N., Epps, J., Bailey, J.: Information theoretic measures for clusterings comparison: Variants, properties, normalization and correction for chance. J. Mach. Learn. Res. **11**(1), 2837–2854 (2010)
17. Fowlkes, E.B., Mallows, C.L.: A Method for Comparing Two Hierarchical Clusterings. J. Am. Stat. Assoc. **78**(383), 553–569 (1983)
18. Jain, A.K., Law, M.H.: Data clustering: a user's dilemma. In: Proceedings of the First International Conference on Pattern Recognition and Machine Intelligence, pp.1–10.Heidelberg, Springer (2005)
19. Chang, H., Yeung, D.Y.: Robust path-based spectral clustering. Pattern Recogn. **41**(1), 191–203 (2008)
20. Gionis, A., Mannila, H., Tsaparas, P.: Clustering aggregation. ACM Trans. Knowl. Discov. Data **1**(1), 1–30 (2007)
21. Fu, L., Medico, E.: FLAME, a novel fuzzy clustering method for the analysis of DNA microarray data. BMC Bioinform. **8**(1), 3 (2007)
22. Veenman, C.J., Reinders, M.J.T., Backer, E.: A maximum variance cluster algorithm. IEEE Trans. Pattern Anal. Mach. Intell. **24**(9), 1273–1280 (2002)
23. Frnti, P., Virmajoki, O.: Iterative shrinking method for clustering problems. Pattern Recogn. **39**(5), 761–775 (2006)
24. UCI repository of machine learning database. http://archive.ics.uci.edu/ml/index.html
25. Street, W.N., Wolberg, W.H., Mangasarian, O.L.: Nuclear feature extraction for breast tumor diagnosis. In: Proceedings of the IS&T/SPIE International Symposium on Electronic Imaging:Science and Technology, pp. 861–870 (1993)
26. Charytanowicz, M., Niewczas, J., Kulczycki, P., et al.: Complete gradient clustering algorithm for features analysis of x-ray images. Adv. Intell. Soft Comput. **69**, 15–24 (2010)
27. Sigillito, V.G., Wing, S.P., Hutton, L.V., et al.: Classification of radar returns from the ionosphere using neural networks. J. Hopkins APL Tech. Dig. **10**(3), 262–266 (1989)

Exponential Fine-Tuning Harmony Search Algorithm

Lipu Zhang and Xuewen Shen

Abstract This paper gives one kind of new harmony search algorithm–exponential fine-tuning harmony search algorithm. The specific improvement is to use an exponential function to change the generation mode of fine-tuning probability, and thus gives a new harmony. The new algorithm is tested by using the standard test function of cec2005, and compared with other four existing harmony search algorithms. The numerical results show that the new algorithm is very competitive with the other harmony search algorithms.

1 Introduction

In 2001, GEEM et al. [1] imitated the process of music improvisation, and then proposed a new nature-inspired algorithm, i.e., harmony search (HS) algorithm. The HS algorithm compares the assignment of each variable to a note. Thus, several notes constitute a harmony feasible solution, and several solutions can be considered as a harmony memory. Improvisation is equivalent to constructing a new harmony. The key points of the HS can be described as: several harmonies are randomly generated to form a harmony memory bank, and the harmony in the memory bank is judged according to the corresponding function value to select the best harmony and the worst harmony. Then a new harmony is generated according to a certain probability, and then compared with the worst harmony in the memory. If the former is better, it is replaced. The algorithm runs according to this process until it reaches a certain termination condition. HS algorithm has good engineering application, and is quickly applied in many fields, such as music synthesis, Sudoku, ecological protection, structural design, aquifer parameter identification, etc. Interested readers can refer to [2–10] for further understanding.

L. Zhang (✉) · X. Shen
College of Media Engineering, Communication University of Zhejiang, Hangzhou, China
e-mail: zhanglipu@cuz.edu.cn

X. Shen
e-mail: cuzmath@126.com

The optimization ability of HS algorithm is affected by the method of generating new harmony, and which is controlled by two parameters: harmony memory control rate (HMCR) and fine-tuning probability (FTP). The specific steps of producing new harmony are as follows:

1. Generating the random numbers of 0 to 1;

    (1) If the number is greater than HMCR, a new harmony is generated randomly in the whole feasible region;
    (2) If the number is less than HMCR, a new harmony is generated according to certain rules in the harmony memory as follows: A new harmony is randomly formed and a random number of 0–1 is generated.
    ① If the number is less than FTP, the new harmony will not be changed;
    ② If the number is greater than FTP, the new harmony will be fine-tuned within the value range.

The HS algorithm performs very well in global search, but not so good in local search. Mahdavi et al. [11] solved the problem of local search by modifying the FTP generation method. Wang and Huang [12] used the dynamic change of harmony memory to automatically adjust FTP and other parameters and constructed an adaptive harmony search algorithm. Fesanghary et al. [13] used continuous quadratic programming to improve the speed of local search. Omran and Mahdavi [14] proposed a global optimal harmony search algorithm based on swarm intelligence to improve the performance of the algorithm. Pan et al. [15] obtained better information from the current optimal solution to form a new harmony and constructed an optimal harmony search algorithm. In [16], they divided the whole harmony memory into many small sub harmony memory banks, and optimized each sub harmony memory independently, and proposed a local optimal harmony search algorithm of dynamic subpopulation. Zhang and Xu [17] generated new harmony from the information of the current global optimal solution and suboptimal solution and obtained the elite decision harmony search algorithm.

Considering that the parameter FTP represents the probability of fine tuning of new harmony, when the search approaches the optimal value, the harmony in the harmony memory gradually approaches the optimal solution. If FTP is constructed to increase the number of searches, the new harmony will be generated in the current harmony memory, which will be more conducive to local search. In this paper, we improve the generation of FTP by introducing an exponential function., and then give a new HS algorithm, named as PDMHS. We use the new algorithm to test some cec2005 [18] test functions. By comparing the numerical results with the other four harmony search algorithms, it can be seen that the new algorithm has better effectiveness and robustness.

This paper consists of the following parts: in Sect. 2, the harmony search algorithm and its improvement are reviewed; in Sect. 3, the PDMHS algorithm is introduced and compared; in Sect. 4, we give some conclusions.

2 Harmony Search Algorithm

The optimization problem studied in this paper is in the following form

$$Minimize\ f(x),$$

$$x_i \in X_i,\ i = 1, 2, ..., N, \tag{1}$$

where $f(x)$ is the objective function, x is the set of each decision variable (x_i), N is the number of decision variables, X_i is the feasible region of each decision variable value, that is $X_i \in [x_i^L, x_i^U]$, where x_i^L, x_i^U is the upper limit and lower limit of each decision variable.

Classic HS algorithm

The classical HS algorithm needs the following parameters:
Harmony memory (HM), harmony memory size (HMS), bandwidth vector (bw).
The procedure of harmony search includes 1–4 steps.
Step 1: initialize a harmony randomly, and the composition is as follows:

$$HM = \begin{pmatrix} x_1^1 & x_2^1 & \cdots & x_N^1 \\ x_1^2 & x_2^2 & \cdots & x_N^2 \\ \cdots & \cdots & \cdots & \cdots \\ x_1^{HMS} & x_2^{HMS} & \cdots & x_N^{HMS} \end{pmatrix}$$

Step 2: constructing a new harmony–a random creation performed by generating a new harmony vector $x' = (x_1', x_2', ..., x_N')$ after definition of HM. Each element of the new harmony vector is generated according to the following principle:

$$x_i' \begin{cases} x_i' \in HM(:, i) & HMCR, \\ x_i' \in X_i, & 1 - HMCR. \end{cases}$$

Among them HMCR refers to the probability of selecting a harmony from HM; 1-HMCR refers to the probability of randomly generating a harmony from the feasible region of the value. Fine-tune the x_i' according to the probability of each FTP obtained from HM, that is:

$$x_i' \leftarrow \begin{cases} x_i' \pm rand[0,1] \times bw & FTP, \\ x_i', & 1 - FTP. \end{cases}$$

where $rand[01]$ is the random number between in 0 and 1.
Step 3: updating the HM—if the new harmony is better than the current worst harmony, then add the new harmony to HM and remove the current worst harmony.

Step 4: repeating step 2 and step 3, until the termination condition is reached.

Modified HS algorithm

In order to eliminate the shortcoming of FTP and bw, which are fixed values, Mahdavi et al. [11] gave a modified harmony search (MHS) algorithm. They used variable FTP and bw in generating the new harmony:

$$\text{FTP}(gn) = \text{FTP}_{min} + \frac{\text{FTP}_{max} - \text{FTP}_{min}}{\text{MaxItr}} \times gn \qquad (2)$$

and

$$\text{bw}(gn) = \text{bw}_{max} e^{c \times gn} \qquad (3)$$

$$c = \frac{\log\left(\frac{bw_{min}}{bw_{max}}\right)}{\text{MaxIter}}$$

where FTP(gn) is the fine tuning probability, FTP_{min} and FTP_{max} are the minimum and maximum of the fine tuning probability, MaxItr and gn are the maximum and current search times, respectively.

Global best harmony search (GHS) algorithm

Different from HS algorithm, Omran and Mahdavi [14] obtained GHS algorithm by modifying the method of constructing new harmony. They obtained the new x' by using the best HM

$$x^{best} = \{x_1^{best}, x_2^{best}, ... x_n^{best}\}$$

The rules are as follows:

$$x'_j = x_k^{best}, \qquad (4)$$

where k is a randomly integer number between 1 and n.

A self adaptive global optimal HS (SGHS) algorithm

In 2010, Pan et al. [15] proposed a SGHS algorithm. They use the current best harmony to construct new harmony, i.e.,

$$x'_j = x_j^{best}, \qquad (5)$$

where $j = 1, ..., n$.

Elite decision harmony search (EDMHS) algorithm

In 2012, Zhang and Xu [17] used the principle of elitist strategy to construct a new harmony, i.e., using the best harmony and the second-best harmony

$$x'_i = \begin{cases} x'_i \in [HM(s,i), HM(b,i)] \\ x'_i \in X_i \end{cases} \quad (6)$$

where HM(s,i) and HM(b,i) refer to the second best and the best harmony in the current HM, respectively.

3 Exponential Fine Tuning Harmony Search Algorithm

The difference between the EFTHS algorithm presented in this paper with the previous algorithms is the method of generating FTP, which is generated by the exponential function. That is:

$$\text{FTP}(gn) = \text{FTP}_{min} + \frac{\text{FTP}_{max} - \text{FTP}_{min}}{\text{MaxItr}} \times e^{gn}$$

Numerical Results

In this chapter, we use cec2005 standard test functions to study the numerical performance of the new algorithm and the other four HS algorithms, including five simple standard test functions and six 30 dimensional standard test functions. For simple functions, the convergence curves and calculation results of each algorithm are drawn, respectively. For 30 dimensional functions, the mean and variance of each algorithm are calculated by searching 50,000 times, respectively.

The parameters of various HS algorithms are given as follows

$$\text{HMS} = 20, \text{HMCR} = 0.90, \text{FTP}_{min} = 0.4$$

and
$$\text{FTP}_{max} = 0.9, bw_{min} = 0.0001, bw_{max} = 1.0.$$

- **Test functions**

We use five famous testing functions, such as Rosenbrock function, Goldstein function, Eason and Fenton function, Wood function and Powell quartic function.

For the five test functions, Tables 1, 2, 3, 4 and 5 show the numerical results of five HS algorithms after 50,000 times, 800 times, 800 times, 70,000 times and 50,000 times of searching, respectively. Figures 1, 2, 3, 4 and 5 show their typical convergence images, respectively. From the numerical results and convergence images, it can be seen that the new algorithm is not inferior to the existing HS algorithm in numerical performance.

Table 1 The numerical results for Rosenbrock function with five HS algorithm

Var	MHS	GHS	SGHS	EDMHS	EFTHS
x_1	1.000001045623	1.092521580337	0.999990060619	1.000000641339	0.999998725065
x_2	1.000001108166	1.092521580337	0.999990049928	1.000000675705	0.999998518983
$f(x)$	0.000000000042	0.769345178147	0.000000004277	0.000000000050	0.000000000058

Table 2 The numerical results for Goldstein function with five HS algorithm

Var	MHS	GHS	SGHS	EDMHS	EFTHS
x_1	−0.000926236	−0.000557584	−0.000000206	−0.000000034	−0.000000271
x_2	−1.000812452	−1.000314622	−1.000000254	−1.000000087	−1.000000251
$f(x)$	3.010488740	3.0202241793	3.000000001	3.000000000	3.000000001

Table 3 The numerical results for Eason and Fenton function with

Var	MHS	GHS	SGHS	EDMHS	EFTHS
x_1	1.74775319976	1.77773319978	1.74344876821	1.7451847176!	1.74345946040
x_2	2.02408447493	2.01840464872	2.02969809400	2.02907854695	2.02960508565
$f(x)$	1.74445789937	1.74884060442	1.74415200593	1.74419165455	1.74415205100

Table 4 The numerical results for Wood function with five HS algorithm

Var	MHS	GHS	SGHS	EDMHS	EFTHS
x_1	1.000020703889	1.048504873163	1.000119098916	1.000008748268	1.000008638413
x_2	1.000039352450	1.165016383518	1.000237337315	1.000015140987	1.000017523632
x_3	0.333338373637	−0.04009927746	−0.13323078788	0.133330376588	0.133325997545
x_4	0.999968672904	0.838627625177	0.999770083868	0.999988351279	0.999978808689
$f(x)$	0.000000090750	0.592532857141	0.000004945118	0.000000032006	0.000000028079

Table 5 The numerical results for Powell function with five HS algorithm

Var	MHS	GHS	SGHS	EDMHS	EFTHS
x_1	−0.002060203	0.006362840	0.001415599	0.008112968	−0.001774928
x_2	0.000204589	−0.000713995	−0.000141303	−0.000811212	0.000176694
x_3	−0.000983926	−0.005359167	0.000967532	0.003678456	−0.001483761
x_4	−0.000984959	−0.005294787	0.000967635	0.003684867	−0.001481943
$f(x)$	6.80835E-07	0.001914874	2.96396E-07	5.54913E-07	1.77721E-07

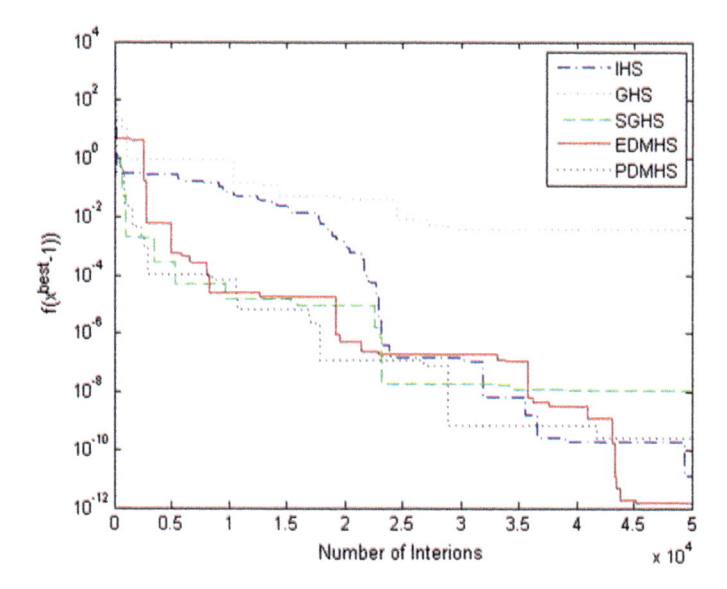

Fig. 1 Rosenbrock function

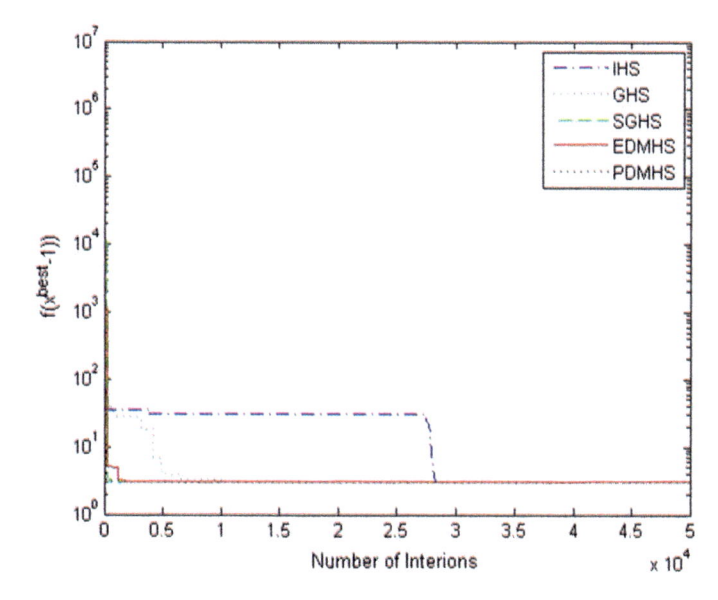

Fig. 2 Goldstein function

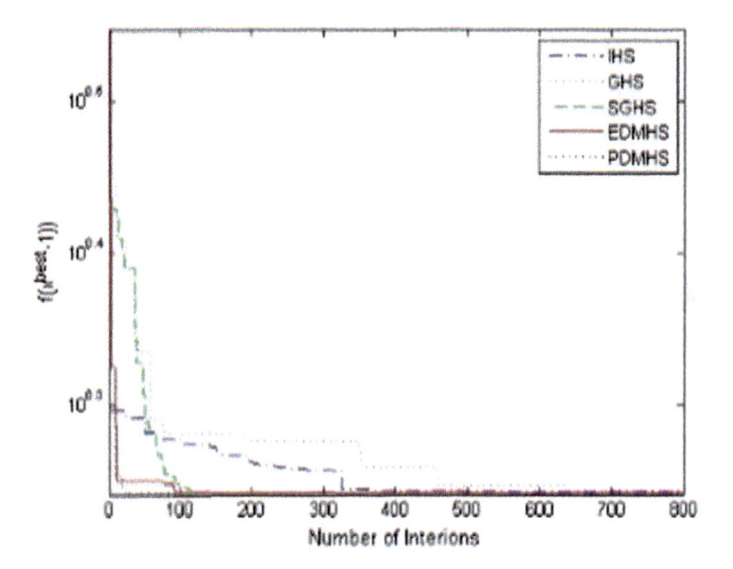

Fig. 3 Eason and Fenton function

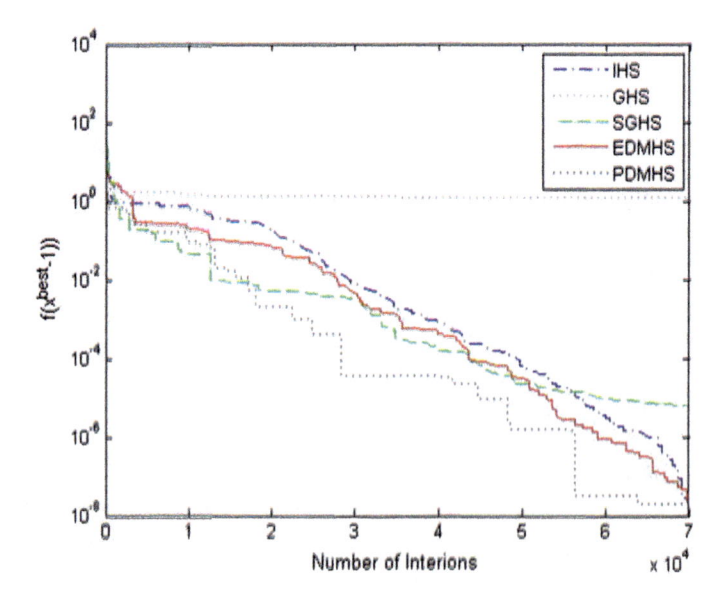

Fig. 4 Wood function

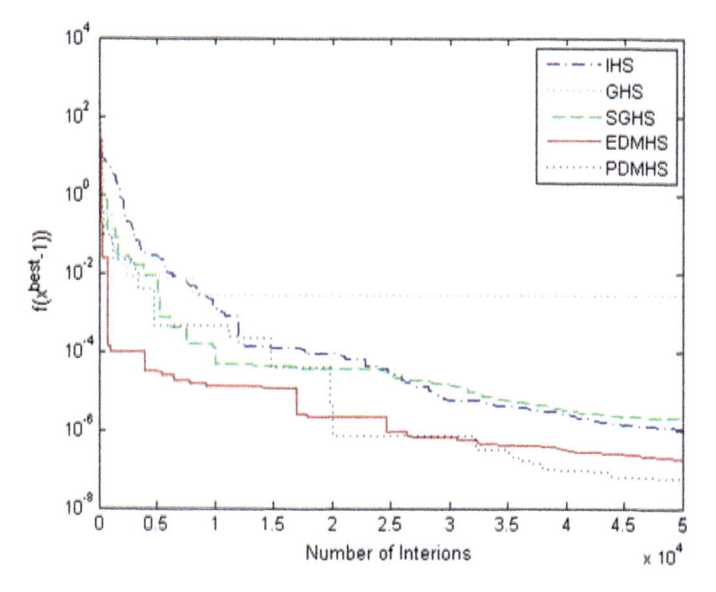

Fig. 5 Powell quartic equation function

4 Conclusion

The new algorithm constructs a new harmony generation method by modifying the fine-tuning probability. Compared with the existing harmony algorithms, the new algorithm has good search ability. The application of the new algorithm in engineering problems is our next research direction.

References

1. Geem, Z., Kim, J., et al.: A new heuristic optimization algorithm: harmony search. SIMULATION **76**, 60–68 (2001)
2. Geem, Z., Choi, J.: Music composition using harmony search algorithm. Appl. Evolut. Comput., 593–600 (2007)
3. Geem, Z.: Harmony search algorithm for solving Sudoku. In: Knowledge-Based Intelligent Information and Engineering Systems, pp. 371–378. Springer
4. Lee, K., Geem, Z.: A new structural optimization method based on the harmony search algorithm. Comput. Struct. **82**, 781–798 (2004)
5. Saka, M.: Optimum geometry design of geodesic domes using harmony search algorithm. Adv. Struct. Eng. **10**, 595–606 (2007)
6. Geem, Z., Williams, J.: Ecological optimization using harmony search. In: Proceedings of American Conference on Applied Mathematics, pp. 24–26
7. Ayvaz, M.: Simultaneous determination of aquifer parameters and zone structures with fuzzy c-means clusteringand meta-heuristic harmony search algorithm. Adv. Water Resour. **30**, 2326–2338 (2007)

8. Geem, Z.: Harmony search applications in industry. Soft Comput. Appl. Industry, 117–134 (2008)
9. Geem, Z.: Music-inspired harmony search algorithm: theory and applications, vol. 191, Springer (2009)
10. Ingram, G., Zhang, T.: Overview of applications and developments in the harmony search algorithm. Music-Inspired Harmony Search Algorithm, pp. 15–37 (2009)
11. Mahdavi, M., Fesanghary, M., Damangir, E.: An improved harmony search algorithm for solving optimization problems. Appl. Math. Comput. **188**, 1567–1579 (2007)
12. Wang, C., Huang, Y.: Self-adaptive harmony search algorithm for optimization. Expert Syst. Appl. **37**, 2826–2837 (2010)
13. Fesanghary, M., Mahdavi, M., Minary-Jolandan, M., Alizadeh, Y.: Hybridizing harmony search algorithm with sequential quadratic programming for engineering optimization problems. Comput. Methods Appl. Mech. Eng. **197**, 3080–3091 (2008)
14. Omran, M., Mahdavi, M.: Global-best harmony search. Appl. Math. Comput. **198**, 643–656 (2008)
15. Pan, Q., Suganthan, P., Tasgetiren, M., Liang, J.: A self-adaptive global best harmony search algorithm for continuous optimization problems. Appl. Math. Comput. **216**, 830–848 (2010)
16. Pan, Q., Suganthan, P., Liang, J., Tasgetiren, M.: A local-best harmony search algorithm with dynamic subpopulations. Eng. Optim. **42**, 101–117 (2010)
17. Xu, Zhang Lipu, Yinghong.: An Elite-Decision-Making harmony search algorithm for optimization problem. J. Appl. Math. **2012**, 1–15 (2012)
18. Suganthan, P., Hansen, N., Liang, J., Deb, K., Chen, Y., Auger, A., Tiwari, S.: Problem definitions and evaluation criteria for the CEC 2005 special session on real-parameter optimization, Nanyang Technological University, Singapore, Tech. Rep 2005005 (2005)

Computational Intelligences

A Location Gradient Induced Sorting Approach for Multi-objective Optimization

Lingping Kong, Václav Snášel, Swagatam Das, and Jeng-Shyang Pan

Abstract In the field of population-based multi-objective optimization, a non-dominated sorting approach amounts to sort a set of candidate solutions with multiple objective function values, based on their dominance relations, and to find out solutions distributed into the first front set, second front set, and so on. A fast non-dominated sorting approach used within the framework of Non-dominated Sorting Genetic Algorithm (NSGA-II) reaches a $\mathcal{O}(MN^2)$ time complexity (N is the population size, M is the number of the objectives). In this paper, we show an approach based on the Location Gradient (LG) number (LG sorting). Our LG sorting method is especially efficient in dealing with duplicate solutions, which only costs $\mathcal{O}(N)$ in front assignment process when all the solutions are duplicate. Except that, in many cases, the LG sorting method can reach $\mathcal{O}(N \log N)$ time complexity. But in the worst case, it still costs $\mathcal{O}(MN^2)$. We demonstrate the efficacy of the LG-based sorting method comparison against several existing non-dominated sorting procedures.

1 Introduction

Due to the population-based nature, evolutionary algorithms (EAs) can approximate the desired solution of a multi-objective Optimization Problem (MOP) in a single run. Researchers have developed EAs to deal with MOPs under the tag of multi-objective evolutionary algorithms (MOEAs).

Many existing MOEAs are designed to give rise to a set of promising solutions [10, 19]. Among these promising solutions, each solution cannot be said to be better than any other solution, which is said to be non-dominated. In the process of finding

---

L. Kong · V. Snášel · S. Das (✉) · J.-S. Pan
Faculty of Electrical Engineering and Computer Science, VSB -Technical University of Ostrava, Ostrava, Czech Republic
e-mail: swagatam.das@isical.ac.in

L. Kong
e-mail: konglingping2007@163.com

V. Snášel
e-mail: vaclav.snasel@vsb.cz

the Pareto optimal solutions, most EAs use temporary dominance relation to decide the orientation of search. Non-dominated sorting is used to separate the solutions into different sets.

Definition 1 Dominance relation [3, 6], a solution $\mathbf{x_1}$ is said to dominate the other solution $\mathbf{x_2}$ when both conditions 1 and 2 are true:

1. The solution $\mathbf{x_1}$ is no worse than solution $\mathbf{x_2}$ in all objectives, or $f_i(\mathbf{x_1}) \leq f_i(\mathbf{x_2})$ for all $i = 1, 2, \ldots, M$.
2. Meanwhile the solution $\mathbf{x_1}$ is strictly better than $\mathbf{x_2}$ in at least one objective, or $f_j(\mathbf{x_1}) < f_j(\mathbf{x_2})$ for at least one $j \in \{1, 2, \ldots, M\}$.

In the following, the notation $s_j \prec s_i$ is used to indicate that s_i is dominated by s_j and $s_j \not\prec s_i$ to indicate that s_i is not dominated by s_j. In this paper, we use $\mathcal{F}$ label as the set of fronts where $\mathcal{F} = \{\mathcal{F}_1, \mathcal{F}_2, \mathcal{F}_3, \ldots\}$, the digits of subscript $\{1, 2, 3, \ldots\}$ is called the front number. $\mathcal{F}_1$ denotes the best front, in which are Pareto front solutions. The notation $\mathcal{F}_k \prec s_i$ is used to indicate that a solution s_i is dominated by some solution from $\mathcal{F}_k$, where $k \in \{1, 2, \ldots\}$.

Definition 2 Non-dominated Sorting [15], among a set of solutions P, the ones that are not dominated by each other compose a non-dominated set. Non-dominated Sorting divides a population into different front sets where the fronts are arranged in decreasing order of their dominance relation. The best non-dominated set, usually called Pareto optimal set, is labeled as $\mathcal{F}_1$. For a solution p in $\mathcal{F}_{k+1}$, exist at least one solution from $\mathcal{F}_k$ which dominates solution p, where $k \in \{1, 2, 3, \ldots\}$.

In the primary approaches, to assign the front number to all solutions, it requires a $\mathcal{O}(MN^3)$ computation complexity at the worst case and a $\mathcal{O}(MN^2)$ computation complexity in the best case. Then, a fast non-dominated sorting approach is presented within the framework of NSGA-II [4, 7]. This sorting approach only requires $\mathcal{O}(MN^2)$ comparisons. In the following years, many additional strategies based on niching, tree structures, label reduction, and archiving were proposed to overcome the high computational cost. The idea of HNDS [1], ENS-SS, ENS-BS [20], and Corner sorting [12, 18] is to use the presorting to make the population ordered based on the first objective values, then assign each candidate a front number according to the presorted list. Jensen [8], DCNC [14, 15], Maxim [2], and KUNG [11] utilized the divide and conquer concept to solve the sorting problems.

1.1 Contribution of the Paper

1. We present a non-dominated sorting algorithm based on the location gradient information. It takes $\mathcal{O}(N)$ space, the time complexity can reach $\mathcal{O}(MN \log N)$ in many of cases (data sets).
2. We design an algorithm for getting a special LG number (a value for a solution), which reduces the difficulty of getting the dominance relation between solutions.

With this LG number, it takes $\mathcal{O}(1)$ time to get the non-dominated relations among solutions with the same LG number and among duplicate solutions.
3 LG number is shown to be important to a population with two objectives. For a population without duplicate solutions, if all solutions are non-dominated, their LG number has to be the same.
4 We introduce three different front assignment processes that can be embedded in the LG sorting method, which is proposed by other algorithms.

The rest of the paper is organized as follows. Section 2 provides the LG number introduction, firstly, it explains what is LG number, followed by the features of it. Then the details of the LG sorting algorithm are presented in Sect. 3. Section 4 describes the experiments in which our LG sorting method is compared with other existing sorting methods. Section 5 concludes the paper.

2 Location Gradient Number

For the sake of simplicity, unless otherwise specified, all the objectives function values are to be minimized in this paper. It is worthy to note that the Location Gradient sorting approach holds equally well for minimization and maximization problem.

Definition 3 The Location Gradient (LG) number of a solution is defined as a summation number of better objectives compared to each solution.

Every solution in a population can be assigned with its LG number. Suppose there is a population with N solutions and each solution has one objective, then the location gradient number of a candidate solution will vary in $LG[i] \in \{0, 1, \ldots, N-1\}$, where $i \in \{1, 2, \ldots, N\}$.

When only one objective domain is being evaluated, and if solution s has the best objective value compared to the rest of $(N-1)$ solutions, the LG number of solution s is $LG[s] = N - 1$. Similarly, if solution s has the worst objective value in population, its LG number will be zero $LG[s] = 0$. In other words, the LG number of solution s presents the number of solutions that have worse objective value than solution s in a specific objective. When each solution has many objectives, the LG number of a solution will be the summation of the gradient numbers in all objectives, in other words, **the LG number of solution s presents a number which counts the worse objective values of the population than solution s**. It can be expressed as $LG[i] = LG[i]_{obj1} + \cdots + LG[i]_{objj} + \cdots + LG[i]_{objM}$, where $i \in \{1, 2, \ldots, N\}$ and $j \in \{1, 2, \ldots, M\}$. The LG definition is similar with the concept of 'solution rank' in paper [5]. However, in paper [5], the 'solution rank' is used as a selection criteria, has not been extended to an algorithm for non-dominated sorting process. Meanwhile, we propose a fast algorithm for getting our LG number.

From this definition, we can find four interesting properties of the LG number:

1. The solution which gets the largest LG number and is distributed into the best front set $\mathcal{F}_1$.
2. The solution which gets the lowest LG number and is distributed into the worst front set.
3. The most important one is, a solution has a larger LG number than the value of its dominating solution has.

The third property is, if there exist $q \prec p$, then q has a larger LG number than p, which works for any population with any number of objectives. Especially, if all objective values are distinct in solutions, the LG gap between two solutions p and q would be at least M. The last property is indicated as, for any solution p with $LG[p] = v$, where v is a random non-negative integer, the solution q which dominates solution p must have an LG number satisfying $LG[q] \geq v + 1$.

Based on these properties of LG number, LG non-dominating sorting approach is proposed. The front number is assigned to solutions with larger LG number first, then to the solutions with smaller LG number. Thus, the front number of a dominated solution cannot exceed its dominating solution's front level. This will be explained further in the next section.

LG number can be obtained by sorting solutions in each objective and by using the well-known merge sort method Merge sort [9], which takes $\mathcal{O}(N \log N)$ time.

Lemma 1 *In a population, if two solutions p and q hold the same LG number, then solutions p and q are non-dominated with respect to each other.*

Proof Suppose solution p dominates q. Based on *Definition* 1, solution p must hold at least one better objective value than solution q. Hence, assuming the LG number of q is $LG[q] = LG[q]_{obj1} + \cdots + LG[q]_{obji} + \cdots + LG[q]_{objM}$. Then there is $LG[p] = LG[q]_{obj1} + \cdots + LG[q]_{obji} + \cdots + LG[q]_{objM} + k$ where $i = \{2, 3, \ldots, M - 1\}$, $LG[p] = LG[q] + k$, where $k > 0$, which is bigger than $LG[q]$. It contradicts the condition where p and q hold the same LG number, so they are non-dominated to each other.

We can draw two other inferences here. A solution s with $LG[s] = k$ definitely cannot dominate a solution t with $LG[t] = k'$ if $k < k'$. On the contrary, a solution s with $LG[s] = k$ may or may not dominate a solution t with $LG[s] = k'$, where $k > k'$. Here k, k' are random non-negative integer.

3 The LG Number Based Sorting Method

LG (Location gradient) sorting method amounts to sort a population of N solutions and each solution is associated with M objective values into the different fronts. All the solutions in a particular front are non-dominated to each other. If a solution s is distributed in $\mathcal{F}_k$, there must be at least one solution in $\mathcal{F}_{k-1}$ which dominates s.

The annotations of all symbols are: P is the population with N solutions $P = \{p_1, p_2, \ldots\}$, where the i_{th} solution is labeled as p_i. objective number is M. $B_{obj}[i]$

and $W_{obj}[i]$ record the best and worst objective label of solution i, which also means solution i has the biggest, lowest location gradient order in this specific objective, respectively. $\mathcal{F}$ is front level, first front labels as $\mathcal{F}_1$. We use $\cup$ for adding a element to a set, $\mathcal{F}_1 = \mathcal{F}_1 \cup p_t$ means adding solution p_t into $\mathcal{F}_1$. $LG[]$ is a array and LG number is a summation of all objectives' position standing for one solution. The i_{th} solution's LG is labeled as $LG[i]$. Sort($P, LG[]$) Sort the population based on LG number in descending order. *rnindex* is a flag records the running front index (right now index). $p_t[i]$ notes i_{th} objective value of solution p_t, $i \in \{1, 2, \ldots, M\}$. Algorithm 1 is the main loop of the LG sorting approach.

LG sorting method is composed of two parts: part one is used for creating the LG numbers for all solutions, and sort the population based on LG number. Part two is used to assign the front number to each solution in the sequence of ordered solutions. The process of LG sorting remains the same for both cases $M > 2$ or $M = 2$, but their time complexities can differ when and only when all the solutions are in a single front set, which will be illustrated in the time complexity section.

Algorithm 1: Main loop of LG sorting

Input: the population, P; population size, N; objective number, M
Output: $\mathcal{F}$ set
Initialize $LG[\ ]$, $B_{obj}[\ ]$, $W_{obj}[\ ]$;
$P \leftarrow sort\ P\ based\ on\ LG[\ ]$; rnindex $\leftarrow 1$;
$\mathcal{F}_1 \leftarrow \mathcal{F}_1 \cup p_1$;
for $k = 2 \rightarrow N$ do
    if $LG[1]$ equals to $LG[k]$ then
        $\mathcal{F}_1 \leftarrow \mathcal{F}_1 \cup p_k$;
    else
        break;
    end
end
for $t = k \rightarrow N$ do
    if $-1 \neq LG[t]$ then
        Fass(P,t, rnindex, LG);
    else
        rnindex ++ ;
        for $z = t \rightarrow N$ do
            $\mathcal{F}_{rnindex} \leftarrow \mathcal{F}_{rnindex} \cup p_z$;
        end
        break;
    end
end

In Algorithm 1,[1] the first ranking solution in LG number is surely belongs to $\mathcal{F}_1$. The ones with the biggest LG number as the p_1 are distributed into $\mathcal{F}_1$ front without any dominance comparison, as in line [4–10]. This is because of the *Lemma 1*. In line [13], we add the infeasible solution checking process. If a solution's LG

---

[1] Front Assignment process, three assignment methods are available [14, 15].

number is −1, it is an infeasible solution. Many problems are with constraints, it is very common to generate some infeasible solutions that might be non-dominated, but they are not under valid bounds. In line [16, 18], those infeasible solutions are divided into the same worst front set. Note that a solution can have a zero LG number, but it is still a feasible solution. In line [14], the important step is to assign a front number to the solutions. Here we provide three different front assignment strategies: Down-to-top front assignment, Top-to-down front assignment, and binary search front assignment (detailed in the supplementary document). For simplicity, we labeled these strategies as *LG-TD*, *LG-DT* and *LG-BS*, respectively. And those three front assignment processes can be found in paper [13, 14].

During the LG sorting process, once the sorted LG number is set, solutions with the biggest LG values can be front assigned directly without any dominance check. A more specific case, **if LG values are equal for all solutions no matter when it is $M = 2$ or $M > 2$, then all the solutions will end up in the same first front, more importantly, there is no dominance comparison needed.**

Algorithm 2: $dominance\_compare(s, p_t)$

Input: M, B_{obj}, W_{obj}; two solutions, s, P_t
Output: dominance relation, true for s dominates p_t

1 **if** $LG[s]$ $equals$ to $LG[p_t]$ **then**
2   return false;
3 **else**
4   **if** $p_t[B_{obj}[p_t]] \prec s[B_{obj}[p_t]]$ **then**
5   return false;
6   **end**
7   **if** $p_t[W_{obj}[s]] \prec s[W_{obj}[s]]$ **then**
8   return false;
9   **end**
10   **for** $each$ $z \in \{1, 2, \ldots, M\}, z \notin [B_{obj}[p_t], W_{obj}[s]]$ **do**
11   **if** $p_t[z] \prec s[z]$ **then**
12   return false;
13   **end**
14   **end**
15   return true;
16 **end**

Dominance_compare(s, p_t) operates the dominance comparison process between solution s and p_t, to find out whether solution s dominates solution p_t or not. There are only two dominance relations of this operation, either solution s dominates solution p_t or solution s does not dominate solution p_t. The solution s is front assigned before solution p_t, so solution s holds a larger LG number to solution p_t (if p_t has an equal LG number to s, they are non-dominated). This means solution p_t can not dominate s. Therefore, based on the dominance definition, if solution p_t holds any one objective value which is better than this particular objective value s holds, then solution s cannot dominate p_t, hence a $false$ flag return as in line

[5, 8] Algorithm 2.[2] To reduce objective comparison, the first two objective values compared are: solution p_t holds the best objective and solution s holds the worst objective. These two objective values are much more representative than other objectives in these two solutions. But if these two objective comparisons are not enough to get the dominance relation, then a full-objective comparison needs to be operated.

Lemma 2 *The proposed LG algorithm finds the correct front number for any solution $s \in P$ from the population set P.*

Proof The LG sorting process starts with a sorted population according to their LG number, hence the solution with bigger LG is always front-assigned first, its front assignment is not affected by the rest of the solutions with smaller LG number. A solution from the succeeding front is the assigned after its dominating solution (with a bigger LG number). This procedure gives each solution a correct front number.

4 Experiment and Results

In this section, we undertake a comparative study between the LG non-dominated sorting algorithm and seven other non-dominated sorting algorithms, which are Deductive sorting [12], Corner sorting [18], ENS-BS, ENS-SS [20], BOS, GBOS [13], and BBOS [16]. Among the algorithms, GBOS is operated in a sequence-search way and it works with an objective reduction operation. BOS is the original best order sorting method, while in this paper, the dominance comparison among individual solutions follows BBOS. LG-TD, LG-DT, LG-BS are three patterns of LG method with top-to-down(TD), down-to-top (DT), and binary search (BS) front assignment process. We test two performance indicators: the number of objective comparisons and the total running time. All the comparing algorithms are implemented in Java Development Kit 1.6 and run in Toshiba computer with 1.7 GHz Intel Core i3 and 64 bit Windows 10 machine [17].

Three datasets are tested by all simulated methods. The procedure for generating the datasets is from the algorithm (given in [18]) of uniformly distributed numbers. The upper and lower limits are set to 1.0 and 0.0, respectively, for the ZDT3 problem to obtain M variables. The first dataset consists of a population of size 10000 with a varied objective number from 2 to 30 as indicated in Fig. 1. The second dataset includes 500–10000 solutions with 500 intervals for 10, 15, 20, and 25 objectives as shown in Fig. 3.

In Fig. 1, LG-TD gives the number of objective comparisons cost by LG non-dominated sorting method with a top-to-down front assignment. In LG sorting method, there are three objective reduction operations, which avoid the unnecessary objective comparison between solutions. All the objective comparison happened in line [4, 5] are counted as a first-comparison operation. A symbol 'LG-' represents

---

[2] $M, B_{obj}, Wobj$ parameters are initialized in Main loop; two solutions, s, P_t.

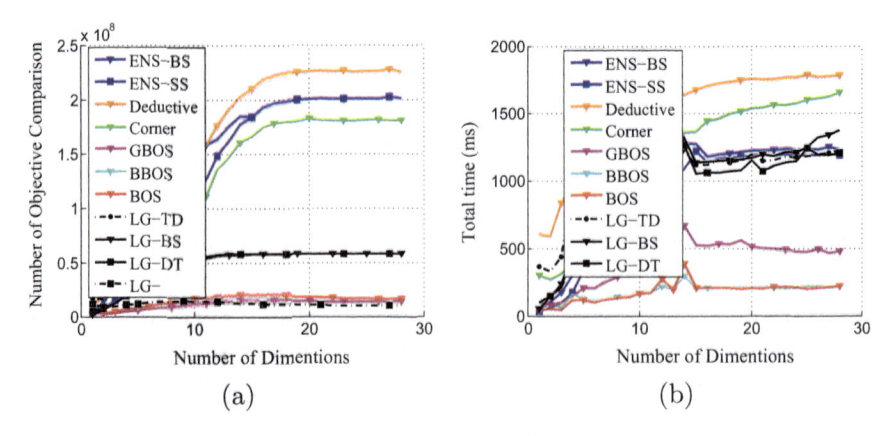

Fig. 1 Number of comparisons **a** and total runtime **b** of population size 1 0000 for increasing number of objectives 2–29

the total number of objective comparison of LG-TD left after eliminating the first-comparison number. Here, we define the first-comparison number because the M sorted lists based algorithms (such as BOS, BBOS, GBOS) take advantage of more space to avoid those first comparisons. Note that single sorted list based methods (such as Corner and Deductive, ENS, LG) will naturally operate these first comparisons which increase the total comparison number.

This experiment tests 2 datasets on 8 sorting methods, which are broadly divided into two groups: single sorted list sorting method and M sorted lists based sorting methods. As in Fig. 1, ENS, Corner, and Deductive methods perform worse than BOS, GBOS and BBOS methods in objective comparison number indicator. At the same time, the running time of these compared methods is also divided into two same groups. To LG sorting methods, it operates based on a single sorted list, while it reserves the best-sorted objective for early termination of the dominance relation comparison. In other words, for all single sorted list based sorting methods, they save space consumption in return for increasing more dominance comparison. On the contrary, M sorted lists based methods waste the space for decreasing the dominance comparison. This is proved by the result of LG-, where all first-comparison numbers are deducted. Besides, the GBOS method performs worse than BBOS and BOS, which is caused by its objective reduction operation in each solution of traversal (Fig. 2).

Figure 3 show that a single sorted list based method performs worse than the M sorted lists based methods. In these two group figures, four figures from the first-row list the dominance comparison number results, while the second row shows the running time, and the third row gives a clear view of the four curves. Also, BOS and BBOS are better than GBOS in running time for most of the figures. More importantly, the LG requires a lesser number of comparisons than most of the methods in all cases.

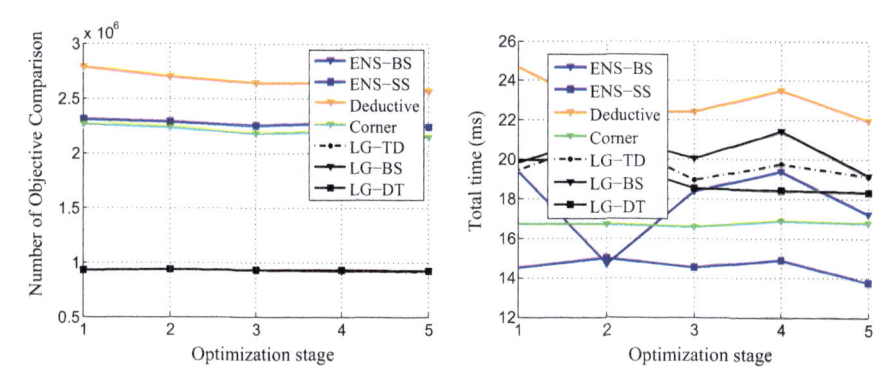

Fig. 2 MOEAs dataset of WFG problem

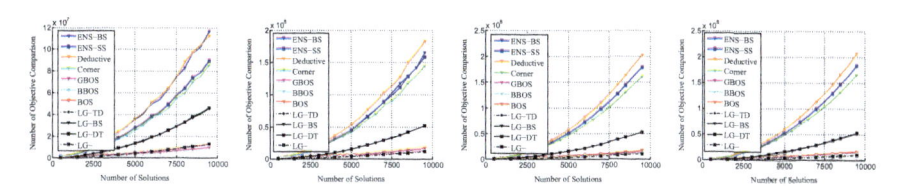

Fig. 3 Number of objective comparisons with increasing population size in objectives 10, 15, 20, and 25

5 Conclusion

In this paper, we present a location gradient-based non-dominated sorting algorithm to speed up the dominance-based multi-objective optimizers. Location gradient number, also called LG number, counts a solution's number of predominant objective values in other solutions, which are worse than this solution. It is proved that two solutions with the same LG number are non-dominated, which decrease the unnecessary dominance comparison. Furthermore, three different front assignment patterns, Top-to-down pattern, Down-to-top pattern, and binary search pattern, can be embedded into this sort. The results show that the LG number based sorting can reduce the objective value comparison between solutions and thus can be an attractive alternative technique to be integrated with dominance-based multi-objective evolutionary algorithms.

Acknowledgements This work was supported by the ESF in "Science without borders" project, $reg.nr.CZ.02.2.69/0.0/0.0/16\_027/0008463$ within the Operational Programme Research, Development and Education, and by the Ministry of Education, Youth and Sports of the Czech Republic in project "Metaheuristics Framework for Multi-objective Combinatorial Optimization Problems (META MO-COP)", reg.no.LTAIN19176.

References

1. Bao, C., Xu, L., Goodman, E.D., Cao, L.: A novel non-dominated sorting algorithm for evolutionary multi-objective optimization. J. Comput. Sci. **23**, 31–43 (2017)
2. Buzdalov, M., Shalyto, A.: A provably asymptotically fast version of the generalized jensen algorithm for non-dominated sorting. In: International Conference on Parallel Problem Solving from Nature, pp. 528–537. Springer (2014)
3. Deb, K.: Multi-objective Optimization Using Evolutionary Algorithms, vol. 16. Wiley, New York (2001)
4. Deb, K., Pratap, A., Agarwal, S., Meyarivan, T.: A fast and elitist multiobjective genetic algorithm: Nsga-ii. IEEE Trans. Evol. Comput. **6**(2), 182–197 (2002)
5. D'Souza, R.G., Sekaran, K.C., Kandasamy, A.: Improved nsga-ii based on a novel ranking scheme (2010). arXiv preprint arXiv:1002.4005
6. Emmerich, M.T., Deutz, A.H.: A tutorial on multiobjective optimization: fundamentals and evolutionary methods. Nat. Comput. **17**(3), 585–609 (2018)
7. Jain, H., Deb, K.: An evolutionary many-objective optimization algorithm using reference-point based nondominated sorting approach, part ii: handling constraints and extending to an adaptive approach. IEEE Trans. Evol. Comput. **18**(4), 602–622 (2013)
8. Jensen, M.T.: Reducing the run-time complexity of multiobjective eas: the nsga-ii and other algorithms. IEEE Trans. Evol. Comput. **7**(5), 503–515 (2003)
9. Knuth, D.: Sorting and searching. Art Comput. Program. **3**, 513 (1998)
10. Kong, L., Pan, J.S., Sung, T.W., Tsai, P.W., Snášel, V.: An energy balancing strategy based on hilbert curve and genetic algorithm for wireless sensor networks. Wireless Commun. Mobile Comput. **2017** (2017)
11. Kung, H.T., Luccio, F., Preparata, F.P.: On finding the maxima of a set of vectors. J. ACM (JACM) **22**(4), 469–476 (1975)
12. McClymont, K., Keedwell, E.: Deductive sort and climbing sort: new methods for non-dominated sorting. Evol. Comput. **20**(1), 1–26 (2012)
13. Mishra, S., Mondal, S., Saha, S., Coello, C.A.C.: Gbos: generalized best order sort algorithm for non-dominated sorting. Swarm Evol. Comput. **43**, 244–264 (2018)
14. Mishra, S., Saha, S., Mondal, S.: Divide and conquer based non-dominated sorting for parallel environment. In: 2016 IEEE Congress on Evolutionary Computation (CEC). pp. 4297–4304. IEEE (2016)
15. Mishra, S., Saha, S., Mondal, S., Coello, C.A.C.: A divide-and-conquer based efficient non-dominated sorting approach. Swarm Evol. Comput. **44**, 748–773 (2019)
16. Roy, P.C., Deb, K., Islam, M.M.: An efficient nondominated sorting algorithm for large number of fronts. IEEE Trans. Cybern. **49**(3), 859–869 (2018)
17. Sun, H.M., Wang, H., Wang, K.H., Chen, C.M.: A native apis protection mechanism in the kernel mode against malicious code. IEEE Trans. Comput. **60**(6), 813–823 (2011)
18. Wang, H., Yao, X.: Corner sort for pareto-based many-objective optimization. IEEE Trans. Cybern. **44**(1), 92–102 (2013)
19. Wang, K., Xu, P., Chen, C.M., Kumari, S., Shojafar, M., Alazab, M.: Neural architecture search for robust networks in 6g-enabled massive iot domain. IEEE Internet J. (2020)
20. Zhang, X., Tian, Y., Cheng, R., Jin, Y.: An efficient approach to nondominated sorting for evolutionary multiobjective optimization. IEEE Trans. Evol. Comput. **19**(2), 201–213 (2014)

Production Line Balance Optimization Based on Improved Imperial Competition Algorithm

Xue-Hua Yang and Chih-Hung Hsu

Abstract In order to solve the common manufacturing enterprises, line balancing rate optimization to maximize production and minimize production line balancing rate smoothness index as the optimization goal, a new improved algorithm is proposed, to improve imperial competition algorithm as a global optimization algorithm, on the basis of the standard empire competition algorithm, crossover and mutation operators of genetic algorithm for reference to take the place of the assimilation process of imperialist competitive algorithm, using a steel coil packaging line instance competition algorithm to improve empire effective verification, verify the improved empire competition the validity of the algorithm to solve the problem of line balancing optimization has better convergence speed and better global optimal solution.

1 Introduction

In today's manufacturing industry, the optimization of production line balance rate is one of the important issues in production management, because the balance rate of the production line will directly affect the efficiency of the whole production line and the efficiency of the manufacturing system. With the concept of continuous improvement, manufacturing companies realize that improving the balance rate of production line is crucial to improve their efficiency. In this paper, an improved imperial competition algorithm is used to optimize the production line of a certain type of package.

X.-H. Yang (✉) · C.-H. Hsu
Institute of Industrial Engineering, School of Transportation, Fujian University of Technology, University Town, No3 Xueyuan Road, Minhou, Fuzhou City, Fujian, China
e-mail: yxuehua0320@163.com

C.-H. Hsu
e-mail: chhsu886@fjut.edu.cn

2 Overview of Production Line Balancing Concepts

CT(Cycle Time), Refers to the time interval between the successive completion of two identical products.

Production line balance rate is used to evaluate the workload balance of each station index. The higher the balance ratio of production line, the higher the balance degree of production line. Its calculation formula is as follows:

$$\eta = \frac{\sum_{i=1}^{m} T(S_i)}{CT \times m} \times 100\% \qquad (1)$$

The production line smoothness index SI is used to evaluate the discreteness of the operation time distribution at each station. The smaller the smoothness index, the higher the balance of the production line. Its calculation formula is as follows:

$$SI = \sqrt{\frac{\sum_{i=1}^{m} [CT - T(S_i)]^2}{m}} \qquad (2)$$

where, η—the balance rate, M – the number of stations, CT –Cycle Time, and T (Si)—the working time of the i station.

3 Examples of a Steel Coil Packaging Production Line

Taking a steel coil package [1] as the research object, the packaging of it is born optimized design of production line. The operation content of each procedure in the packaging production line, Job prioritization and job time is shown in Table 1.

4 Production Line Balance Model Construction

4.1 Model Building

This study on the basis of the production line workstation is known, to balance η and SI as the index to evaluate the production line balance, require the production line to meet operation element precedence relation, under the condition of the production process and reasonable assigned to each workstation, make production line to minimize the beat theory, maximizing production line balancing rate and the highest level of balance, to establish a balance model for production line:

Table 1 Steel coil packaging sequence

Work number	Operation content	Immediate successor activity	Operation time/S
1	Confirmation	2, 3, 4	60
2	Steel coil loading	5, 6	60
3	Coil weighting	5, 6	30
4	Measure dimension and deviation and label	5, 6	40
5	Place the inner core of anti-rust paper	7	60
6	Place the outer rust proof paper	7	50
7	Anti-rust paper is completely closed and pasted	8	50
8	Pre-set plastic sleeve	9	50
9	Plastic bag enclosed	10, 11	120
10	Paper wrap inner corner guard	12, 13	40
11	Paper wrap outer corner protection	12, 13	60
12	Wrap the surface of flat paper	14	100
13	Packing of peripheral cladding plate	14	120
14	Install inner circumferential guard plate	15, 16	40
15	Install round guard sheet paper	17, 18	60
16	Install iron round guard plate	17, 18	50
17	Place iron inner corner guards	19	30
18	Place iron outer corner guards	19	30
19	Circumferential baling	20	120
20	Radial baling	21	100
21	Manual labelling	22	40
22	Send coils, meaning hoisted into storage		70

$$\text{maxf}(x) = A \cdot \frac{\sum_{i=1}^{m} T(S_i)}{CT \times m} + B \cdot \sqrt{\frac{m}{\sum_{i=1}^{m} [CT - T(S_i)]^2}} \qquad (2)$$

The closer the balance ratio of the production line is to 1, the better the production state is. The smaller the smoothing index is, the more balanced the workstation load is. A and B are weights assigned to the objective function. In this paper, A + B = 1 is set to make the balance rate and smoothness index of the production line of the same importance for the balance problem.

Based on the specific situation of the coil production line, the proposed constraint conditions are as follows [2, 3]:

(1) The work unit is the smallest work unit that can not be divided, that is, the same work unit can not be assigned to different workstations at the same time:

$$S_i \cap S_j = \emptyset (i \neq j; i, j = 1, 2, \ldots, m) \quad (4)$$

(2) Each job unit shall be finally assigned to the workstation:

$$U_{i=1}^n S_i = E \quad (5)$$

(3) The operating time of each workstation must be less than or equal to the production tempo of the production line:

$$T(S_i) \leq CT (= 1, 2, \ldots, m) \quad (6)$$

(4) The 0–1 matrix that needs to satisfy the priority relation for assignment of job units:

$$A = (a_{ij})_{n \times n} \quad (7)$$

where, if represents that operation unit I is not the immediate pre-operation of operation unit j; If aij = 1, it means that operation unit I is the immediate pre-operation of operation unit j.

5 Balance Design of Coil Production Line Based on Imperial Competition Algorithm

5.1 Algorithm is Introduced

The imperialist competitive algorithm is an intelligent optimization algorithm based on the competition mechanism between different countries in 2007 [4]. Like other algorithms, imperial competition algorithm starts with a series of population problems, where each individual represents a country and is represented by a vector or an ordered sequence of numbers. All the countries were divided into two groups according to their initial power. The stronger countries were treated as imperialists and the rest as colonies. An independent imperialist country and its affiliated colonial countries constitute a new imperial group. The power of the imperial group consists of the power of the imperialist countries and the power of the colonial countries with a certain weight. When the imperial group was established, the colonial countries began to move closer to the imperialist countries to which they belonged, that is, to move to the position of the imperialist countries in the search space. The mode of movement is described in the following article. When all the imperialist blocs were formed, there was competition between the different imperialist blocs. Similar to the evolution of history and society, each empire group wanted to increase its power by occupying other empires. If a country cannot win the competition or stop the decline of its power, it will be eliminated. In the process of empire group competition, the powerful empires will become stronger and stronger, while the weak empires will weaken and eventually perish.

5.2 Algorithm Steps

Step1: Set the number of initial countries as N and randomly generate N initial countries, that is, there are N balance schemes of the production line. Turn to Step1.

Step2: According to the randomly generated countries and the number of countries, the objective function is calculated according to the balance model. As the value function of the algorithm, it is transferred to Step3.

Step3: Select M countries with relatively large value function as imperialist countries, and the remaining countries as colonial countries. Determine the number of colonies in each imperialist country. A corresponding number of colonial countries are randomly assigned to each imperialist country to form M imperialist groups Step4.

Step4: Asassimilate colonial countries in each empire group according to the method described, and then calculate the upper objective function of newly created colonies, and transfer to Step5.

Step5: Select the maximum value function of these new colonies and compare it with the value function of the imperialist countries. If it is greater than the value function of the imperialist countries, then exchange the positions of the colonial countries and imperialist countries to which the maximum value function belongs. If the value function is smaller than that of the imperialist countries, then keep the position unchanged and go to Step6.

Step6: After the imperialist and colonial countries of each imperial group are established, calculate the total power of each imperial group and turn to Step7.

Step7: Determine whether the maximum number of iterations is reached or only one empire group is left. If so, the algorithm is finished.

Table 2 The product line balancing results of steel coil packaging

Workstation number	Workstation element	Workstation time/S	Free time/S
1	1, 4	100	20
2	2, 3	90	30
3	5, 6	110	10
4	7, 8	100	20
5	9	120	0
6	10, 11	100	20
7	12	100	20
8	13	120	0
9	14, 15	100	20
10	16, 17, 18	110	10
11	19	120	0
12	20	100	20
13	21, 22	110	10

6 Effect Evaluation After Production Line Optimization

The process distribution after improving the balance of the production line by improving the empire competition algorithm is shown in Table 2.

The balance ratio of the production line is calculated as follows: $\eta = 88.46\%$, The number of production line stations is 13.

7 Conclusion

In this paper, the improved empire competition algorithm is used to study the production line balance problem, through the mathematical modeling of the production line balance problem, the optimal solution is obtained, so as to get the highest balance degree of the production line scheme, improve the efficiency of the production line to 88.46%. To a certain extent, the model and algorithm are proved to be of practical significance for improving the balance degree of production line.

Acknowledgments This paper was supported by Natural Science Foundation of Fujian Province of China (Grant No. 2019J01790).

References

1. Hai-jiang, L.I.U.: The Steel Product Packaging Technology Report of Bao steel Development[R]. Institute of Modern Manufacturing Technology of Tongji University, Shanghai (2012)
2. Triki, H., Mellouli, A., Hachicha, W., et al.: A hybrid genetic algorithm approach for solving an extension of assembly line balancing problem[J]. Int. J. Comput. Integr. Manuf. **29**(5), 504–519 (2016)
3. Suresh, G., Sahu, S.: stochastic assembly line balancing using simulated annealing[J]. Int. J. Prod. Res. **32**(8), 1801–1810 (1994)
4. Atashpaz, G.: LUCASC. imperialist competitive algorithm: an algorithm for optimization inspired by imperialistic competition. In: Proceedings of the 2007 IEEE Congress on Evolutionary Computation. Piscataway, pp. 4661–4667. IEEE Press (2007)

Research on the Key Resilience Indexes of Logistics Enterprises in Response to Supply Chain Disruption Risks

Xu-He, Chih-Hung Hsu, and Xian-Tuo Xiao

Abstract With the development of economic globalization and enterprise globalization, the enterprise supply chain system in all industries presents an expanding development trend, which also means that the enterprise will face more and more unpredictable risks, which leads to the interruption of the enterprise supply chain and brings huge losses to the enterprise. This study takes the supply chain of a logistics company as the research object. Through field investigation and expert interview, focus group interview method, Focus Group Interview and Failure Mode Effects Analysis (FMEA) and Fuzzy Delphi method are applied to construct the house of quality model. Research results show that the logistics enterprise can through supply chain security monitoring, early warning/maintenance and maintenance of supply chain carry out quality control and reduce defects as well as supplier exploitation/management/development and so on resilient enhance criteria to further improve and develop supply chain resiliency, so that enterprises can avoid or overcome potential or in the supply chain system is facing the risk, eventually make the enterprise to achieve a steady business result.

1 Introduction

Over the past decade or so, supply disruption risk has become a unique supply chain topic. And supply disruptions are considered to be "unexpected" disruptions in the supply chain that disrupt normal logistics, and thus have an operational and financial impact on businesses in the supply chain. Supply chain risks also are described

---

Xu-He (✉) · C.-H. Hsu · X.-T. Xiao
Institute of Industrial Engineering, School of Transportation, Fujian University of Technology, University Town, No3 Xueyuan Road, Minhou, Fuzhou City, Fujian Province, China
e-mail: 1358538994@qq.com

C.-H. Hsu
e-mail: chhsu886@fjut.edu.cn

X.-T. Xiao
e-mail: 384520970@qq.com

as "adverse and unexpected events that may occur that directly or indirectly cause disruption to the supply chain". In particular, the 2020 Covid-19 caused significant supply chain disruptions last year [1].

Enterprises are now no longer limited to traditional supply chain risk factors, but are exposed to a large number of risks (quality, safety, product, leadership, labor and environment), which makes it difficult for managers to develop effective risk management strategies [2]. As a result, scholars have sought to better understand the various types of disruption risks faced by enterprises and the response mechanisms that organizations have developed to mitigate the impact of these disruption events [3].

In conclusion, this paper takes some logistics enterprise as an example, by adopting the combination of qualitative research and quantitative research of mixed research methods, from the empirical research, this paper discusses the research key resilience indexes of supply chain disruption risks, and based on this, advances the resilient enhance criteria, according to the results identify the key rule of resilient enhance criteria, for enterprise decision-makers to provide prevention or strengthening of a risk management strategy. The purpose of this study is as follows:

1. Find out the risk factors of supply chain interruption in logistics enterprises and establish the relationship between them and the resilience indexes and the resilient enhance criteria.
2. Qualitative and quantitative methods were used to determine the correlation between each factor and decision-making.
3. Can provide the reference basis for the decision-makers of logistics enterprise to prevent or strengthen risk management strategies.

2 Literature Review

The supply chain, as a network of suppliers, manufacturers, warehouses and retailers, aims to minimize total costs and meet service level requirements by producing and distributing the right quantity of goods at the right time [5]. In today's business environment with fierce global competition, the most important element of supply chain system is to reduce the total cost; Optimize delivery time; It also has a reliable distribution network. Therefore, when the supply chain is disrupted due to risk factors, it will affect not only the operations of a few enterprises, but all enterprises included in the supply chain.

In the literature, Chopra and Sodhi [6] summarized various supply chain risks, including interruption, delay, information, prediction, intellectual property, purchase, accounts receivable, inventory and production capacity, as shown in Table 1.

Scholars at home and abroad have made abundant discussions on the research and analysis of resilient enhance criteria. Among them, Liu Haohua, a domestic scholar, gave a detailed explanation on how to build an elastic supply chain from the following points [4]:

Table 1 Supply chain risk and its causes

Risk categories	Risk factors
Interrupt	Natural disasters Labor dispute Supplier bankruptcy War and terror Poor productivity and agility of backup suppliers while relying on a single supplier source
Delay	High supplier utilization Vendor agility is poor Poor product quality or supply failure Too many processing links in transit or transit
System	Information infrastructure collapses System integration or network is too complex The electronic commerce
Forecast	Long lead time, seasonal factors, product diversity, short life cycle, promotion, incentive, supply chain lack of visibility, whip effect
Intellectual property rights	Vertical integration of supply chain Global outsourcing and global markets
Procurement	Currency fluctuations Percentage of major components and raw materials that depend on a single source of supply Production utilization rate of the industry Long term and short term contracts
Accounts receivable	Number of customers The financial strength of the customer
Inventory	Scrap rate Product inventory carrying cost
Production capacity	Cost of production Productivity elasticity

1. Maintain appropriate redundancy;
2. Enhance supply chain flexibility:

    (1) redesign products and standardize processes;
    (2) Concurrent engineering development and production;
    (3) Reduce the types of parts and components;
    (4) To conclude flexible contracts;
    (5) Flexible use of various strategies according to the specific situation.

3. "Embed" Resilience in the design;
4. Improve supply chain agility;
5. Establish a full-depth and multi-level Resilient defense system.

3 Research Method

In this study, based on quality function deployment (QFD), two houses of quality (HOQ) were established to construct correlations between risk, resilience indexes and resilient enhance criteria. First disruption of the supply chain risk factors for related literature collection and sorting, secondly, organizational focus group interviews, and focus on case the key risk factors, logistics company will be the result of the interview to import the failure mode and effects analysis (FMEA) to investigate the failure causes and the possible impact of the risk factors and key are calculated and analyzed. House of Quality (HOQ) was used to combine risk factors with resilience indexes. Finally, the ranking and weight values of resilience indexes were calculated by using the Vikor sorting method, and the results of the first stage of the house of quality were obtained.

In order to improve the resilience of supply chain by literature review, this study compiled related criteria improve enterprise resilience, and use the Fuzzy Delphi Method, Fuzzy Delphi Method(FDM) selected corresponding enhance criteria, will be the first HOQ as a result, together with the resilience indexes, higher resilience indexes into the second HOQ, finally re-use VIKOR sorting method of analysis and calculation, find out the key of resilient enhance criteria, the conclusion and suggestions are put forward. The two HOQ building compositions of this study are shown below: (Fig. 1).

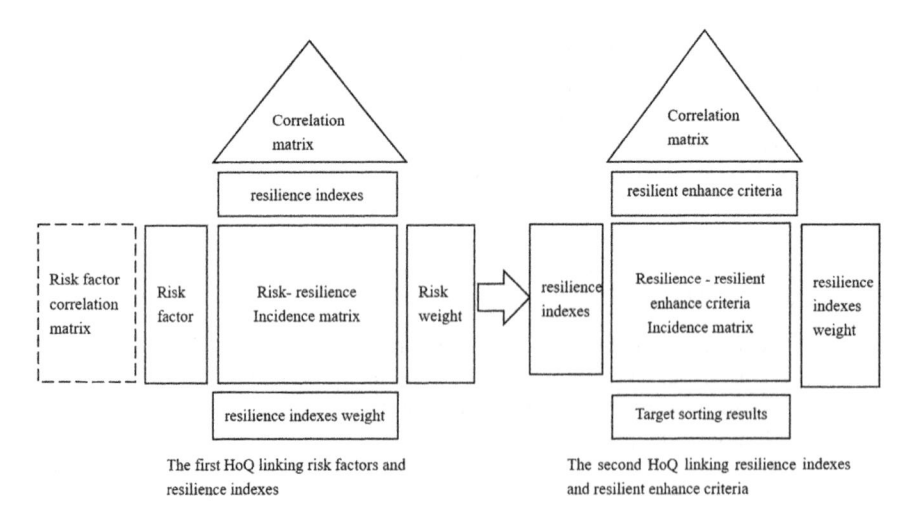

Fig. 1 The two HOQ building compositions of this study

4 Case Analysis

After the above comprehensive discussion, this chapter will begin to carry out empirical analysis on the case company. The case company (Guangdong Supply Chain Technology Co., Ltd.) is a logistics sub-group formally established in 2017 by the largest self-operated e-commerce enterprise in China. The aim is to better output the professional logistics ability of the enterprise to the whole society, and help the upstream and downstream partners of the industrial chain to reduce the logistics cost of the supply chain, improve the circulation efficiency and create the ultimate customer experience together.

As mentioned above, this study focuses on the case company and collects data in the form of questionnaire. Among them, the respondents are the employees of the case company who have more than three years of experience, and the departments they belong to are all involved in the relevant business in each link of the company's supply chain. Nine questionnaires were sent out and nine were recovered with a recovery rate of 100%.

Step 1. Focus Group Interview

According to the 9 interviewed employees listed above, including the production department, procurement department, sales department and quality assurance department, a focus group was formed to conduct the focus group interview.

After screening common risk factors through literature review, 30 possible risk factors conforming to the actual situation of the company in the case were established,

Step 2 Failure Mode Effects Analysis

According to the records of the focus group interview, the failure impact and possible causes of each risk are sorted out, and the FMEA questionnaire is designed. Each member scores each risk factor according to the previous measurement criteria, which is quantified into a numerical value for easy analysis, so as to obtain the risk priority value RPN of each risk factor. Among them, the failure effects and possible causes of each risk factor were summarized through focus group interviews in the previous stage. Once in the process of interview, some people think that a risk has impact on the company's supply chain, will continue to cross-examine the causes and effects, coupled with the failure mode and effects analysis above, each member in view of the failure O, fault inspection degrees D, fault severity rating 1–5 points, respectively, S final statistics 9 valid questionnaire responses received.

Step 3 Risk factor-resilience indexes correlation matrix analysis

In this part, the VIKOR sorting method is mainly used for calculation and analysis, so as to construct the first stage of the house of quality, and take the calculation result as the weight value of the second stage of the house of quality. Before the analysis of VIKOR method, the correlation between risk factors, the correlation between resilience indexes and the correlation between risk factors and resilience indexes should be fully considered, so as to obtain the initial matrix, so as to carry out the calculation and analysis of VIKOR steps. At this point, the analysis of the first-stage HOQ has been completed.

According to the results of the first-stage HoQ analysis, and according to the results of the first-stage HQ analysis, "E3 redundancy", "E10 market positioning" and "E12 security " of the resilience indexes are the most priority, indicating that their group utility is the highest and individual regret is the lowest, and they are the most effective indicators to measure the capacity of the supply chain system and mitigate different risks ." Redundancy" refers to the strategic allocation of additional capacity and/or inventory at potential "bottlenecks". By providing "margin" in the supply chain to deal with the surge effect, managers can build reasonable redundant inventory and capacity by establishing system-wide or redundant IT systems. Other risk factors, such as "E10 market positioning" and "E12 security ", should not be ignored.

Next, we will design a big data system, use focus group interview method and FMEA analysis, screen out representative factors. According to the actual situation of the case company, the resilience indexes are put forward. Based on the combination of qualitative analysis and quantitative analysis, VIKOR method and correlation analysis are used to construct the house of quality (HoQ) model. Finally, key resilient enhance criteria are selected, which can be provided as the reference basis for enterprise decision-makers to prevent or strengthen the risk management strategy.

5 Conclusions

The main purpose of this study is to establish the identification method of enterprise supply chain risk, evaluate and seek the mitigation plan to deal with the risk, so as to improve the elasticity of enterprise supply chain and avoid the occurrence of interruption risk. Taking a supply chain technology co., LTD in Guangdong as a case company, this paper constructs the actual risk factors of the case company by using the focus group interview method, and then selects the most representative key risk factors by FMEA analysis. In order to enhance the ability to resist risks, resilience is taken as an important measure index standard, and key resilience indexes are selected to objectively measure the resilience of supply chain system. Collected at the same time, a series of resilient enhance criteria, using Fuzzy Delphi method to the key resilient enhance criteria unity awareness of screening. Finally, the VIKOR method is adopted to rank the resilient enhance criteria. This study adopted the structure of quality function deployment (QFD), enables the "risk factor, resilience indexes, resilient enhance criteria" to close union, by correlation analysis method to establish three aspects: the number of links between each other, so that the enterprise can avoid or overcome the potential or current interruption risks in the supply chain system, and finally achieve the effect of sustainable and stable operation.

Acknowledgments This paper was supported by Natural Science Foundation of Fujian Province of China (Grant No. 2019J01790).

References

1. Remko, v.H.: Research opportunities for a more resilient post-COVID-19 supply chain–closing the gap between research findings and industry practice. Int. J. Oper. Prod. Manag. **40**(4), 341–355 (2020)
2. Marucheck, A., Greis, N., Mena, C., Cai, L.: Product safety and security in the global supply chain: issues, challenges and research opportunities. J. Oper. Manag. **29**(7–8), 707–720 (2011)
3. Sawik, T.: Integrated supply, production and distribution scheduling under disruption risks. Omega (United Kingdom) **62**(7), 131–144 (2016)
4. Haohua, L.: Building Elastic Supply Chain. J. Central Univ. Financ. Econ. **5**, 63–68 (2007)
5. Simchi-Levi, D., Kaminsky, P., Simchi-Levi, E. et al.: Designing and Managing the Supply Chain: Concepts, Strategies And Case Studies. Tata McGraw-Hill Education (2008)
6. Chopra, S., Sodhi, M.S. : Managing Risk to Avoid Supply-Chain Breakdown. MIT Sloan Manage-ment Review (2004)

Optimization of Resource Service Composition in Cloud Manufacture Based on Improved Genetic and Ant Colony Algorithm

Wang Zhengcheng

Abstract Aiming at resource service composition optimization under cloud manufacturing, a service composition and optimization objective function model for cloud manufacturing resource based on quality of service was established. An improved genetic and ant colony algorithm to solve the model was also proposed. The hybrid algorithm combined the advantages of local optimization of ant colony algorithm and global search of genetic algorithm. The improved algorithm can solve slow convergence speed and easy to fall into local optimum existed in ant colony algorithm, also can solve local search ability poor and easy to premature convergence existed genetic algorithm. Simulation results showed that the algorithm contributed to reducing problem search space and time, and can achieve identifying and matching of resource services quickly and accurately. The improved algorithm can solve the optimization problem of cloud manufacturing resource services composition more effectively.

1 Introduction

The goal of the development of advanced manufacturing models has always been around how to share efficient information and optimally allocate manufacturing resources. In 2010, academician Li Bohu and others took the lead in proposing the concept of cloud manufacturing, which is defined as "a kind of network and cloud manufacturing service platform that organizes online manufacturing resources (manufacturing cloud) according to user needs, and provides users with various on-demand manufacturing services, which defined as a new model of networked manufacturing" [1]. Cloud manufacturing resource service optimization configuration mainly includes key links, such as manufacturing resource and task modeling, manufacturing task decomposition, manufacturing resource virtual packaging, resource service retrieval and matching, service portfolio optimization, service monitoring and scheduling, and feedback evaluation [2].

W. Zhengcheng (✉)
Zhejiang Institute of Mechanical and Electrical Engineering, Hangzhou 310053, China
e-mail: achengwang@163.com

Cloud manufacturing is committed to realizing the new resource sharing and collaboration model of "centralized use of decentralized resources" and "centralized resources and decentralized services". Due to the features of resource services in the cloud manufacturing model, which are interconnected, intelligent, socialized, and globalized, it is very important for users to construct and choose the best cloud manufacturing resource optimization configuration according to user's personalized task requirements. Currently, resource service combination and optimization configuration mainly focus on two aspects: model strategy and solution algorithm. In terms of model strategies, there are mainly business flow-based combination methods [3], multi-agent based service combination [4], Petri net based service combination [5], and artificial intelligence Planning (Artificial Intelligence Planning) based service composition [6], graph based service composition [7], and QoS based service composition [8–10], etc. In terms of solving algorithms, there are mainly ant colony algorithm [11], genetic algorithm [12], immune optimization algorithm [13], particle swarm algorithm [14], and so on.

On the basis of the above research, this paper considers the high target dimension, dynamic change of search space, NP-hard complexity, and multiple quality of service (QoS) constraints for cloud manufacturing resource service optimization and configuration issues, and constructs based on quality of service (QoS) constraints. The multi-dimension cloud manufacturing service optimization configuration model of the cloud manufacturing service, and the advantages of ant colony algorithm and genetic algorithm are combined to solve the model, so as to ensure the solution effect and performance of the algorithm under the multi-objective background and dynamic environment of the cloud manufacturing service optimization configuration.

2 The Question Raised

The optimal allocation of cloud manufacturing resources is essentially a service composition optimization problem driven by multitask and multi-objective constraints. This paper proposes to solve the problem in two stages. The first stage is cloud manufacturing resource service selection driven by single-atom task constraints, and the second stage is cloud manufacturing resource service optimization configuration based on multi-objective constraints, as shown in Fig. 1.

First, the total task of cloud manufacturing T and its constraints QT is decomposed into individual atomic tasks and their constraints with the support of the service platform's case library, product process library, model library, etc., which are formalized as $T = \{T1, T2, T3, \cdots, Tn\}$ and $QT = \{QT1, QT2, \cdots, QTn\}$, where n is the number of decomposed atomic tasks and their corresponding sub-constraints, the specific decomposition process and algorithm displayed in reference [2]. Secondly, according to the semantic ontology-based resource, task and constraint description template provided by the cloud manufacturing service platform, use the ontology description language to describe cloud manufacturing resources, tasks and constraints and

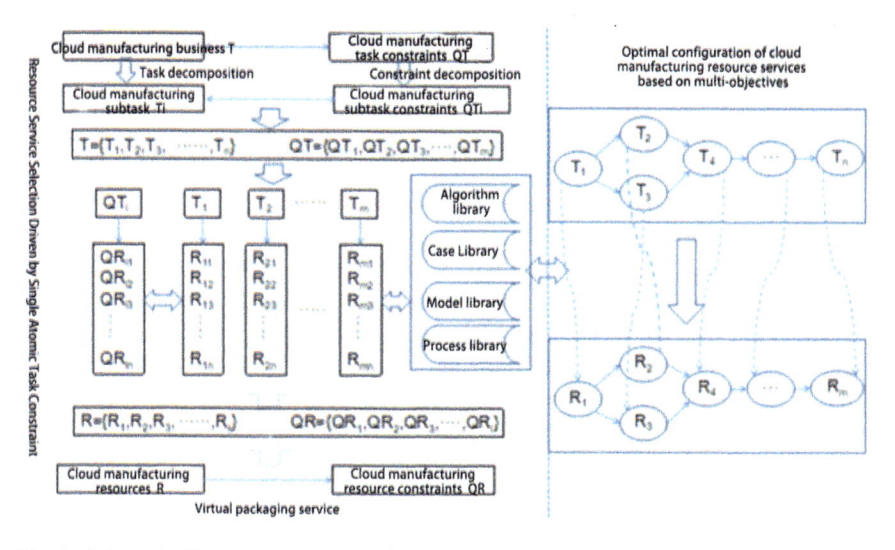

Fig. 1 Schematic diagram of cloud manufacturing resource optimization configuration process

complete platform release registration, cloud manufacturing resource virtual packaging and service deployment also shown in reference [2]. Each cloud manufacturing resource service and its constraints are formalized as R = {R1,R2,···,Rm}, QR = {QR1,QR2,···,QRm}, m is the cloud platform manufacturing resource services and the number of constraints. Then based on service quality constraints, the selection of candidate resource services that can complete a single atomic task is realized. The process only implements the selection of candidate resource services Rij based on the constraints QTi of atomic tasks Ti and the constraints QRj of cloud manufacturing resource services Rj. Rij represents the atomic task Ti that can be completed by cloud manufacturing resource service Rj. Finally, according to the total task constraint of cloud manufacturing and the constraint between resource services, QRij, with the support of case library and process library, the optimization configuration of cloud manufacturing resource services based on multi-objective and multi-constraint is completed. QRij indicates that there are connection constraints between cloud manufacturing resource services Ri and Rj.

3 Model Building

3.1 Service Quality Model

Quality of Service (QoS) is a non-functional evaluation index system that evaluates the service capability or level of cloud manufacturing resources. It is used as a measure of the pros and cons of atomic services and combined services. It is the basis of QoS-based cloud manufacturing resource service composition and optimization technology. Based on the actual characteristics of cloud manufacturing resources, this paper selects six QoS indicator attributes that are universal, can reflect user needs, and are representative of the field: service price (Price), execution time (Time), quality level (Quality), Reliability, Value, and Capability. When considering the selection of candidate resource services based on the quality of service atomic tasks, some indicators are strong constraints, and some indicators are weak constraints. This paper uses 1 to indicate strong constraints on the corresponding indicators, 0 indicates weak constraints on the corresponding indicators. Strong constraint is a mandatory index that must be met when the candidate resource service is matched. Through strong objective constraints, the problem-solving space can be greatly reduced, so as to obtain the optimal solution of the problem space quickly. Therefore, the service quality models of cloud manufacturing tasks and cloud resource services are constructed as shown in formula (1) and formula (2), respectively.

$$QT = \begin{bmatrix} QTC_1, 0\,or\,1 & QTT_1, 0\,or\,1 & QTR_1, 0\,or\,1 & QTV_1, 0\,or\,1 & & QTP_1, 0\,or\,1 \\ QTC_2, 0\,or\,1 & QTT_2, 0\,or\,1 & QTR_2, 0\,or\,1 & QTV_2, 0\,or\,1 & \cdots & QTP_2, 0\,or\,1 \\ \bullet & \bullet & \bullet & \bullet & & \bullet \\ QTC_n, 0\,or\,1 & QTT_n, 0\,or\,1 & QTR_n, 0\,or\,1 & QTV_n, 0\,or\,1 & & QTP_n, 0\,or\,1 \end{bmatrix}$$
(1)

$$QR = \begin{bmatrix} QRC_1, 0\,or\,1 & QRT_1, 0\,or\,1 & QRR_1, 0\,or\,1 & QRV_1, 0\,or\,1 & & QRP_1, 0\,or\,1 \\ QRC_2, 0\,or\,1 & QRT_2, 0\,or\,1 & QRR_2, 0\,or\,1 & QRV_2, 0\,or\,1 & \cdots & QRP_2, 0\,1 \\ \bullet & \bullet & \bullet & \bullet & & \bullet \\ QRC_m, 0\,or\,1 & QRT_m, 0\,or\,1 & QRR_m, 0\,or\,1 & QRV_m, 0\,or\,1 & & QRP_m, 0\,or\,1 \end{bmatrix}$$
(2)

In formula (1), QT is the total task constraint of cloud manufacturing, and QTC, QTT, QTR, QTV, QTQ, and QTP represent cloud manufacturing task price, time, reliability, credibility, quality level, and capacity constraints, respectively. The constraints of sub-task QT_i are shown in formula (3), where n is the number of sub-tasks after the total task is decomposed.

$$QT_i = \{\, QTC_i,\ QTT_i,\ QTR_i, QTV_i, QTQ_i, QTP_i \,\} \cdot i \in [1,n] \quad (3)$$

In formula (2), QR is the total constraint of cloud manufacturing resource service, QRC, QRT, QRR, QRV, QRQ, QRP, respectively, represent the cloud manufacturing task price, time, reliability, reputation, quality level and capacity constraints. The constraints of service QR_j are shown in formula (4), where m is the number of cloud manufacturing resource services.

$$QR_j = \{\, QRC_j,\ QRT_j, QRR_j, QRV_j, QRQ_j, QRP_j \,\} \cdot j \in [1,m] \quad (4)$$

3.2 Treatment of Constraint Indicators

Because the dimensions of each constraint index are different, it needs to be normalized. According to the six service quality constraint indicators described in 1.1, cost and time are quantitative indicators; quality, reputation and capability are qualitative indicators. The normalized value of reliability indicators is calculated according to formula (5), and $N_{j-success}$ is the number of times that the cloud manufacturing platform monitors the success of the service resource R_j, $N_{j-total}$ is the total number of call executions of the service resource R_j monitored by the cloud manufacturing platform.

$$QRR_j^* = \frac{N_{j-sucess}}{N_{j-total}} \tag{5}$$

Without losing generality, this article proposes the qualitative domain S = {excellent, good, medium, poor, poor} of three indicators of quality, reputation and ability, and the corresponding quantitative domains {1.0–0.8, 0.8–0.6, 0.6–0.4, 0.4–0.2, 0.2–0}. The three indicators of quality, reputation, and ability are normalized as formula (5): QRQ_{ij}^*, QRV_{ij}^*, QRP_{ij}^*.

The two quantitative indicators of cost and time are the smaller the better. The price includes the quotation of resource services and the logistics cost between resource services, and the time includes the request time, service execution time and logistics time between resource services. The standardization of the two indicators is proposed as shown in formula (6) and formula (7). QRC_{ij}^* and QRT_{ij}^*, respectively, represent the resource service QR_j cost and time standardized value corresponding to the cloud manufacturing task QT_i, QRC_{i-max} and QRC_{i-min}, respectively, represent the maximum and minimum value of the resource service concentration cost of the cloud manufacturing task candidate QT_i. QRC_{ij} is the cost of the cloud manufacturing task QT_i corresponding to the service candidate resource QR_j. QRT_{i-max} and QRT_{i-min}, respectively, represent the maximum and minimum service time of the cloud manufacturing task QT_i according to its candidate resource. QRT_{ij} represents the candidate resource QR_j service time corresponding to the cloud manufacturing task QT_i.

$$QRC_{ij}^* = \frac{QRC_{i-max} - QRC_{ij}}{QRC_{i-max} - QRC_{i-min}} \tag{6}$$

$$QRT_{ij}^* = \frac{QRT_{i-max} - QRT_{ij}}{QRT_{i-max} - QRT_{i-min}} \tag{7}$$

QRC_{ij}^* and QRT_{ij}^*, respectively, represent the normalized value of service cost and time of the candidate resource QR_j corresponding to the cloud manufacturing task QT_i, and $QRC_{ij}^* \in [0, 1]$, $QRT_{ij}^* \in [0, 1][0, 1]$.

3.3 Objective Function Model

Based on the constraint indicators of manufacturing tasks and resource services, with time, cost, quality, reputation, capability and reliability as the optimization goals, the optimal resource service of single-atom tasks and the optimal resource driven by task combination can be obtained through the matching of constraint indicators. Different manufacturing tasks have different emphasis and preference on resource service index, so the two-stage indicator weighting method is adopted to transform the multi-objective problem into a single-objective planning problem, and the weight coefficient is used to express the preference of the manufacturing task to each constraint indicator. The first stage selects the candidate resource service set based on the single-atom task constraint preference. And the second stage forms the optimal service combination set based on the timing constraint combination matching of multi-atomic tasks. In the second stage, the optimal service composition set is formed based on the temporal constraint combination matching of multi-atomic tasks.

The first stage: the objective function of the candidate resource service QR_j selection for the atomic task QT_i.

$$\max Q_i^* = w_T {}^1\!/\!QRT_j^* + w_c {}^1\!/\!QRC_j^* + w_q QRQ_j^* + w_v QRV_j^* + w_p QRP_j^* \\ + w_r QRR_j^* \quad j \in [1, m] \tag{8}$$

$$w_T + w_C + w_Q + w_V + w_P + w_R = 1$$

w_T, w_C, w_Q, w_V, w_P and w_R are, respectively, expressed as the constraint index time, cost, quality, reputation, ability and reliability weights, and the weight coefficients are determined by the analytic hierarchy process. If a certain index of manufacturing task QT_i is a strong constraint 1, and the corresponding index value of resource service.

QR_j is less than the index value of QT_i, then resource service QR_j is not a candidate resource service for manufacturing task.

The second stage: Formation of optimal service composition objective function based on multi-atomic task timing constraint composition matching

$$\min P = w_c \sum_{i=1}^{n-1} \sum_{j=1}^{m} (QRC_{(i,i+1),j} + QRC_{ij} + QRC_{i+1,j}) \\ + w_t \sum_{i=1}^{n-1} \sum_{j=1}^{m} (QRT_{(i,i+1),j} + QRT_{ij} + QRT_{i+1,j}) \tag{9}$$

w_c is the cost weight, w_t is the time weight and $w_c + w_t = 1$. $QRC_{(i,i+1),j}$ is the connection cost between the two candidate resource services. QRC_{ij} and $QRC_{i+1,j}$ are the processing and manufacturing costs of the two candidate resource services. $QRC_{i+1,j}$ is the connection time between the two candidate resource services, QRT_{ij} and $QRT_{i+1,j}$ are the processing and manufacturing time of the two candidate resource services.

4 Algorithm Design

4.1 Basic Principles

Genetic algorithm (Genetic Algorithm, GA) starts from a population that may potentially solve the problem. A population consists of a certain number of individuals that have been genetically coded. After the initial population is generated, according to the principle of survival of the fittest, it gradually evolves to produce the better approximate solution. In each generation, individuals are selected according to their fitness in the problem domain, and crossover and mutation are carried out with the help of genetic operators of natural genetics to produce a population that represents a new solution set. This process will lead to the natural evolution of the population, and the offspring population will be more adaptive to the environment than the previous generation. The optimal individual in the last generation population can be decoded and used as the approximate optimal solution to the problem. The basic ant colony algorithm expression is shown in formula (10):

$$GA = (C, f, P_0, M, \varphi, \Gamma, \psi, T) \qquad (10)$$

C is the population individual coding method, f is the individual fitness function, P_0 is the initial population, M is the population size, φ is the selection operator, Γ is the crossover operator, ψ is the mutation operator and T is the termination operating condition.

Ant Colony Algorithm (ACT) is an intelligent algorithm that simulates biological activities. Its basic principle is that ants will leave some chemical pheromones in the places they pass by when they are looking for food or looking for the way back to the nest. These substances can be felt by the ants arriving in the same ant colony and choosing the path containing the most material pheromone, so that within a certain period of time, the shorter the path will be accessed by more ants, and the more information will be accumulated. This process will continue until all ants take the shortest path. Ant colony algorithm has the characteristics of self-organization ability, positive feedback mechanism, easy to combine with other algorithms and parallel search ability. The model of the basic ant colony algorithm is mainly composed of movement rules and pheromone update rules. The movement rules are shown in Eq. (11), and the pheromone update rules are shown in Eq. (12).

$$P_{ij}^k(t) = \begin{cases} \dfrac{[\tau_{ij}(t)]^\alpha [\eta_{ij}(t)]^\beta}{\sum\limits_{s \in allowed_k} [\tau_{is}(t)]^\alpha [\eta_{is}(t)]^\beta} & j, s \in allowed_k \\ 0 & j, s \notin allowed_k \end{cases} \quad (11)$$

$$\eta_{ij}(t) = 1/d_{ij}$$

$$\tau_{ij}(t+1) = \rho \tau_{ij}(t) + \Delta \tau_{ij} \quad \rho \in (0, 1)$$

$$\Delta \tau_{ij} = \sum_{k=1}^{m} \Delta \tau_{ij}^k \quad (12)$$

$$\Delta \tau_{ij}^k = \begin{cases} Q/L_k & \text{The ant k passes through node i and node j in this cycle} \\ 0 & \end{cases}$$

In formula (11), $P_{ij}^k(t)$ is the transfer rule of the ant k from node i to node j at time t, and $\tau_{ij}(t)$ represents the amount of information remaining on the edge connecting node i and node j at time t. $\eta_{ij}(t)$ is the heuristic of the edge connecting node i and node j at time t. d_{ij} is the distance between node i and node j, and C is a constant number. α is the pheromone heuristic factor, reflecting the importance of the residual information on the edge connecting node i and node j, and β is the expected value heuristic factor, reflecting the importance of the heuristic information on the edge connecting node i and node j. $allowed_k$ is the set of nodes that the ant is allowed to select in the next step.

In formula (12), m represents the number of ants in the population, ρ represents the residual degree of information on the edge connecting node i and node j over time, and Q is a normal number, which represents the total amount of pheromones released by ants in a cycle or a process, L_k represents the length of the path taken by the ant k in this cycle, and $\Delta \tau_{ij}^k$ represents the increment in the information left by the ant k on the edge connecting node i and node j in this cycle, $\Delta \tau_{ij}$ represents the information increment of all ants connecting node i and node j in this cycle.

4.2 Algorithm Improvement Strategy

4.2.1 Improvement of Genetic Algorithm

The key to the efficiency of the genetic algorithm is the selection operators: the crossover operator and the mutation operator. Traditional genetic algorithm only selects individuals with high to low adaptability, but does not consider the changes of individual living environment. This paper proposes an individual selection strategy based on the evolutionary characteristics of the population as shown in formula (13):

$$P_i t = \frac{f_i(t) - f_{\min}(t)}{\sum_{k=1}^{m}(f_k(t) - f_{\min}(t))} \quad (13)$$

$P_i(t)$ is the probability that individual i is selected in the generation of population evolution t, $f_i(t)$ is the fitness value of individual i in the generation of population evolution t, $f_k(t)$ and $f_{\min}(t)$ are the fitness value and minimum value of individual k in the generation of population evolution t, and m is the number of population individuals and the fitness value of population individuals are calculated by formula (8).

Crossover probability P_c and mutation probability P_m directly affect the convergence of genetic algorithm. The larger the P_c, the faster the generation of new individuals, but the greater the damage to the high fitness individuals. The smaller the P_c, the slower the convergence rate of the algorithm will stagnate. If P_m is too small, it is not easy to produce a new individual structure. If P_m is too large, it will become a random search algorithm that affects convergence. Therefore, according to the idea of Srinvivas adaptive genetic algorithm, this paper proposes the improvement strategies of crossover operator and mutation operator as shown in formula (13) and formula (14), respectively.

$$P_{ci}(t) = \begin{cases} P_{c2} + \frac{(P_{c1} - P_{c2})(f_{\max}(t) - f_i(t))}{f_{\max}(t) - f_{avg}(t)} & f_i(t) \geq f_{avg}(t) \\ P_{c1} & f_i(t) < f_{avg}(t) \end{cases} \quad (14)$$

$$P_{mi}(t) = \begin{cases} P_{m2} + \frac{(P_{m1} - P_{m2})(f_{\max}(t) - f_i(t))}{f_{\max}(t) - f_{avg}(t)} & f_i(t) \geq f_{avg}(t) \\ P_{m1} & f_i(t) < f_{avg}(t) \end{cases} \quad (15)$$

Take $P_{c1} = 0.9$, $P_{c2} = 0.5$; $P_{m1} = 0.1$, $P_{m1} = 0.1 f_{\max}(t)$ as the maximum fitness of the t-generation population, and $f_{avg}(t)$ is the average fitness of the t-generation population.

4.2.2 Improvement of Ant Colony Algorithm

(1) Improvement of information retention

The degree of information residue ρ is directly related to the global search ability of the ant colony algorithm and its convergence speed. If ρ is large, the path that has not been searched will be selected with low probability; if ρ is small, it will affect the convergence speed of the algorithm. Therefore, in the early stage of the algorithm, hoping to find the optimal solution as soon as possible to obtain a constant and larger value of ρ. In the later stage of the algorithm, to prevent the search from falling into the local optimal stagnation, if the optimal solution is not significantly improved, the information residual degree ρ value is adaptively adjusted according to formula (14), where C is a normal number, $C \in (0, 1)$。

$$\rho(t+1) = \begin{cases} 0.9\rho(t) & if \left| Nf(t) - \sum_{i=1}^{N} f(t-i) \right| \leq CN \sum_{i=1}^{N} f(t-i) \\ \rho_{min} & Otherwise \end{cases} \quad (16)$$

(2) Improvement of heuristic factors α and β

The pheromone heuristic factor α reflects the relative importance of the amount of information accumulated by the ants in the movement process $\tau_{ij}(t)$ in the ant colony search, and the expected value heuristic factor β reflects the relative importance of the ants inspiring information during the movement process $\eta_{ij}(t)$ in the ant colony search [2]. The larger the value of β, the faster the convergence speed of the algorithm search, but the randomness in the search process is weakened. The larger the α value, the greater the possibility that the ant will choose the path it has traveled before, and it is easy to fall into the local optimum. Therefore, the paper sets the constants α and β in the basic ant colony algorithm as the time-varying parameters $\alpha(t)$ and $\beta(t)$ to enhance the algorithm's adaptive anti-interference ability.

$$\alpha(t) = \alpha_0 + k_\alpha t \quad \beta(t) = \beta_0 - k_\beta(t) \quad (17)$$

4.3 Algorithm Flow

Combine a variety of optimization algorithms to overcome the limitations of a single algorithm, to ensure that the problem space has a good global search ability, and the solution can quickly converge and find an accurate solution. This paper combines the advantages of genetic algorithm and ant colony algorithm and proposes an improved strategy. In the stage of selecting candidate resource services for single-atom tasks, due to the large number of resource services, the solution space is large. Improved genetic algorithm is used to solve the problem, making full use of its convergence and global search capabilities. The advantages of the solution space, and can quickly search out all the candidate resource service solutions in the solution space. In the stage of forming the optimal service combination based on the combination of multi-atomic task timing constraints and matching, due to the relatively small number of solution spaces, in order to solve the crossover of the genetic algorithm's local search capabilities, it is easy to prematurely converge and fall into the local optimal problem. The improved ant colony algorithm is used to solve the problem. Take advantage of its strong local optimization ability and high ability to find accurate solutions. The steps of algorithm design are as follows, and the flowchart is shown in Fig. 2.

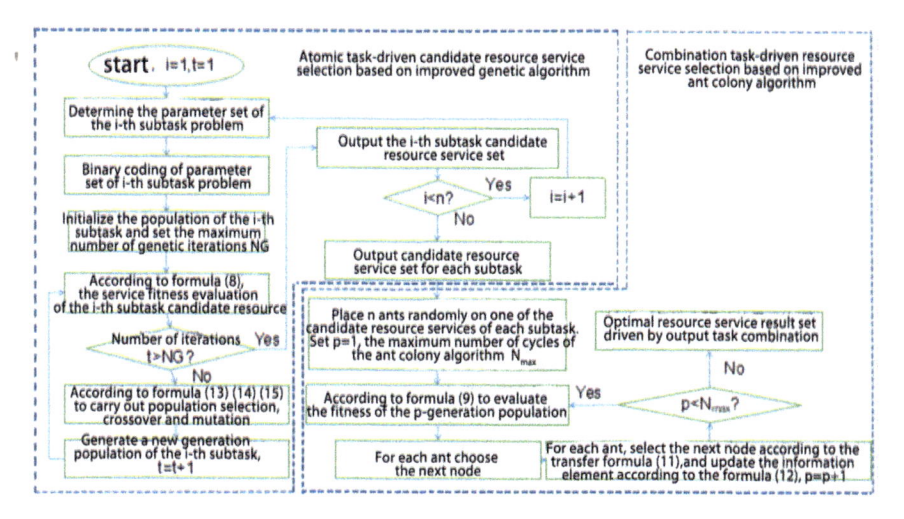

Fig. 2 Flow chart of hybrid algorithm

(1) The first stage takes advantage of the improved genetic algorithm's strong global search capability, and quickly converges to output a set of candidate resources driven by atomic tasks.
(2) The second stage takes advantage of the strong local search ability of the improved ant colony algorithm, and quickly converges to output a resource service combination scheme driven by the combination of sequential atomic tasks.

5 Simulation Examples

Taking automobile engine cloud manufacturing as an example, the model and algorithm mentioned in this article are verified. Suppose an engine manufacturing company requires the completion of a certain specification engine assembly manufacturing task within 30 days, and the company's engine manufacturing requires cloud manufacturing platform collaboration after task decomposition. The process of completing the main sub-tasks is shown in Fig. 3. Mainly include structural design (QT1), process analysis and planning (QT2), numerical control programming (QT3), simulation processing (QT4), cam processing (QT5), connecting rod processing (QT6), piston processing (QT7), cylinder processing (QT8) and machining (QT9) 9 sub-tasks.

Establish the task matrix according to formula (1):

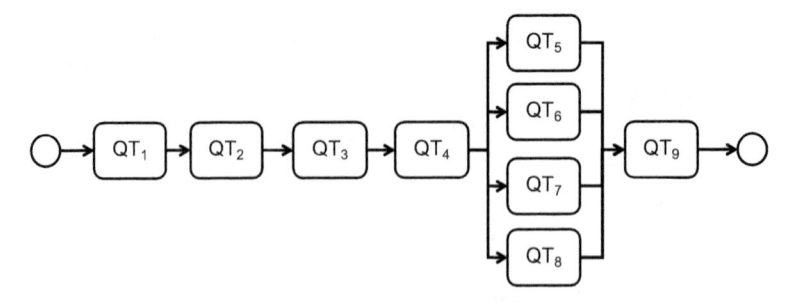

Fig. 3 Schematic diagram of automobile engine manufacturing process

$$QT = \begin{vmatrix} QT_1 \\ QT_2 \\ QT_3 \\ QT_4 \\ QT_5 \\ QT_6 \\ QT_7 \\ QT_8 \\ QT_9 \end{vmatrix} = \begin{vmatrix} \text{cost} & \text{time} & \text{reliability} & \text{reputation} & \text{quality} & \text{capability} \\ 10, 1 & 2, 1 & 0.85, 0 & 0.72, 0 & 0.90, 0 & 0.75, 0 \\ 12, 1 & 3, 0 & 0.90, 0 & 0.68, 0 & 0.82, 0 & 0.76, 0 \\ 10, 1 & 1, 1 & 0.95, 1 & 0.84, 0 & 0.78, 0 & 0.80, 0 \\ 10, 0 & 1, 0 & 0.86, 0 & 0.75, 0 & 0.72, 0 & 0.78, 0 \\ 35, 1 & 3, 1 & 0.86, 0 & 0.70, 0 & 0.74, 0 & 0.82, 0 \\ 20, 1 & 3, 1 & 0.78, 0 & 0.76, 0 & 0.75, 0 & 0.80, 0 \\ 22, 1 & 4, 1 & 0.82, 1 & 0.78, 0 & 0.76, 0 & 0.82, 0 \\ 40, 1 & 6, 1 & 0.92, 1 & 0.85, 1 & 0.84, 0 & 0.84, 0 \\ 25, 1 & 5, 1 & 0.95, 0 & 0.86, 1 & 0.82, 0 & 0, 80, 0 \end{vmatrix}$$

Set cost weight $w_c = 0.35$, time weight $w_t = 0.2$, reliability weight $w_r = 0.15$, reputation weight $w_v = 0.1$, quality weight $w_q = 0.1$ and capability weight $w_p = 0.1$. According to the genetic algorithm proposed in this paper, combined with the candidate resource service information that has been registered and collected on the platform, the initial value of the genetic algorithm is set to 50, the ending evolution algebra is 100, the crossover probability is 0.9 and the mutation probability is 0.1. The output service attribute value of each sub-task candidate resource is shown in Table 1, and the objective function value is calculated according to formula (8).

According to the manufacturing process flow chart of Fig. 3, the logical sequential network topology structure of each subtask candidate resource service based on connection time and connection cost is shown in Fig. 4.

The connection time and connection cost between the cloud manufacturing task initiating S, ending E and each candidate resource service are shown in Table 2.

Set the initial value of the ant colony algorithm parameters, the number of ants is 50, the information intensity is 100, the information residual degree is 0.85, the expected heuristic factor is 3.8 and the heuristic factor is 1.24, time weight $w_t = 0.4$, cost weight $w_c = 0.6$. According to the improved ant colony algorithm and objective function formula (9) proposed in this paper, the solution result is shown in Fig. 5. The total manufacturing cost is 1.673 million yuan, the total logistics cost is 21,800 yuan, the total manufacturing time is 16.2 days and the total logistics time is 12.4 days.

Optimization of Resource Service Composition in Cloud ...

Table 1 Service attribute values of candidate resources for each sub-task

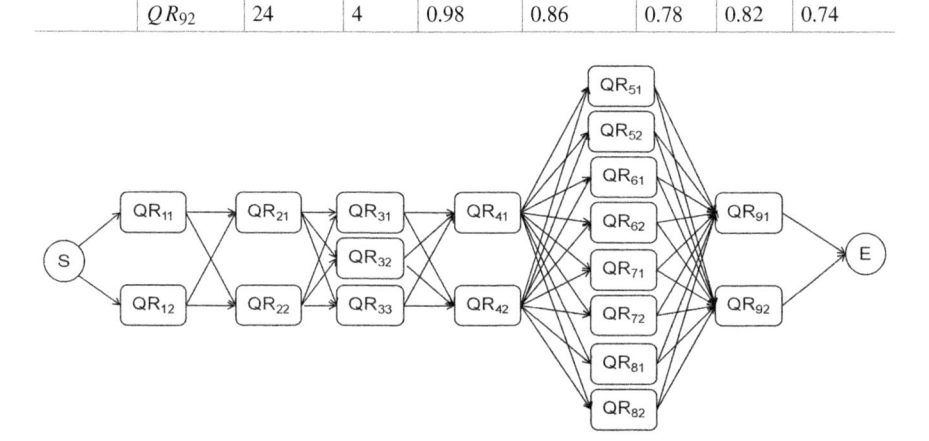

Sub-task name and weight	Candidate resource service name	Cost (ten thousand yuan)	Time (days)	reliability	Credibility	quality	ability	Objective function value
$QT_{1(0.12)}$	QR_{11}	9.4	1.8	0.88	0.70	0.85	0.78	0.715
	QR_{12}	9.6	1.5	0.82	0.85	0.80	0.82	0.57
$QT_{2(0.07)}$	QR_{21}	11.5	2.5	0.86	0.72	0.84	0.78	0.563
	QR_{22}	11.3	2.7	0.92	0.81	0.80	0.84	0.733
$QT_{3(0.06)}$	QR_{31}	9.5	0.8	0.95	0.86	0.88	0.76	0.7
	QR_{32}	9.6	0.7	0.96	0.85	0.80	0.80	0.74
	QR_{33}	9.4	1	0.95	0.80	0.75	0.72	0.72
$QT_{4(0.06)}$	QR_{41}	9.6	0.8	0.86	0.80	0.78	0.76	0.563
	QR_{42}	9.4	1	0.90	0.86	0.82	0.80	0.733
$QT_{5(0.15)}$	QR_{51}	36	2.8	0.87	0.75	0.82	0.86	0.57
	QR_{52}	33	3	0.85	0.84	0.88	0.80	0.73
$QT_{6(0.12)}$	QR_{61}	19	2.8	0.80	0.86	0.90	0.84	0.58
	QR_{62}	18	3	0.82	0.85	0.82	0.86	0.726
$QT_{7(0.12)}$	QR_{71}	22	3.5	0.82	0.82	0.85	0.80	0.568
	QR_{72}	20	4	0.83	0.84	0.88	0.82	0.73
$QT_{8(0.09)}$	QR_{81}	40	5.5	0.92	0.86	0.80	0.82	0.586
	QR_{82}	38	6	0.93	0.85	0.78	0.80	0.73
$QT_{9(0.15)}$	QR_{91}	25	3	0.96	0.88	0.80	0.84	0.6
	QR_{92}	24	4	0.98	0.86	0.78	0.82	0.74

Fig. 4 Schematic diagram of the connection topology of each candidate resource service

Table 2 The connection time and connection cost of each candidate resource service

Connection name	time (天)	cost (元)	Connection name	time (天)	cost (元)	Connection name	time (天)	cost (元)
S-QR$_{11}$	1.2	2300	QR$_{41}$ – QR$_{51}$	1.4	1200	QR$_{61}$ – QR$_{91}$	1.5	1200
S-QR$_{12}$	1.8	2500	QR$_{41}$ – QR$_{52}$	1.6	1500	QR$_{62}$ – QR$_{91}$	2.0	1800
QR$_{11}$ – QR$_{21}$	1.6	1500	QR$_{41}$ – QR$_{61}$	2.1	1800	QR$_{71}$ – QR$_{91}$	2.4	2200
QR$_{11}$ – QR$_{22}$	1.3	1200	QR$_{41}$ – QR$_{62}$	1.6	1500	QR$_{72}$ – QR$_{91}$	1.6	1400
QR$_{12}$ – QR$_{21}$	1.5	1600	QR$_{41}$ – QR$_{71}$	2.5	2300	QR$_{81}$ – QR$_{91}$	1.8	1600
QR$_{12}$ – QR$_{22}$	1.7	1800	QR$_{41}$ – QR$_{72}$	1.8	1600	QR$_{82}$ – QR$_{91}$	2.2	2000
QR$_{21}$ – QR$_{31}$	2.0	1800	QR$_{41}$ – QR$_{81}$	2.0	1800	QR$_{51}$ – QR$_{92}$	2.4	2100
QR$_{21}$ – QR$_{32}$	1.8	1200	QR$_{41}$ – QR$_{82}$	2.2	2000	QR$_{52}$ – QR$_{92}$	1.8	1600
QR$_{21}$ – QR$_{33}$	2.4	2600	QR$_{42}$ – QR$_{51}$	1.8	1500	QR$_{61}$ – QR$_{92}$	1.5	1300
QR$_{22}$ – QR$_{31}$	2.2	2300	QR$_{42}$ – QR$_{52}$	1.6	1300	QR$_{62}$ – QR$_{92}$	2.4	2200
QR$_{22}$ – QR$_{32}$	2.5	2400	QR$_{42}$ – QR$_{61}$	2.4	2200	QR$_{71}$ – QR$_{92}$	2.0	1800
QR$_{22}$ – QR$_{33}$	1.8	1700	QR$_{42}$ – QR$_{62}$	1.6	1400	QR$_{72}$ – QR$_{92}$	1.6	1500
QR$_{31}$ – QR$_{41}$	2.3	2400	QR$_{42}$ – QR$_{71}$	1.5	1300	QR$_{81}$ – QR$_{92}$	2.5	2300
QR$_{31}$ – QR$_{42}$	1.6	1800	QR$_{42}$ – QR$_{72}$	2.2	1800	QR$_{82}$ – QR$_{92}$	2.3	2100
QR$_{32}$ – QR$_{41}$	2.5	2300	QR$_{42}$ – QR$_{81}$	2.5	2300	QR$_{91}$-E	1.5	1300
QR$_{32}$ – QR$_{42}$	2.2	2000	QR$_{42}$ – QR$_{82}$	1.6	1400	QR$_{92}$-E	1.8	1600
QR$_{33}$ – QR$_{41}$	2.0	1800	QR$_{51}$ – QR$_{91}$	1.8	1600			
QR$_{33}$ – QR$_{42}$	1.6	1500	QR$_{52}$ – QR$_{91}$	2.0	1800			

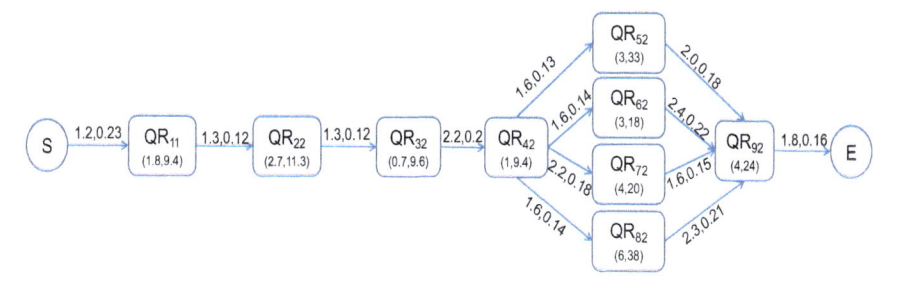

Fig. 5 Service composition structure based on time cost constraints

6 Conclusion

Aiming at the shortcomings and existing problems of the task-driven resource service combination in the cloud manufacturing mode, this paper first proposes a flow chart for the optimal configuration of cloud resource service combination in the cloud manufacturing mode and explains the task-driven cloud resource service optimization configuration process. Secondly, a service quality model for cloud resource combination is constructed, and a normalization method for constraint indicators is proposed. Based on the optimized configuration flow chart and the service quality model, a two-stage cloud resource service combination objective function is proposed again, and an improved genetic algorithm and an ant colony algorithm are designed to solve the objective function. A simulation example proves that the proposed hybrid algorithm can quickly and effectively find the optimal solution of the problem space. In the next step, we will conduct in-depth research on the synergy of resource service portfolio driven by multiple concurrent tasks in the cloud manufacturing model. Different multi-task portfolio structures will seek more solving algorithms to maximize the comprehensive benefits of each concurrent task and maximize the benefits of cloud resource service sharing.

References

1. Bohu, L.I., Lin, Z., Shilong, W. et al.: Cloud manufacturing: a new service-oriented networked manufacturing model. Comput. Integr. Manuf. Syst.16(1):1–8(in Chinese) (2010)
2. Zhengcheng, W.: Study on Several Key Problems for Networked Manufacturing Resources Integration Platform. Zhejiang University (2009)
3. Tianyang, L.: Research on key technologies of mass customization service by manufacturing cloud. Harbin Institute of Technology (2018)
4. Longfei, Z., Lin, Z., Yongkui, L.: Survey on scheduling problem in cloud manufacturing. Comput. Integr. Manuf. Syst. 23(6), 1147–1166 (2017)
5. Min, H., Guoqing, S., Danchen, Z. et al.: Test method to quality of service composition based on time-varying petri net. J. Softw. 30(8), 2453–2468 (2019)
6. Ming , G.: Modeling,service planing and service composition in knowledge intensive collaborative work flows. DONGBEI University of Finance & Economic (2013)

7. Li, M., Zhiyang, Q., Yanping, C. et al.: Semantic web service selection based on QoS. Comput. Sci. **44**(3), 226–230, 246 (2017)
8. Chenghua, L., Jisong, K.: Multi-attribute decision making and adaptive genetic algorithm for solving QoS optimization of web service composition. Comput. Sci. **46**(2), 187–195 (2017)
9. Chen, F., Jindong, W., Hengwei, Z. et al.: Multi-constraint service selection based on decomposition of global QoS. J. Syst. Simul. **30**(10), 3893–3902 (2018)
10. Zhang, Z.J., Zhang, Y.M., Xu, X.S., et al.: Manufacturing service composition self-adaptive approach based on dynamic matching network. Ruan Jian Xue Bao/J. Softw. **29**(11), 3355–3373 (2018)
11. Zhengcheng, W.A.N.G., Xiaohong, P.A.N., Xuwei, P.A.N.: Resource service chain construction for networked manufacturing based on ant colony algorithm. Comput. Int. Manuf. Syst. **16**(1), 174–181 (2010)
12. Wenan, T., Yao, Z.: Web service composition based on chaos genetic algorithm. Comput. Integr. Manuf. Syst. 24(7), 1822–1829 (2018)
13. Yuanfeng, M.A., Angru, L.I., Huimin, Y.U. et al.: Dynamic crowding distance-based hybrid immune algorithm for multi-objective optimization problem. Comput. Sci. **45**(6A), 63–68 (2018)
14. Zhengcheng, W., Da, X.: Research on inter-organizational resource chain construction based on improved PSA. China Mech. Eng. **24**(9), 1186–1190.1194 (2013)

Real-Time Multi-person Multi-camera Tracking Based on Improved Matching Cascade

Yundong Guo, Xinjie Wang, Hao Luo, Huijie Pu, Zhenyu Liu, and Jianrong Tan

Abstract In a small-scale distributed multi-camera system like video surveillance system of a museum, shopping mall, plaza, etc., or advanced driving assistance system, real-time multi-person tracking is essential for public and pedestrian safety consideration in the smart security system. In this paper, a real-time multi-person multi-camera tracking framework is presented, which is compatible with both overlapping and non-overlapping views. Since cameras have different orientations and exposures, false matching occurs frequently when people cross the camera boundaries or reenter the same camera. To deal with this challenge, an improved multi-person multi-camera matching cascade scheme is proposed, which can increase the accuracy of inter-camera person re-identification (Re-ID) by taking advantage of association priorities of targets and features. Besides, the proposed method can deal with the occlusion of people and variation of appearance features. Experiments are implemented with overlapping and non-overlapping videos, and results show that the proposed method has robust performance in different situations.

Y. Guo · Z. Liu · J. Tan
School of Mechanical Engineering, Zhejiang University, Hangzhou 310027, China
e-mail: yundong88@zju.edu.cn

Z. Liu
e-mail: tree1118220@126.com

J. Tan
e-mail: 58151161@qq.com

X. Wang
College of Information Science and Engineering, Ocean University of China, Qingdao 266100, China
e-mail: wangxinjie@ouc.edu.cn

H. Luo (✉)
School of Aeronautics and Astronautics, Zhejiang University, Hangzhou 310027, China
e-mail: luohao@zju.edu.cn

H. Pu
Pan Asia Technical Automotive Center Co., Ltd. (PATAC), Shanghai 201206, China

1 Introduction

With the consideration of safety or supervision, video surveillance systems are deployed at many public and private places and continuously monitored by security guards. Since surveillance systems keep running all the time, some hidden dangers are occasionally neglected due to the fatigue or carelessness of human beings. To reducing the workload of artificial monitoring and analysis, or even supervising without relying on any constant human participation, the demands of automatically real-time multi-person multi-camera tracking, rapidly increase in recent years. Besides, it is also a key technology in advanced driving assistance systems.

It is a challenging problem, especially when there are both overlapping and non-overlapping views among cameras. And observations of the same person differ obviously in different cameras, which have various angles or exposure conditions. Not to mention there are strict requirements of real-time tracking performance in driving assistance scenarios. Generally, real-time multi-person multi-camera tracking can be broken down into two modules, i.e., multi-object tracking in a single camera and person Re-ID across cameras. Specifically, techniques like pedestrian detection, feature extraction, person Re-ID, and data association are included.

Tracking multiple objects in a single camera is called Single Camera Tracking (SCT) or Multiple Object Tracking (MOT). The most critical challenge is to distinguish individuals when occlusion frequently occurs. There has been a lot of research to deal with this task [1–6], but most approaches are offline methods and focused on global optimization. Thanks to the significant advances of object detection [7–11], tracking-by-detection becomes a popular and powerful strategy, which makes real-time SCT come true.

Compared to SCT, associating the same person among different cameras is the characteristic of multi-camera tracking (MCT). Person Re-ID based on individual features is one of the most effective methods to achieve MCT. Because the features are generated by a trained feature extractor [12–14], appearance features based on convolutional neural networks (CNN) could perfectly work with pedestrian detection framework. By formulating the data association problem, which means determining each detection of person correspond to which identity or individual, the whole pipeline of MCT could be established.

In this paper, a new framework for a real-time multi-person multi-camera tracking system is presented. The main contributions of this paper are as follows:

- A real-time multi-person multi-camera tracking framework for both overlapping and non-overlapping views, in which pedestrian detection, motion model, feature extraction, and matching cascade are included.
- An improved multi-person matching cascade scheme for inter-camera, which can increase the accuracy of inter-camera person Re-ID by taking advantage of association priorities of targets and features.

The rest part of this paper is organized as follows. Section 2 gives a brief review of the related work including MCT, person Re-ID, and data association. Section 3 extensively describes the proposed method and theory. Section 4 demonstrates the experimental results. Finally, conclusions are drawn, and the associated future work is given in Sect. 5.

2 Related Work

Due to the illumination difference, viewpoints variance, and blind areas among cameras, real-time multi-person multi-camera tracking is challenging work. There has been a lot of works aiming at MCT in recent years [15–18], which are mostly based on tracking-by-detection strategy and designed to deal with videos captured by surveillance systems. The main idea is to obtain bounding boxes of targets firstly and then estimate motion formulation or extract appearance features of all targets simultaneously. At last, tracklets across cameras are linked using association methods. These frameworks usually focus on either overlapping [15] or non-overlapping [16, 18] multi-camera systems, and could not deal with those having both views robustly. Besides, MCT methods aiming at online frame-to-frame matching are hardly published.

In frameworks introduced above, techniques like person Re-ID and data association have the most influence on the performance of MCT, so optimization and innovation are mainly focused on these fields. In 2014, a deep metric learning method by using siamese CNN is proposed, which is the first work to apply deep learning in person Re-ID problem [12]. With the availability of large datasets like Market-1501 [19] and MARS [20], person Re-ID is usually addressed within a metric learning framework in recent years, and several CNN architectures for end-to-end learning have been designed to deal with pose variations and viewpoint changes [14]. Besides, an attribute-person recognition network is presented, which learns a discriminative CNN embedding for both person Re-ID and attributes recognition [21].

In the field of data association, the Hungarian algorithm [22] could solve the bipartite matching problem efficiently and is proved to work well in a frame-to-frame association of SCT [23]. But it only considered consecutive frames, rather than all pairwise evidence, and may result in more identity switches and lower quality trajectories when occlusion occurs. A graph-based formulation of the data association problem is presented by introducing a novel local pairwise feature based on local appearance matching, which is robust to partial occlusion and camera motion [3]. A continuous evaluation metric for association problems is formulated, which addresses the difficulty of association about people crossing camera boundaries with appearance, biometric, and location information [24]. Besides, a matching cascade that assigns higher priorities to more frequently seen targets is introduced, which could deal with the occlusion situation well [14]. Methods above for data association are performed on either SCT or MCT, and a unified scheme for both overlapping and non-overlapping views of the multi-camera system is not developed yet.

3 Proposed Method

3.1 Overview

In this section, a new real-time multi-person multi-camera tracking framework is presented, which has robust performances for both overlapping and non-overlapping views. The global multi-person matching cascade scheme is improved by establishing a shared set with multiple features from confirmed targets. Furthermore, with different association priorities of targets and features, the accuracy of person Re-ID across cameras is increased.

Overview of the proposed method is shown in Fig. 1, which can be divided into three parts: Multi-camera System, Matching Cascade, and Tracking Result. A multi-camera system with three cameras having overlapping views is demonstrated as an example. Each camera works independently, which is responsible for capturing an image from a particular perspective. Then, the captured image is processed by a real-time detector like SSD [9] or YOLOv3 [10] to obtain detections. At last, the location and appearance features of each detection are collected for the matching cascade. Once the tracking pipeline is working, the proposed tracking framework would maintain a shared set with multiple features from confirmed targets. During the phase of matching cascade, a shared set is used to perform bipartite matching between detections and confirmed targets based on appearance features. Then, unmatched detections and targets are attempted to be matched based on intersection over union (IOU) of bounding boxes. Finally, appearance features of matched targets are saved to the shared set, and detections are assigned with corresponding labels according to matching results. The motion state of each confirmed target is updated based on a fusion of its motion model and matching results.

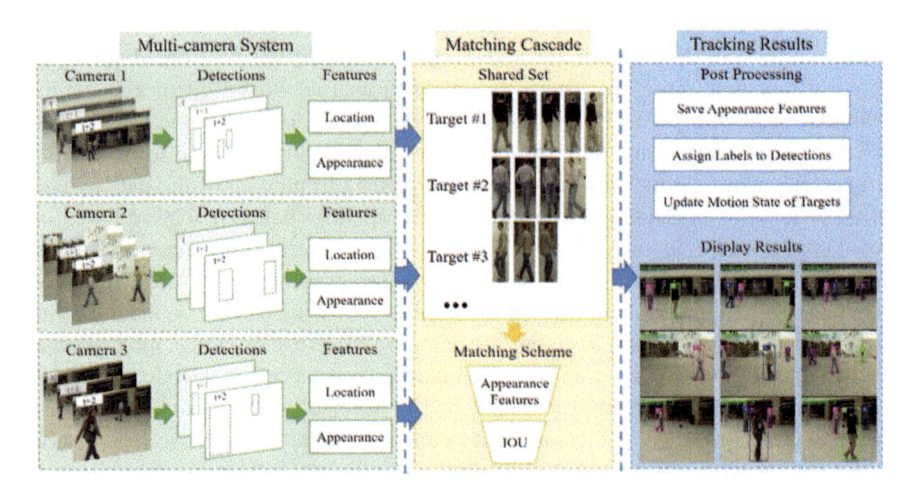

Fig. 1 Overview of the proposed method

During the pipeline of proposed real-time multi-person multi-camera tracking framework, pedestrian detection, motion model, feature extraction, and matching cascade are the most important technologies, which will be illustrated below in detail.

3.2 Pedestrian Detection

The widely used tracking-by-detection strategy is built upon an object detector, which provides detection results to drive the tracking procedure. But it depends heavily on the quality of detectors, and an identical approach would produce tracking results with significant performance differences when using different detectors. Besides, some approaches perform well in specific video sequences because their detector is trained in specific videos and does not generalize well in other video sequences [6].

In this paper, YOLOv3 is used to compromise between the accuracy of detections and running speed. The training dataset is collected and organized based on multiple well-known pedestrian datasets like Caltech Pedestrian Dataset [25], CUHK Occlusion Dataset [26], DukeMTMC Data Set [27]. Like anchor boxes obtained by k-means clusters on the COCO dataset, the bounding box priors for pedestrian detection are set with reasonable dimensions based on the training dataset.

In real-time multi-person multi-camera tracking, the tolerance of false negative is higher than false positive. So, the loss function of YOLOv3 is defined as

$$L_{YOLOv3} = \varphi_{coord} \sum L_{coord} + \varphi_{conf} \sum L_{conf} + \sum L_{cls} \quad (1)$$

where L_{YOLOv3} is the total loss, L_{coord} is location loss, L_{conf} is confidence loss, L_{cls} is classification loss, φ_{coord} and φ_{conf} are weight of location loss and confidence loss. In the situation of multi-person multi-camera tracking, φ_{conf} should be larger than φ_{coord} to achieve less false positive.

3.3 Motion Model

The main task of tracking is localizing every target at any time in videos, image sequences, or live video streams. Building a motion model is a simple way to achieve it.

In this paper, a Kalman Filter with constant velocity motion and linear observation model is built to deal with state estimation of each target, which is proved to be functional [14]. The motion state of each target is defined as a column vector

$$\mathbf{S} = (x, y, w, h, \dot{x}, \dot{y}, \dot{w}, \dot{h})^T \quad (2)$$

where **S** is the vector of motion state. The center coordinate of bounding box is (x, y), with width w and height h. Their velocities are $(\dot{x}, \dot{y}, \dot{w}, \dot{h})$, respectively, which could be positive or negative.

3.4 Feature Extraction

Lots of research shows that appearance features obtained by deep metric learning have a good performance for both MCT and Re-ID [12–14]. Since state estimation based on the motion model is untrustworthy occasionally, the CNN-based appearance features are used to distinguish individuals primarily in this paper.

In general, a wide residual network is employed to extract features from images of detections. A feature vector with 128 dimensions is computed for each detection and normalized to a unit hypersphere to perform a cosine metric in the matching phase. This CNN network is trained using a large-scale person Re-ID dataset called MARS [20], which contains 1,261 identities and around 20,000 video sequences. The trained network is validated on DukeMTMC Data Set [27], which shows a good generalization performance.

3.5 Matching Cascade

In SCT, frequent occlusion is the most difficult problem to deal with. Meanwhile, person Re-ID when people crossing camera boundaries or reentering the same camera is the trickiest one in MCT. An improved matching cascade scheme is proposed to cope with occlusion problems and increase the accuracy of person Re-ID across cameras by taking advantage of association priorities of targets and features. There are three key roles in this scheme, which are priorities, metrics, and bipartite matching method.

The improved matching cascade scheme is shown in Fig. 2, which can be taken as a funnel-shaped set of bipartite matching procedures based on different metrics. There are four matching procedures totally, and the matching priority of each procedure is marked by a number. Since appearance features based on CNN are more reliable than state estimation based on motion model, detections and targets are matched by appearance features firstly and by IOU secondly. During matching by appearance features, detections are matched to features used to appear in the same camera first. Then, the unmatched detections are matched to features obtained from other cameras. Besides, during matching by IOU, unmatched detections are matched to unmatched targets using IOU in case of a sudden change of appearance features, which often occurs after a long period of occlusion. At last, unmatched detections are matched to tentative targets, which would turn to be confirmed targets if successfully matched five times continuously.

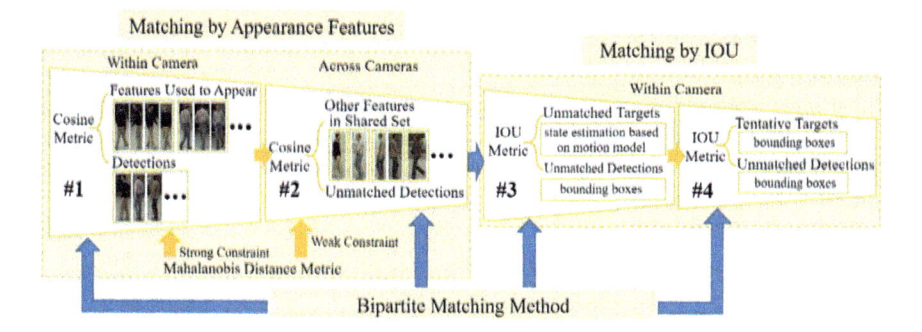

Fig. 2 The improved matching cascade scheme

Cosine metric is used to quantify the distance between two vectors, which is a good way to determine whether two appearance features are from one target. The cosine distance between the i-th detection and j-th target is defined as

$$D_{cosine}(i, j) = 1 - \mathbf{r}_i^T \mathbf{r}_j \tag{3}$$

where $\mathbf{r}_i$ is feature vector of the i-th detection that has 128 dimensions and is normalized to a unit hypersphere, $\mathbf{r}_j$ is feature vector of the j-th target. If a target has more than one feature vector, the minimum $D_{cosine}(i, j)$ calculated would be the real cosine distance.

Mahalanobis Distance is a metric to measure how many standard deviations a detection is away from the mean target location, which is estimated by a Kalman Filter. During matching by appearance features, it is a quantitative factor to avoid bad results caused by false positive detections and eliminate the misunderstanding of two targets with similar appearance features.

IOU Metric is a widely used method to calculate the distance between two bounding boxes. If FPS is high enough like 25, the IOU Metric would be a simple way to match detections and targets when there is no occlusion. IOU of two bounding boxes A and B is calculated by

$$\text{IOU} = \frac{A \cap B}{A \cup B} \tag{4}$$

The bipartite matching method is performed at every matching procedure. It is a classic assignment problem, which could be solved by the Hungarian algorithm perfectly [22], which could achieve convincing results and satisfying speed.

4 Experimental Results

Experiments are implemented with this setup: integrated development environment (IDE) is VS Code, deep learning platforms are TensorFlow for feature extraction and Darknet for YOLOv3, and computational hardware is based on a PC running Windows 10 with Intel i7 CPU and one GeForce RTX 2080Ti GPU installed.

A sequence called "campus" shot outside on the campus of EPFL with three cameras [28] is used to evaluate the proposed real-time multi-person multi-camera tracking method. Three cameras have overlapping views and up to five people are simultaneously walking in front of them. In Fig. 3, three captures with tracking results at 00:08 s, 00:12 s, and 00:16 s of videos are shown, respectively. Images in the same column show simultaneous tracking results of multiple cameras at a certain moment, and images in the same row show tracking results of multiple people in one camera. The results show that people walking around in the watched area could be tracked perfectly. During a long time running, the average FPS is around 25, which can be considered real-time.

To evaluate the performance of the proposed method on non-overlapping sequences, we captured videos in an outdoor environment with three cameras. The field of view (FOV) of each camera is 45°, and the total FOV is nearly 140°.

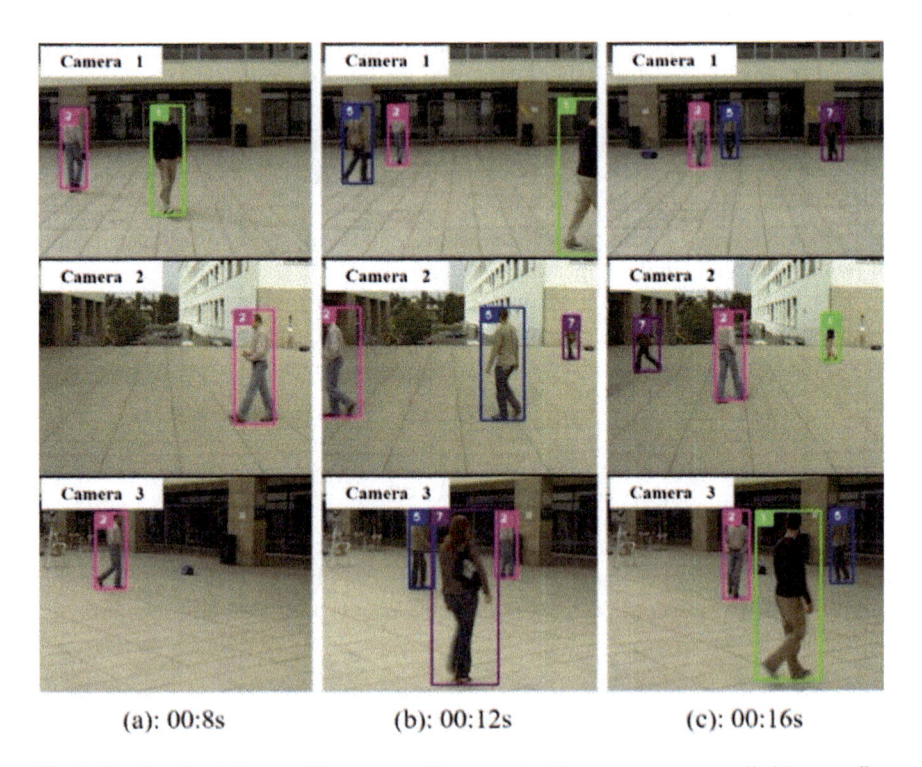

(a): 00:8s   (b): 00:12s   (c): 00:16s

Fig. 3 Results of real-time multi-person multi-camera tracking on a sequence called "campus"

Fig. 4 Results of real-time multi-person multi-camera tracking on an outdoor sequence

Figure 4 shows that three people with different clothes can be tracked perfectly in a non-overlapping layout of cameras. Images in the same row show simultaneous tracking results of multiple cameras at a certain moment, and images in the same column show tracking results of multiple people in one camera. During a long time running, the average FPS is around 25 on both sequences, which can be considered real-time. In the second row, the middle image captured from camera 2 contains an occlusion. Besides, the person with tracking ID 2 turns around when walking across the boundary of camera 1 and camera 2, which makes the appearance features change. Figure 4 shows proposed method can deal with these two situations well.

5 Conclusions

A real-time multi-person multi-camera tracking framework is presented, which works perfectly for both overlapping and non-overlapping views. With an improved multi-person matching cascade scheme for inter-camera, the tracking IDs remain the same when people walk across the camera boundaries. Besides, the proposed method can deal with the occlusion of people and variation of appearance features.

A continuously adaptive strategy for adjusting constraints of matching cascade will be studied in the future. Meanwhile, the proposed method will be deployed on embedded systems like NVIDIA Jetson.

Acknowledgments This paper is supported by Zhejiang Province Basic Public Welfare Research Program (LGG19F020021), Shanghai Automotive Industry Science and Technology Development Foundation (1815).

References

1. Andriyenko, A., Schindler, K., Roth, S.: Discrete-continuous optimization for multi-target tracking. In: 2012 IEEE Conference on Computer Vision and Pattern Recognition, pp. 1926–1933. IEEE (2012).
2. Milan, A., Schindler, K., Roth, S.: Multi-target tracking by discrete-continuous energy minimization. IEEE Trans. Pattern Anal. Mach. Intell. **38**(10), 2054–2068 (2015)
3. Tang, S., Andres, B., Andriluka, M., Schiele, B.: Multi-person tracking by multicut and deep matching. In: European Conference on Computer Vision, pp. 100–111. Springer, Cham (2016).
4. Son, J., Baek, M., Cho, M., Han, B.: Multi-object tracking with quadruplet convolutional neural networks. In: Proceedings of the IEEE Conference on Computer Vision and Pattern Recognition, pp. 5620–5629. (2017).
5. Feng, W., Hu, Z., Wu, W., Yan, J., Ouyang, W.: Multi-object tracking with multiple cues and switcher-aware classification. arXiv:1901.06129. (2019).
6. Luo, W., Xing, J., Milan, A., Zhang, X., Liu, W., Kim, T. K.: Multiple object tracking: A literature review. Artif. Intell., 103448. (2020).
7. Girshick, R.: Fast r-cnn. In Proceedings of the IEEE International Conference on Computer Vision, pp. 1440–1448. (2015).
8. Redmon, J., Divvala, S., Girshick, R., Farhadi, A.: You only look once: Unified, real-time object detection. In: Proceedings of the IEEE conference on computer vision and pattern recognition, pp. 779–788. (2016).
9. Liu, W., Anguelov, D., Erhan, D., Szegedy, C., Reed, S., Fu, C.Y., Berg, A.C.: Ssd: Single shot multibox detector. In: European Conference on Computer Vision, pp. 21–37. Springer, Cham (2016).
10. Redmon, J., Farhadi, A.: Yolov3: An incremental improvement. arXiv:1804.02767. (2018).
11. Bochkovskiy, A., Wang, C.Y., Liao, H.Y.M.: Yolov4: Optimal speed and accuracy of object detection. arXiv:2004.10934. (2020).
12. Yi, D., Lei, Z., Liao, S., Li, S. Z.: Deep metric learning for person re-identification. In: 2014 22nd International Conference on Pattern Recognition, pp. 34–39. IEEE (2014).
13. Ristani, E., Tomasi, C.: Features for multi-target multi-camera tracking and re-identification. In: Proceedings of the IEEE Conference on Computer Vision and Pattern Recognition, pp. 6036–6046. (2018).
14. Wojke, N., Bewley, A.: Deep cosine metric learning for person re-identification. In: 2018 IEEE Winter Conference on Applications of Computer Vision (WACV), pp. 748–756. IEEE (2018).
15. Liem, M.C., Gavrila, D.M.: Joint multi-person detection and tracking from overlapping cameras. Comput. Vis. Image Underst. **128**, 36–50 (2014)
16. Lee, Y.G., Tang, Z., Hwang, J.N.: Online-learning-based human tracking across non-overlapping cameras. IEEE Trans. Circuits Syst. Video Technol. **28**(10), 2870–2883 (2017)
17. Ristani, E.: People Tracking and Re-Identification from Multiple Cameras (Doctoral dissertation, Duke University). (2018).
18. Yoon, K., Song, Y.M., Jeon, M.: Multiple hypothesis tracking algorithm for multi-target multi-camera tracking with disjoint views. IET Image Proc. **12**(7), 1175–1184 (2018)

19. Zheng, L., Shen, L., Tian, L., Wang, S., Wang, J., Tian, Q.: Scalable person re-identification: A benchmark. In: Proceedings of the IEEE International Conference on Computer Vision, pp. 1116–1124. (2015).
20. Zheng, L., Bie, Z., Sun, Y., Wang, J., Su, C., Wang, S., Tian, Q.: Mars: A video benchmark for large-scale person re-identification. In: European Conference on Computer Vision, pp. 868–884. Springer, Cham (2016).
21. Lin, Y., Zheng, L., Zheng, Z., Wu, Y., Hu, Z., Yan, C., Yang, Y.: Improving person re-identification by attribute and identity learning. Pattern Recogn. **95**, 151–161 (2019)
22. Kuhn, H.W.: The Hungarian method for the assignment problem. Naval Res. Logist. Q. **2**(1–2), 83–97 (1955)
23. Wojke, N., Bewley, A., Paulus, D.: Simple online and realtime tracking with a deep association metric. In: 2017 IEEE International Conference on Image Processing (ICIP), pp. 3645–3649. IEEE (2017).
24. Narayan, N., Sankaran, N., Arpit, D., Dantu, K., Setlur, S., Govindaraju, V.: Person re-identification for improved multi-person multi-camera tracking by continuous entity association. In: Proceedings of the IEEE Conference on Computer Vision and Pattern Recognition Workshops, pp. 64–70. (2017).
25. Dollar, P., Wojek, C., Schiele, B., Perona, P.: Pedestrian detection: An evaluation of the state of the art. IEEE Trans. Pattern Anal. Mach. Intell. **34**(4), 743–761 (2011)
26. Ouyang, W., Wang, X.: A discriminative deep model for pedestrian detection with occlusion handling. In: 2012 IEEE Conference on Computer Vision and Pattern Recognition, pp. 3258–3265. IEEE (2012).
27. Ristani, E., Solera, F., Zou, R., Cucchiara, R., Tomasi, C.: Performance measures and a data set for multi-target, multi-camera tracking. In: European conference on computer vision, pp. 17–35. Springer, Cham (2014).
28. Fleuret, F., Berclaz, J., Lengagne, R., Fua, P.: Multicamera people tracking with a probabilistic occupancy map. IEEE Trans. Pattern Anal. Mach. Intell. **30**(2), 267–282 (2007)

Design and Optimization of the Seat of the Elderly Scooter Based on Solar Energy

Ya-Zheng Zhao and Yi-Jui Chiu

Abstract Based on ergonomics, this research carried out a humanized design of the backrest and cushion curve of the seat of the elderly scooter, and rationally selected design parameters such as the angle between the backrest and the cushion, the height of the armrest, and the width of the seat. Finally, the geometric model and finite element model of the human body and the seat are established, and the stress distribution simulation results are analyzed. Since the stress distribution simulation analysis can better simulate the interaction between the driver and the seat, the comfort of the seat can be better studied. The simulation results show that the seat has a reasonable body pressure distribution.

1 Introduction

The car's good driving comfort has increasingly become one of the important criteria for people to buy a car, and factors such as vehicle design and lightweight have a great influence on car comfort. Seats, steering wheels, pedals, etc., are all key components that affect the comfort of the vehicle. Among them, the good design performance of the seat is very important. With the increasing popularity of automobiles, the market for the elderly has become larger due to the advent of an aging society, and the elder's requirements for products such as seats have also increased.

The design of the seat must meet several conditions: the arrangement of the seat in the entire compartment should be reasonable; the shape of the seat should meet the ergonomic requirements; the seat must have sufficient rigidity; the seat should be reliably locked structure. There are some researches in this field, such as Li et al. [1]. In order to find a way to evaluate the comfort of the seat, they studied the relationship between the seat pressure distribution and comfort. The specific experiment is to simulate the driving environment, collect the distribution of seat pressure, let the experimenters drive for a specified time according to their

---

Y.-Z. Zhao · Y.-J. Chiu (✉)
School of Mechanical and Automotive Engineering, Fujian Province, Xiamen University of Technology, No. 600, Ligong Rd, Xiamen 361024, China
e-mail: chiuyijui@xmut.edu.cn

© The Author(s), under exclusive license to Springer Nature Singapore Pte Ltd. 2022
J.-F. Zhang et al. (eds.), *Advances in Intelligent Systems and Computing*, Smart Innovation, Systems and Technologies 268, https://doi.org/10.1007/978-981-16-8048-9_20

comfortable driving state, and analyze the collected pressure data to conclude that the seat pressure distribution can be used to evaluate comfort degree. Jiang et al. [2] studied the driving posture of the driver. Long-term driving causes physical and psychological fatigue of the driver, and the driving posture is the main cause of physical fatigue of the driver. The main purpose of this research is to explore driving postures that can reduce physical fatigue and improve driving comfort. Considering that there is little information about drivers of different body types in anthropometrics in the past, the researchers established corresponding three-dimensional human body and seat models based on the sitting postures of nearly 300 men and nearly 420 women, and passed effective pressure simulation experimental data is used to evaluate the degree of cooperation between the seat back and a variety of different human models, and then in turn guide the seat design based on the data analysis results. In order to reduce the error of collecting pressure distribution data, Bordignon et al. [3] researched an effective method and analyzed factors such as seat stiffness and driving posture comfort to create a seat comfort model. Hassan et al. [4] designed a pressure distribution simulation experiment, which established geometric and finite element models, and obtained the stress distribution of the hips and legs. Wang et al. [5] studied the influence of lumbar support on comfort. The study established a finite element model of the marginal human body. The study showed that for drivers of different ages and body types, reasonable lumbar support can maintain the natural curvature of the spine. Reiko et al. [6] used MATLAB and other software to draw graphs related to pressure distribution, and studied the pressure distribution on the body surface when a person maintains a sitting posture. This research is a breakthrough in pressure distribution simulation experiments.

The original design designed the seat back and cushion to be vertical and horizontal. This design makes the lumbar spine in the middle of the spine bear most of the force of the upper body, and the force of the intervertebral disc is uneven. At the same time, the back needs to rely on muscles to keep the waist bent, which is prone to lumbar muscle strain. The flat design of the seat cushion prevents the hips and thighs from contacting the seat well, and the force is uneven. After experiments, the design factors such as the armrest height and seat width and depth of the seat are also unreasonably selected. Combining the above unreasonable factors, this study designed the seat elements such as the seat back cushion curve and armrest height of the elderly scooter, and applied the simulation of the seat human body stress distribution to evaluate the car comfort.

2 Body Size of the Elderly

Establishing effective ergonomic dimensions is essential in the process of seat design. The human body dimensions that need to be used in the design of the elderly scooter seat mainly include sitting knee height, sitting depth, hip width, and elbow height. Considering that the height of the elderly will be shorter than when they were young,

Table 1 Elderly body size

Model	P5	P50	P90
knee height/mm	447	484.6	513
Sitting depth/mm	413	449	476.8
Sitting hip width/mm	292	314	337
Sitting elbow height/mm	225.6	259	285.5

it is necessary to have appropriate correction data to select the size of the mannequin. Since among the elderly, women are about 4% shorter than normal adults in body size, while men are about 2%, so the average 3% is used as the correction data for the height decline of the elderly. The revised human body data is shown in Table 1.

The correction formula for the size of the elderly is as follows:

$$a = b(100 - 3)\% \tag{1}$$

where a is the revised size of the elderly, b is adult standard vertebral size.

$$h = b - a \tag{2}$$

where h is height reduction data.

3 Structural Design

3.1 Curved Design of Backrest and Cushion

The design of the curve shape follows the "point-line-surface" idea. The seat cushion and backrest are meshed with the concave and convex points of the spinal cord and hip curve as the boundary, the regional nodes are marked with coordinates, and the boundary value is determined by the initial parameters. The shape of the surface is controlled by coordinates. Control the shape of the surface. Use MATLAB cubic spine interpolation to obtain the center contour line of the backrest and cushion.

Figure 1 shows the fitting diagram of the center contour line of the backrest. x is the vertical distance from a point on the human spine to the origin, and y is the horizontal distance from a point on the human spine to the origin.

Backrest center contour line formula:

$$y = A_0 + A_1 x + A_2 x^2 + A_3 x^3 + A_4 x^4 + A_5 x^5 \tag{3}$$

Fig. 1 Backrest center contour

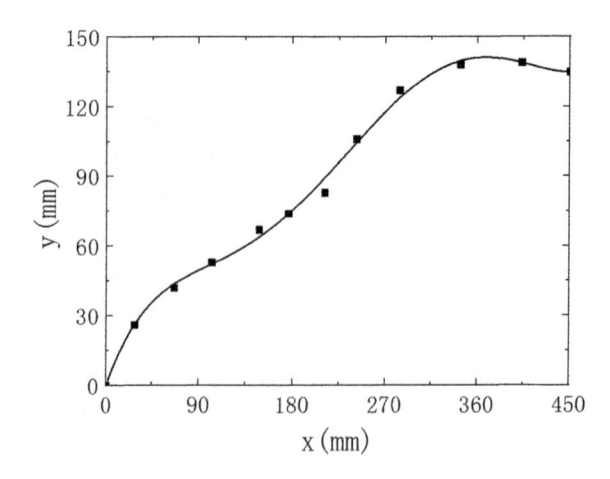

Coefficients:

A0 = −0.370	A1 = 1.269	A2 = −0.013
A3 = 7.513 × 10⁻⁵	A4 = −1.750 × 10⁻⁷	A5 = 1.412 × 10⁻¹⁰

Goodness of fit:

R-square: 0.997 Adjusted R-square: 0.995.

Figure 2 shows the fitting diagram of the center contour line of the cushion. x is the horizontal distance from a point of the human hip and thigh to the origin and y is the vertical distance from the origin of the human hip and thigh to the origin.

Cushion center contour line formula:

$$y = A_0 + A_1 x + A_2 x^2 + A_3 x^3 + A_4 x^4 + A_5 x^5 \tag{4}$$

Fig. 2 Cushion center contour line

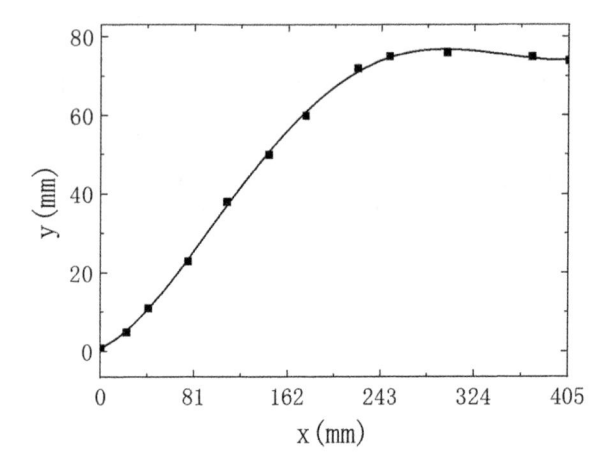

Coefficients:

A0 = 0.925	A1 = 0.131	A2 = 0.003
A3 = -1.612×10^{-5}	A4 = 2.446×10^{-8}	A5 = -1.019×10^{-11}

Goodness of fit:

R-square: 0.999 Adjusted R-square: 0.998.

3.2 Armrest Height

The height of the armrest needs to be matched with the height of the elbow. According to the human body data of the fifty percentile, the design is more reasonable when the distance between the bottom of the hip and the upper surface of the armrest is 240–250 mm.

3.3 Seat Width and Depth

The choice of seat width should match the hip width in the human body size. When sitting in a sitting position, the hip width of men in the 90th percentile is 337 mm. Considering that the driver may wear thick clothes and need posture adjustment when driving, a correction value of 70 mm is appropriately selected. In summary, the seat width parameter is rounded to 410 mm.

The seat depth is designed with the 50th percentile of the human body, which should match the seat depth of the human body. The sitting depth of the elderly person of P50 is 449 mm, and the correction value of 70 mm is selected. In summary, the seat depth parameter is rounded to 520 mm.

4 Finite Element Model

This article needs to study the pressure distribution of the seat back and cushion. Therefore, a full-body human body model needs to be established. Three full-body models of elderly drivers of different sizes are selected, namely short-sized driver A, medium-sized driver B, and tall-sized driver C.

The division of finite element mesh mainly considers the selection of mesh type and size. First use the grid division technology in the workbench to divide, then adjust the overall grid size, and then further adjust the grid size on the surface of the cushion and backrest. The seat adopts a 2 mm tetrahedral grid, the seat cushion and

Fig. 3 Finite element model

backrest surface adopts a 1 mm tetrahedral grid, and the human body model adopts a 2 mm tetrahedral grid. Figure 3 shows the established seat finite element model.

Table 2 shows the detailed mesh data. The total number of seat grids is 376125, the average grid size is 1.78 mm, and the smallest grid is 0.59 mm. The total number of human body grids is 269123, the average grid size is 1.96 mm, and the smallest grid is 1.02 mm.

Finite element model material: considering that polyurethane foam has suitable softness and cushioning properties, polyurethane foam is selected as the material of backrest and cushion. The simulation uses superelastic foam. The strength limit of this material is 178.65 kPa, the elastic modulus is 39.6 MPa, and the Poisson's ratio is 0.42. The selection of human body materials in the simulation is more complicated. In this study, an isotropic linear elastic material is used. The elastic modulus of the material is 0.85 MPa, Poisson's ratio is 0.46, and the density is 1100 kg/m^3.

Appropriate simulation boundary conditions mainly refer to the reasonable type of assembly and contact between the human body and the seat, and the human body and the seat have suitable constraints. This research uses UG to reasonably assemble the three-dimensional model of the human body and the seat, as shown in Fig. 4. In Workbench, the contact type between the driver and the seat is set to frictional contact, and the coefficient is set to 0.3. In order to make the simulation closer to the actual situation, the loading method of the human body model adopts gravity field loading.

Regarding the restraint between the human body and the seat, there are six degrees of freedom to restrain the back of the backrest and the bottom of the seat cushion. Constrain the degree of freedom of the hands and feet of the model mannequin to

Table 2 Finite element mesh

Model	Grid type	Number of grids	Average grid size/mm	Minimum grid size/mm
Seat	Tetrahedral Mesh	376,125	1.78	0.59
Human body	Tetrahedral Mesh	269,123	1.96	1.02

Fig. 4 Geometric model of human body

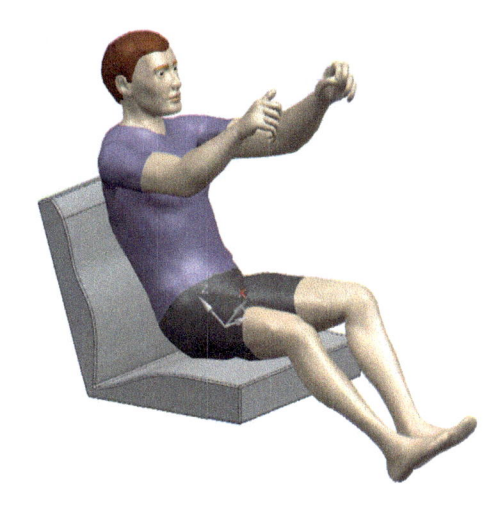

Fig. 5 Simplified seat

simulate the driver's grasp of the steering wheel and pedaling. Part of the boundary condition settings are shown in Fig. 5.

5 Analysis of Simulation Experiment Results

The experiment shows the stress distribution surface of the ischial tuberosity section of the driver A, B, C. A is the 5th percentile driver. B is the 50th percentile driver. C is the 90th percentile driver. The maximum stress is located at the ischial tuberosity; this area is where the human body receives the greatest pressure when maintaining a sitting posture. The maximum stress of the driver is 14.355 kPa.

Figure 6 shows the stress distribution diagram of the seat. (a), (b), (c) are the interaction between the seat and the human body in the three percentiles. Analyzing

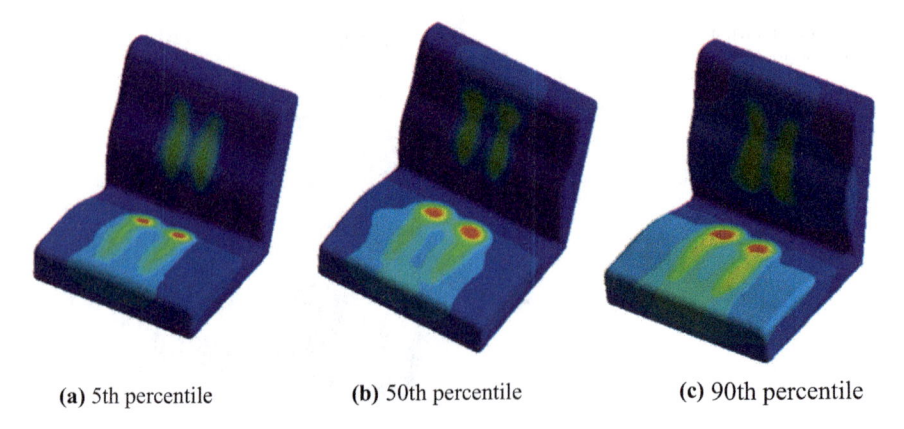

(a) 5th percentile (b) 50th percentile (c) 90th percentile

Fig. 6 Seat stress distribution

Table 3 Simulation index results

Driver	Body parts	Maximum stress /KPa	Contact area/mm^2
Driver A	Backrest	6.590	37,694
	Cushion	9.727	67,100
Driver B	Backrest	5.758	45,356
	Cushion	12.331	75,985
Driver C	Backrest	5.542	59,587
	Cushion	9.677	91,774

seat stress diagram, the maximum stress of the seat cushion appears at the ischial tuberosity like the driver, from which the stress value gradually decreases in a circular distribution, from the hip to the thigh. The value gradually decreases. As shown in Table 3, the contact area between the back of the driver A and the seat back is 37694 mm^2, and the maximum stress is 6.590 kPa. The contact area between the back of the driver B and the seat back is 45356 mm^2, and the maximum stress is 5.758 kPa. The contact area between the back of the driver B and the seat back is 59587 mm^2, and the maximum stress is 5.542 kPa.

Analyzing the stress distribution diagram of the backrest, the stress value gradually decreases outwards with the lumbar spine and the scapula as the center. Due to the different body types of drivers A, B, and C, the maximum stress values are not at the same place. The maximum stress of the seat cushion corresponding to the three drivers is 9.727 kPa, 12.331 kPa, 9.677 kPa. The stress value decreases outward with the lumbar spine and scapula as the center, because the upper body mass is mainly borne by these two places.

6 Conclusion

This paper combines ergonomics to design the seat back and cushion curves, and selects parameters. The armrest height of the seat is 240–250 mm. The seat width and seat depth are 410 and 520 mm, respectively. After the structural design is completed, a simulation analysis of the driver-seat stress distribution is carried out. The total number of seat and human body grids are 376,125 and 269,123. The seat material uses polyurethane foam, and the isotropic material is used in the finite element analysis. The modulus of elasticity is 0.85 MPa, Poisson's ratio is 0.46, and the density is 1100 kg/m^3.

The finite element analysis obtained the following results. The maximum stress of the seat back corresponding to driver A is 6.590 kPa, and the maximum stress of the seat cushion is 9.727 kPa. The maximum stress of the backrest and seat cushion corresponding to driver B is 5.758 and 12.331 kPa. C corresponds to the maximum stress of the backrest and cushion is 5.542 and 9.677 kPa. The results show that the design meets the requirements of body pressure distribution.

References

1. Li, W., Mo, R., Yu, S.H.: The effects of the seat cushion contour and the sitting posture on surface pressure distribution and comfort during seated work. Int. J. Occup. Med. Environ. Health **7**(2), 89–91 (2020)
2. Jiang, Y.X., Duan, J., Deng, S.P.: Sitting posture recognition by body pressure distribution and airbag regulation strategy based on seat comfort evaluation. J. Eng. **2019**(23), 8910–8914 (2019)
3. Bordignon, M., Cutini, M., Bisagli, C.: Evaluation of agricultural tractor seat comfort with a new protocol based on pressure distribution assessment. J. Agric. Saf. Health **24**(1), 13–26 (2018)
4. Hassan, I., Raja, M.W., Muhammad, A.A.: Study of comfort performance of novel car seat design for long drive. Proc. Inst. Mech. Eng. Part D J. Automob. Eng. **234**(2–3), 645–651 (2020)
5. Wang, X.G., Michelle, C., Georges, B.: Effects of seat parameters and sitters' anthropometric dimensions on seat profile and optimal compressed seat pan surface. Appl. Ergon. **73**(6), 13–21 (2018)
6. Reiko, M., Kazuhito, K., Nei, K.: Analysis of body pressure distribution on car seats by using deep learning. Appl. Ergon. **75**(12), 283–287 (2019)

Research on Digital Intelligence Enabled Omnimedia Communication System and Implementation Path in 5G Era

Weilong Chen and Jing Zhang

Abstract In 5G era, AI, big data, cloud computing, AR/VR and blockchain and other digital intelligence technologies are continuously and dynamically deconstructing and reconstructing the Omnimedia communication system. Omnimedia communication is a communication system of "all things are media" after the deep integration of media under the power of digital intelligence. It is an upgrade from the initial simple "superposition" to the deep intelligent "mutual embedding" driven by technical logic. Exploring the theoretical interpretation, implementation path and comprehensive evaluation index system of Omnimedia communication will help to further clarify the Omnimedia communication system with Chinese characteristics. It is a beneficial exploration to further implement the "Omnimedia communication project" proposed in "the 14th Five-Year Plan" and "the Long-Range Objectives Through the Year 2035" and promote the sustainable development of Omnimedia industry.

1 Introduction

With the rapid development of digital technology, especially artificial intelligence technology, digital intelligence enabled Omnimedia research is in full swing. Digital technology has induced the deep transformation of Omnimedia communication order and media ecology, and the original media communication ecosystem has been

---

This work is supported by Zhejiang Federation of Humanities and Social Sciences Circles (ZJFHSSC's Grant No. 2021N86), Zhejiang Province Soft Science Research Plan Project of China (Grant No. 2022C35072), the Humanity and Social Science Foundation of Ministry of Education of China (Grant No. 21YJCZH213).

W. Chen (✉)
Office of Academic Research and Creation, Communication University of Zhejiang, Hangzhou, China
e-mail: 88071998@163.com

J. Zhang (✉)
School of Business Administration, Zhejiang Gongshang University, Hangzhou, China
e-mail: zhangjing0035@163.com

completely reshaped. The "pseudo behavior" of traditional media under the same evolutionary logic leads to the sharp increase of Matthew effect under the vicious competition of niche. The media industry is facing the "double failure" dilemma of "market failure" such as homogeneous competition, internal friction of resources, lack of profit and "government failure" such as endangering cultural security and regulation anomie, as well as the "multi-demand" stress dilemma of media integration and industrial upgrading under the empowerment of digital technology. How to realize the development of strategic transformation and build an Omnimedia communication system, through the establishment of a hierarchical and classified Omnimedia ecological platform, reshape the benign niche pattern of mainstream media, local media and Internet platform, further enhance the media communication power, create media credibility, expand media influence and improve media competitiveness, is the current focus of Omnimedia communication. It is an important mission of system construction and implementation path exploration.

In the digital age, the inducement of technology will induce new ecological species, and the changes from macro to micro dimensions will form new niche and ecological structure. In the 5 g era, technologies such as big data, cloud computing, AI, AR/VR and blockchain will rapidly subvert and reconstruct the Omnimedia ecosystem along the media value chain. The systematic research on Omnimedia communication system and implementation path of logarithmic intelligence empowerment is a beneficial attempt to clarify the current situation and problems of Omnimedia communication, explore the internal mechanism of Omnimedia communication and build a benign ecosystem of Omnimedia communication.

2 Connotation of Omnimedia Communication

Since Omnimedia appeared in the late 1990s, it has remained in the stage of a few words for a long time, and its initial application is mainly in the life service industry. With the gradual deepening of the national "tri-networks integration", the "Omnimedia" appears more and more frequently in the domestic academia and industry. Search with "Omnimedia" as the subject word in CNKI database, as of March 9, 2021, there are 21,058 articles in the field of journalism, communication and publishing. Omnimedia has become a hot issue in the academic circle and has been at a high level for more than ten years. The trend of the annual number of papers in CNKI is shown in Fig. 1.

In 2000, a scholar made a preliminary exploration of Omnimedia from the perspective of news communication. AOL, the Internet giant, announced in New York that it would acquire Time Warner, the world's largest media entertainment company. In this case, the acquisition finally formed a "Omnimedia" which integrates many businesses such as network, newspapers, film and television entertainment, cartoon and so on. At this time, the understanding of the Omnimedia is mainly in the media form

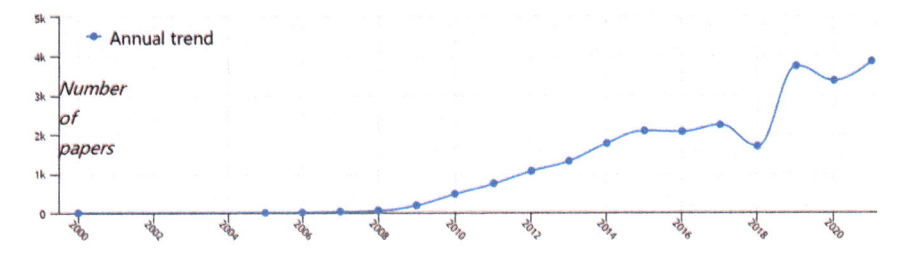

Fig. 1 Annual trend of papers about "Omnimedia"

of convergence and coordination. "Omnimedia" became the high-frequency vocabulary of research that began in 2008, which is in line with the rapid development of Internet media technology and the transformation of media industry in China.

Domestic scholars have done a lot of research on the connotation of Omnimedia. If we say that the early connotation of "Omnimedia" is more the initial speculation of the concept, then since 2008, the understanding of Omnimedia has become more specific, vivid and rich [1]. The industry's understanding of Omnimedia more refers to the comprehensive use of different media forms, similar to the traditional sense of "multimedia". In 2008, Yantai daily media group established the first "Omnimedia news center" in China to build an Omnimedia practice of multi-media platform integration, and the application of Omnimedia practice is gradually emerging in China [2]. With the wide application of the industry, the academia has conducted follow-up research on Omnimedia from different perspectives.

Peng Lan (2009) clearly put forward the concept of "Omnimedia" from the perspective of media operation, which referred to the overall mode and strategy of business operation, that is, to use Omnimedia means and platforms to build a large reporting system. Zhou Yang (2009) held that the concept of Omnimedia came from the application level of the media industry and was the product of "cross media" after the integration of media. From the perspective of media integration, some scholars believed that Omnimedia is a new form of communication that used various forms of expression to display the content of communication and various communication technologies. The integration of digital technology and media also induced the concept of "intelligent media", the overlapping of instrumental rationality and value rationality, which makes intelligent media become a hot topic in academia and industry. The essence of Omnimedia communication is the deep integration of traditional media and intelligent media under the digital intelligence. "People, media and things" are integrated to develop the communication system of "all things are media". It is the thought crystallization of technology enabling media and integration development [1].

On the whole, the academic circles tend to think that Omnimedia is the localization exploration of media integration in the Chinese context, and observe the "Omnimedia" communication from the four dimensions of time, space, subject and efficiency. That is, the "four Omnimedia" theory of "whole process media, holographic media, all staff media, and all effect media". It is the blueprint for the development of media convergence. What's more, the media has evolved from the initial simple "superposition" to the deep integration "mutual embedding" [1].

3 Research on Omnimedia Communication System Under the Digital Background

According to the literature search results of "Omnimedia" as the subject word in CNKI, and further analysis of the distribution of the number of "main and secondary topics" literature, the top keywords are: Omnimedia era, media convergence, media convergence, fusion development, Omnimedia environment, mainstream media, new media, traditional media, influence, Omnimedia transformation, new media platform, credibility, etc. It can be seen that the research on Omnimedia is the research on how to deeply integrate and transform the development of Chinese media. In China, the construction of Omnimedia communication system has been explored, and related research is mainly done from the following three aspects:

A. Theoretical interpretation of Omnimedia communication

In terms of theoretical research, scholars have systematically explained the connotation of Omnimedia communication system, mainly dealing with many relationships, namely, the relationship between intelligent media and traditional media, central media and local media, mainstream media and business platform, social media and professional media.

At the same time, the theoretical interpretation also includes macro, meso and micro levels. At the macro level, the overall research is mainly combined with economic, political, social governance, network security and supervision. At the medium level, it mainly studies the overall linkage mechanism of the internal driving force and external factors of its development from the perspective of industrial planning and development needs, as well as various special construction and social service functions in Omnimedia communication, and the special construction and social service function in all media communication [3]. At the micro level, mainly from the perspective of management and evaluation, it designs the content and implementation path of each element, link, mechanism of the in-depth development of media integration and intelligent construction.

B. Practical exploration of Omnimedia communication

In the aspect of practice and exploration, it mainly constructs the Omnimedia communication ecosystem from the aspects of content, demand, media, technology, function and platform. Innovation and development is the core of Omnimedia communication ecosystem, which can meet the diversified social needs through content innovation, optimize multi-group relationships through structural innovation and build a digital ecosystem through technological innovation, to realize the cluster embedded platform mode of upgrading from digital level to data level, seek the leading position in the media transformation and development, and constantly improve the media "four forces" construction. With the development of digital reform, more and more scholars are paying attention to the research of intelligent Omnimedia communication system. Zhi Tingrong (2019) proposed that the construction of Omnimedia communication system is a multi-dimensional and multi-level composite system, which should be constructed from the aspects of system construction, functional coupling and objective optimization, so as to form a "Trinity" Omnimedia communication system [4].

C. The combination of Omnimedia communication and digital intelligence technology

Driven by "5G + AI", the Omnimedia has formed an intelligent innovation path from the complete communication chain of transmitter, content, media, user and feedback, which provides the intelligent implementation path for the Omnimedia from four dimensions called "four Omnimedia" [5]. AI has the technical characteristics of Internet of things and big data-driven, self-evolution, human–computer interaction and so on. AI is rapidly subverting the way of information collection, content writing, content distribution, content audit and monitoring along the media value chain. Cloud ecology is a new change of media ecological environment in the era of cloud communication [6]. It has three service functions of "connector, sensor and catalyst". The construction of the integrated media industry ecosystem such as "Guangxi Cloud" provides us with a good sample. Digital technologies such as intelligent algorithm, big data, blockchain, VR/AR are deeply influencing the Omnimedia communication system.

4 Digital Intelligence Enabled Omnimedia Communication Implementation Path

With the wide application of 5G technology and the in-depth integration of "5G + " digital intelligence technology, the innovative development and implementation path of Omnimedia communication have taken off. Driven by the technology logic from "overlapping" to "fusion", digital intelligence technology is changing the media environment and media industry dynamically and continuously in the two dimensions of space and time, forming an independent and integrated "technology ecosystem", and reshaping the Omnimedia communication ecosystem in the process of deconstruction and reconstruction.

A. The content drove by 5G

5G will increase the mobile bandwidth by hundreds, which means that it will bring more abundant and faster "Cloud + " information collection, information processing, information transmission and information storage [7]. The high-capacity application scenarios such as high-quality long video and other large-scale applications with limited bandwidth will be brought to spring. The rich application scenarios such as 5G + 4 K/8 K, 5G + AR/VR, 5G + AI and the emerging mass of online content provide an extraordinary new experience for the audience.

B. Participants drove by 5G

In terms of participants, 5G further optimized the intelligent production end and immersive consumer end. PGC (professional production content) and OGC (professional production content) provide users with high-quality professional production content through 5G "multi-screen and multi-terminal" distribution. 5G enables UGC (user integrated content) to provide a more efficient, diversified and rich way of information production through massive personalized "smart and interest" production, and richer and immersive application scenarios will become accessible. The "MTU" (the collective name of MGC, TGC and UGC) inspired by artificial intelligence technology complements UGC, PGC and OGC with human as the core, so that users can feel better interactive effect through immersive experience, thus continuously improving user viscosity.

C. The information connection drove by 5G

"5G + " Internet of things has built a world of "everything is media" for us. Sensors and chips in all kinds of terminal devices connect everything and transmit it to the cloud through 5G to form a big data center. With the help of cloud computing and AI, information multi-screen and multi-terminal real-time intelligent sharing is formed. Information ubiquity makes media everywhere, and all things connect seamlessly. The Omnimedia industry with digital intelligence will cover multiple industrial clusters, and form an industrial ecosystem of synergy, symbiosis, mutual benefit and complementarity.

In the future, AR/VR will subvert the receiving end, 5G will unify the transmission platform, AI will restructure the production end, blockchain will optimize the network governance end, cloud computing will intellectualize the interactive process, in the vast ocean of high-speed mobile Internet, any single media technology system will no longer exist, and the Omnimedia ecology will face a huge transformation [9].

5 Research on the Evaluation System of Omnimedia Communication

According to the theory of "four Omnimedia" and "four forces", the model of communication effect evaluation system is in line with the theoretical system of Omnimedia and has more practical guiding value (Xie Huwei,2020). The evaluation system includes four first-class indexes and 20 s-class indexes of communication power (whole process), guidance power (holography), credibility (full staff) and influence (full effect), and the weight of each first-class index is 25% determined by Delphi method. By building an Omnimedia communication effect evaluation system model with Internet users as the core and social value evaluation as the first, we can explore the shortcomings and problems of the current media itself to a certain extent, and continuously improve the core competitiveness of the media through targeted learning from each other [10].

The Omnimedia communication system is a complex ecosystem under the empowerment of digital intelligence. Based on the perspective of media ecosystem, it will be a new exploration to explore the evaluation system of Omnimedia communication and explore the ecological balance evaluation index system of Omnimedia communication in horizontal and vertical dimensions. The horizontal dimension can include technology departments, content producers, cluster ecological platforms, production and broadcasting service companies and other ecological subjects in the "creation-production-communication-consumption" integrated value chain. The vertical dimension can include multiple subjects such as national central media, provincial media, county-level financial media and Internet platform.

6 Conclusion

In 2021, "the National 14th Five-Year Plan" and "the Long-Range Objectives Through the Year 2035" proposed that "promoting the deep integration of media, implementing the Omnimedia communication project, strengthening the new mainstream media, and building the county-level financial media center" [8]. "5G + digital intelligence technology" provides a strong technical guarantee for the implementation of Omnimedia communication project. According to *the white paper on global 5G industry development* issued by ITU (International Telecommunication Union), 5G technology will be widely used in the three business scenarios of "enhanced mobile broadband, massive Internet of things communication, high reliability and low delay communication" [5]. How to seize the fleeting opportunity of the times under the background of technology iteration, and in the field of deep integration of intelligent technology "empowerment", is a beneficial exploration of great practical significance to build a sustainable Omnimedia ecosystem.

References

1. Peng, D.: Research on the construction of all media communication system under the background of intelligence. China TV **05**, 48–51 (2020)
2. Xin, L.: What is "all media." Chinese Journalist **03**, 82–83 (2010)
3. Duan, P.: On the practical path of the construction of China's intelligent all media communication system: content, framework and mode. Modern Publ. **03**, 11–8 (2020)
4. Tingrong, Z.: Holographic perspective of all media communication system: system construction, function coupling and target optimization. J. Northwest Normal Univ. (SOCIAL SCIENCE EDITION) **06**, 32–39 (2019)
5. Lei, S.: Research on the intelligent construction path of "four all media" driven by "5G+AI." News Lovers **01**, 90–92 (2021)

6. Xie, X., Zhang, J.: AI empowerment: artificial intelligence and reconstruction of media industry chain. Wide Angle Publ. **11**, 26–9 (2020)
7. Zizhong, Z., Hao, G.: Construction of all media communication system from the perspective of Technological Ecology. News Writ. **01**, 12–17 (2021)
8. Jianwu, S.: How to implement the all media communication project. News Writ. **01**, 1 (2021)
9. Xiangzhong, L.: From media convergence to media convergence: the choice and approach of TV people. Modern Commun. (J. Commun. Univ. China) **01**, 1–7 (2020)
10. Huwei, X., Zhu, S., Li, K.: Research on the evaluation system of communication effect of "four Omnimedia". Media **19**, 74–7 (2020)

Analysis of Interaction Patterns of Intercountry Cooperation and Conflict Events Based on Complex Networks

Xiao-Mei Mo and Ding-Guo Yu

Abstract As a highly extraordinary and special year in the post-Cold War period, 2020 is characterized by intricate interactions of cooperation and conflict events between countries and regions, and a comprehensive and timely analysis of the latest intercountry cooperation and conflict event interaction patterns is of great reference value for China's diplomatic development planning. This paper researches the network characteristics of the event data in the Global Data on Events, Location and Tone (GDELT) based on theory of complex network and further analyzes the relations between countries. Firstly, this paper constructs national interaction networks using GDELT, then uses the network features to statistically analyze the cooperation and conflict relations among global countries in 2020; and finally analyzes the characteristics of cooperation and conflict patterns between China and the other 15 countries with the highest percentage of edge strength. This paper provides a new perspective for the exploration of international relations in the era of big data, as well as a reference for the analysis of news media data.

1 Introduction

International relations are the relations between two or more specific country subjects, including two basic types of union and war [1]. Complex international relations can be broken down into a series of events, such as visits, alliances, protests, threats, wars, and so on. By analyzing the event unit data, the international relations between two countries can be judged and predicted. The year 2020, which has just passed, was an extraordinary and exceptional year in the post-Cold War period [2], with a complex interplay of cooperation and conflict events between countries and regions. A comprehensive and timely analysis of the latest patterns of interaction between intercountry

---

X.-M. Mo (✉) · D.-G. Yu
College of Media Engineering, Communication University of Zhejiang, Hangzhou 310018, China
e-mail: moxmi@163.com

D.-G. Yu
e-mail: yudg@cuz.edu.cn

cooperation and conflict events is of important reference value to China's diplomatic development planning. As the world's largest news event database, GDELT (Global Database of Events, Language, and Tone) is comprehensive, granular, and frequently updated [3], making it an excellent data support for international relations research.

Network analysis, especially complex network theory, has rapidly flourished and developed since the end of the twentieth century. The theory and techniques of complex networks are used to model the relationships between entities to construct networks, and the relationships embedded in them can be easily analyzed [4]. GDELT database contains information about national or regional entities on both sides of an event and their attributes such as cooperation and conflict, which can be used to well study the patterns of cooperation and conflict events between different countries using the theory and methods of complex networks.

In recent years, some studies on constructing national interaction networks based on GDELT datasets and exploring network characteristics using the theory and methods of complex networks have started to appear at home and abroad. For example, Kun Qin studied the network of interaction relations between different countries appearing in the same document [5]. Kiran Sharma used a complex network approach to quantitatively analyze ethnic conflicts and human rights violations [6]. There is also a network of "The News Co-occurrence Globe" that presents geographic network structure over the planet akin to "communities" [7]. However, the current construction of intercountry relationship network based on GDELT mainly focuses on the co-occurrence analysis of countries, i.e., whether different countries appear in the same event or the same news report and the number of appearances, which is reflected in the network analysis as a simple network composed of undirected edges and fails to reflect the analysis of the number and direction of the interaction patterns of cooperation and conflict between countries.

Based on the characteristics of the event dataset in GDELT, this paper establishes an interaction network model of cooperation and conflict events between global countries in 2020; explores the network characteristics using the theory and methods of complex networks to analyze the cooperation and conflict relationships between countries and regions. The analysis method proposed in this paper can provide a new perspective for the research and exploration of the interaction model of intercountry cooperation and conflict events in the era of big data.

2 Data and Network Model

2.1 Introduction of GDELT Event

As a major database of GDELT, GDELT Event records are stored in an expanded version of the dyadic CAMEO [8] (Conflict and Mediation Event Observations) format, capturing two actors and the action performed by Actor1 upon Actor2. A wide array of variables break out the raw CAMEO actor codes into their respective fields to make it easier to interact with the data such as the location of the event.

In this paper, we select 3 fields: QuadClass, Actor1Geo_CountryCode, and Actor2Geo_CountryCode to construct network. Numeric codes in QuadClass field map to the Quad Classes as follows: 1 = Verbal Cooperation, 2 = Material Cooperation, 3 = Verbal Conflict, 4 = Material Conflict. Actor1Geo_CountryCode is the 2-character FIPS10-4 country code for the location of actor1, and Actor2Geo_CountryCode is the 2-character FIPS10-4 country code for the location of actor2.

2.2 Cooperation and Conflict Network Construction

According to the principle of complex network construction, both the country where Actor1 is located (the active country) and the country where Actor2 is located (the passive country) can be defined as the basic nodes in the network, and a directed edge is defined from the active country of the event to the passive country (the case where the active and passive parties of the event are the same country is excluded). If it is found in other records that the same directed edge can be created, the existing edge is weighted with the value of the number of directed edges (i.e., events initiated by the active country to the passive country) that appear in the network. And depending on the value of QuadClass, the network is further divided into two networks for intercountry cooperation (value of QuadClass is 1 or 2) and conflict (value of QuadClass is 3 or 4).

According to the above network construction method, for several example data shown in Table 1 that meet the requirements of the cooperative network (QUADCLASS value equals to 1 or 2), the generated intercountry cooperative network is shown in Fig. 1.

Table 1 Example of data in event database to create cooperative network

Actor1Geo_CountryCode	Actor2Geo_CountryCode	QuadClass
US	CH	1
US	CH	2
US	CH	1
CH	US	1
CH	US	2
RS	CH	1
CH	RS	2

Fig. 1 Intercountry cooperation network by example data

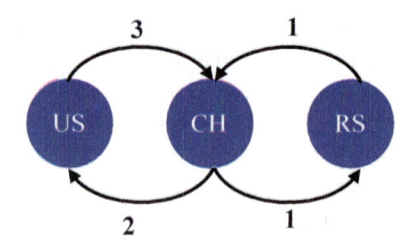

3 Experiment and Analysis

This paper selects data from the first day, the first week, the whole month and the whole year of December 2020 as four-time scales to analyze the basic characteristics of the global network of cooperation and conflict events.

3.1 Network Topology Features

In this paper, the overall topological characteristics of intercountry cooperation and conflict networks are first counted. As can be seen in Table 2, the number N of nodes in these networks is above 200 in nearly all four time scales (274 FIPS country codes in total), except for 1 day scale in which the number N of nodes in intercountry conflict networks is 180. Overall, within each time scale, the number of nodes N and the number of connected edges M in the cooperative network are larger than the results for the conflicting network, indicating that the scope of cooperation and the number of events between countries are larger than the conflicting ones at the global scale.

The average degree K represents the average number of countries with which a country has cooperation or conflict events in a certain period, and the graph density D represents the tightness of connections between countries where cooperation or conflict events occur. In terms of k, on average, a country has cooperation events with nearly 13 countries and conflict events with six to seven countries during a day. In terms of D, the value of the cooperation network is also larger than that of the conflict network at all time scales, again indicating that the connection density of cooperation events between countries globally is greater than that of conflict events.

Both the average clustering coefficient C and the average path length L can reflect the small-world characteristics of the network. In the 2 networks, C is larger while L is smaller, reflecting the small-world characteristics of the cooperation and conflict networks between countries.

Table 2 Topological characteristics of intercountry cooperation and conflict network

Network type	Time scale	N	M	k	D	C	L
Cooperation	1 day	217	2789	12.853	0.06	0.418	2.375
	1 week	237	7079	29.945	0.127	0.533	2
	1 month	245	12,965	52.918	0.217	0.611	1.825
	1 year	255	28,935	113.471	0.447	0.737	1.554
Conflict	1 day	180	1171	6.506	0.036	0.348	2.705
	1 week	224	3366	15.027	0.067	0.448	2.232
	1 month	237	7204	30.397	0.127	0.546	2
	1 year	249	19,337	77.659	0.313	0.671	1.701

Note N represents the number of network nodes; M represents the number of connected edges; k represents the average degree; D represents the graph density; C represents the average clustering coefficient; L represents the average path length

3.2 Network Scale-Free Feature Analysis

The CCDF method has been introduced to estimate distributions and analyze the scale-free feature in the node intensity and single-node edge strength of national interaction network [5]. It is given by: $P(X \geq x) = Cx^{-\alpha}$ and $P(X \geq x) = C(x + \rho)^{-\alpha}$. Here, we use CCDF to investigate the node intensity distributions of cooperative and conflict networks. For intercountry cooperation and conflict networks constructed based on the event database, node intensity is the weighted degree, which represents the total number of cooperation or conflict events between a country and other countries in a certain time period.

Figure 2 a and b show the CCDF of the intercountry cooperation and conflict network in 2020 in double logarithmic coordinates, and Fig. 2 c and d show the distribution obtained by shifting the node intensities, i.e., adding a shifting parameter ρ to each node intensity value. The distribution of node strengths of the cooperative and conflicting networks can be well fitted with the Zipf-Mandelbrot function regardless of the event scale, proving that the distribution is scale-free. That is, in a period of time, the intensity varies greatly between countries. A very small number of countries participate in cooperation or conflict events particularly frequently, while most countries participate very few times.

Figure 3 shows the top 20 countries in terms of node intensity in the intercountry cooperation and conflict network in 2020. As can be seen from Fig. 3, the United States, China, and Russia are firmly top three in the two different nature of networks, and the United States has a much higher node strength than the other countries, with 12% and 13%, respectively, while China has both 7% and Russia has 5% and 6%, respectively. The gap between other countries is not very large, with the percentage of 5% and below. Among the 274 country codes in the event database, the top 20 countries in terms of node strength have a total share of 60% in the cooperative network and 66% in the conflict network. Sixteen of these countries appear in this

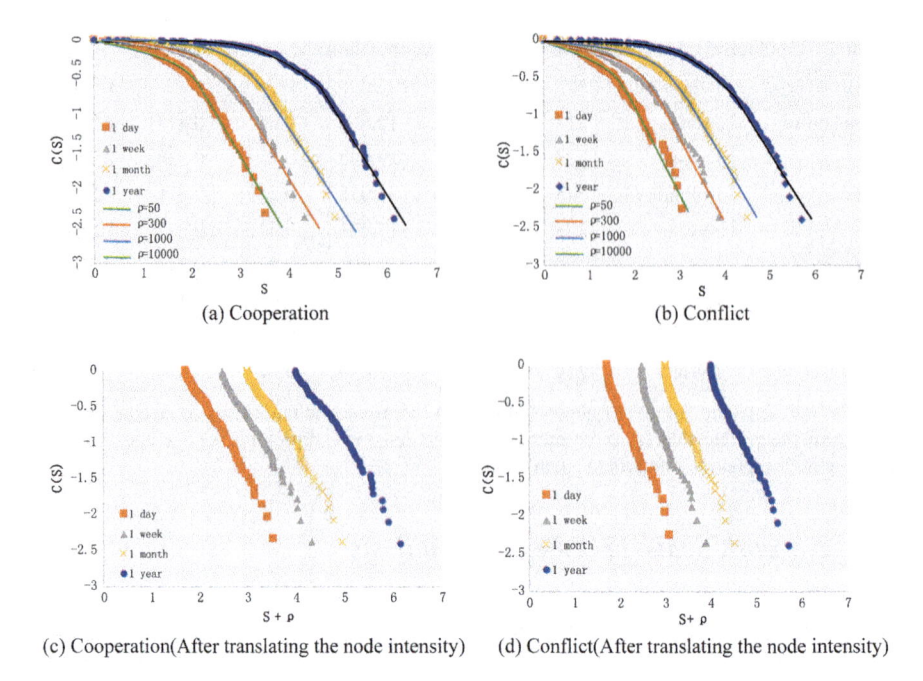

Fig. 2 CCDF of node strength in intercountry cooperation and conflict network under double logarithmic coordinates

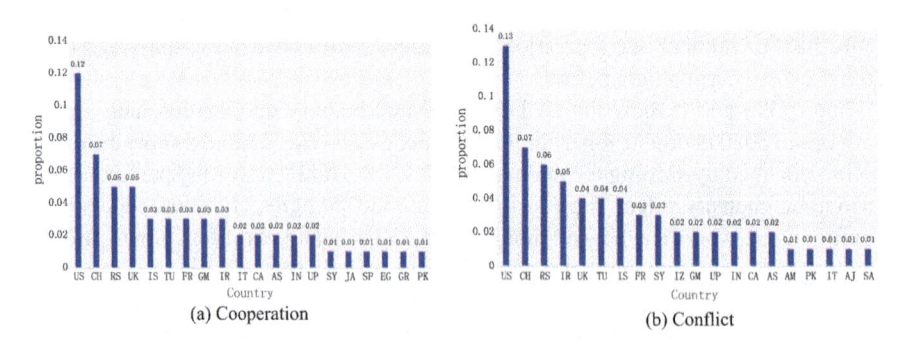

(a) Cooperation
(b) Conflict
(c) Cooperation(After translating the node intensity)
(d) Conflict(After translating the node intensity)

(a) Cooperation
(b) Conflict

Fig. 3 Top 20 countries of node strength in intercountry cooperation and conflict network in the year 2020. *Note* The horizontal coordinate is the top 20 countries in terms of node strength; the vertical coordinate is the proportion of the node strength of that country in the network node strength

range at the same time, and their total share of node strength in the cooperative network is 56% and in the conflict network is 61%. This indicates that the majority of events in both cooperation and conflict are associated with only a handful of dozen countries in 2020.

3.3 Analysis of Interaction Patterns Between China and 15 Countries

Sixteen countries have been identified in the previous section as being among the top 20 countries in terms of node strength in both cooperation and conflict networks. As China ranks second in the node strength of both networks, issues that deserve further attention include: the number of cooperation and conflict events from the other 15 countries toward China as a percentage of all the same types of events from that country abroad, and the number of cooperation and conflict events from China toward these 15 countries as a percentage of the same types of events. To this end, we study and analyze the percentage of the edge strength between the 15 countries and China in the two networks.

Figure 4 shows the percentage of cooperation or conflict edge strength between 15 countries and China in the same type of edge strength toward their foreign countries. As can be seen in Fig. 4.

(1) In 15 countries toward China's percentage of cooperation and conflict edge strength, the United States, Australia, India, the United Kingdom toward China's conflict is significantly higher than that of cooperation, other countries were similar, while the percentage of cooperation edge strength of Italy, France, Germany, Ukraine, Israel, Syria and other countries toward China is higher than that of conflict. This gives us some reference to distinguish the countries that are relatively "friendly" and "unfriendly" to China in 2020.

(2) The percentage of cooperation or conflict edge strength from China toward the 15 countries is basically smaller than that from these countries toward China in the same category. The only exception: the percentage of cooperation edge strength from China toward the United States has a slightly higher than that of opposite direction.

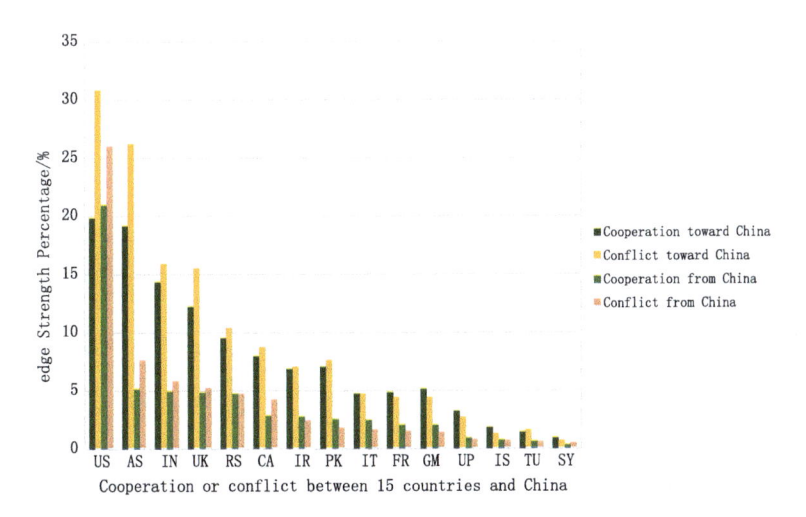

Fig. 4 Percentage of cooperation or conflict edge strength of 15 countries with China

It shows that in addition to these 15 countries, there are other countries with which China has more event concern, and the difference in the percentage of cooperation between the United States and China also reflects the different diplomatic attitudes between the two countries.

Based on the above analysis, the characteristics of percentage of different types of edge strength between countries in the complex network of intercountry cooperation and conflict events provide us with a new and effective way to analyze the patterns of intercountry relations from the perspective of big data mining.

4 Conclusion and Prospect

Based on the characteristics of event data set in GDELT, this paper proposes a method to construct event network of cooperation and conflict between countries by using the theory and method of complex network. The topological characteristics and scale-free characteristics of the global intercountry cooperation and conflict event networks under different time scales are analyzed. Finally, combining the statistics on the percentage of edge strengths of the two types and four directions in 2020, this paper analyzes the characteristics of cooperation and conflict patterns between China and the top 15 countries in terms of node strength.

The work in this paper provides new ideas for analyzing the pattern characteristics of intercountry cooperation and conflict events from the perspective of big data mining. Further research includes:(1) combining other data sets in GDELT to further analyze the characteristics of interaction patterns between countries;(2) studying the spatiotemporal evolution and predictive analysis methods of networked data mining.

Acknowledgments This work was supported by the Key Research and Development Program of Zhejiang Province, China (No.2019C03138).

References

1. Sun, J.-J.: Several basic models of international relations in history. In: Membership Congress of the Chinese Society for Historians of China's foreign Relations, pp. 19–22 (2005). (in Chinese)
2. Qiu, Y.-P.: Review and prospect of international situation. Contemp. Int. Relat. **01**, 1–4+63 (2021). (in Chinese)
3. Leetaru, K., Schrodt, P. A.: GDELT: Global Data on Events, Language, and Tone, 1979–2012. In: International Studies Association Annual Conference. SanDiego, CA (2013)
4. Mo, X.-M.: Analysis of global news flow patterns based on complex networks. J. Southwest Univ. (Nat. Sci. Ed.) **42**(12), 15–24 (2020). (in Chinese)
5. Qin, K.: Networked mining of GDELT and international relations analysis. J. Geo-Inf. Sci. **21**(1), 14–24 (2019). (in Chinese)
6. Sharma, K.: A complex network analysis of ethnic conflicts and human rights violations. Sci. Rep. **7**(1), 8283 (2017)

7. The GDELT project. Mapping Media Geographic Networks: The News Co-occurrence Globe. https://blog.gdeltproject.org/mapping-media-geographic-networks-the-news-co-occurrence-globe/. Accessed 23 Apr 2021
8. Gerner, D.J., et al.: The creation of CAMEO (Conflict and Mediation Event Observations): An event data framework for a post cold war world. In: annual meeting of the American Political Science Association, pp. 29 (2002)

Video, Image, and Others

Application of the Novel Parallel QUasi-Affine TRansformation Evolution in WSN Coverage Optimization

Jeng-Shyang Pan, Geng-Chen Li, Jianpo Li, Min Gao, and Shu-Chuan Chu

Abstract With the development of wireless sensor network (WSN) technology, the coverage optimization of sensor nodes has gradually become the focus of researchers. Therefore, how to maximize the network coverage of WSN is worth studying. In this paper, Parallel QUasi-Affine TRansformation Evolution (P-QUATRE) is proposed based on the traditional QUATRE, which improves the performance of the algorithm. Next, a 3D WSN network layout model is proposed. Finally, the proposed P-QUATRE is applied to the model for simulation experiments to improve the network coverage. Through the analysis of the experimental results, compared with Particle Swarm Optimization (PSO), Gases Brownian Motion Optimization (GBMO) and Tunicate Swarm Algorithm (TSA), P-QUATRE shows better performance in this field.

1 Introduction

Evolutionary algorithm as a new optimization technology, its idea comes from the study of biological group behavior or the evolution process of natural phenomenon. It is an intelligent method which is inspired by the intelligent phenomena of biological groups in nature [1–3]. It has the characteristics of self-organization, self-learning,

J.-S. Pan · S.-C. Chu
College of Computer Science and Engineering, Shandong University of Science and Technology, Qingdao 266590, China
e-mail: jspan@cc.kuas.edu.tw

J.-S. Pan · G.-C. Li (✉) · J. Li · M. Gao
School of Computer Science, Northeast Electric Power University, Jilin 132012, China
e-mail: 605544372@qq.com

J. Li
e-mail: jianpoli@163.com

M. Gao
e-mail: 15822553806@163.com

S.-C. Chu
College of Science and Engineering, Flinders University, Clovelly Park, Adelaide, SA 5042, Australia

self-adaptation and inherent parallelism. Compared with traditional pure mathematical methods, swarm intelligence algorithm solves problems through Darwin's natural law of "survival of the fittest, survival of the fittest" [4–6]. It has no requirements on the continuity and differentiability of problems, has good generality, and is easier to solve complex optimization problems [7]. In the existing research, there are many types of evolutionary algorithms, such as PSO [8, 9] based on the foraging habits of birds, GBMO [10] is proposed from the point of view of Brownian motion and turbulent rotating motion of gas, and TSA [11] is based on the behavior of the deep-sea tunicate swarm searching for food, etc.

QUATRE is a novel evolutionary algorithm proposed by Meng in 2016 [12, 13]. Due to the excellence of the algorithm, many researchers started to study QUATRE and put forward a lot of improved algorithms. In [14], the QUATRE-EAR algorithm proposes an enhanced mutation strategy to solve the problem of DE control variables. This strategy uses a timestamp mechanism. In [15], a new E-QUATRE algorithm is proposed. It evolves utilizing automatic adjustment of control parameters and adaptive generation matrix. This makes it possible to avoid falling into the confusion of local optimum in the process of optimizing. A QUATRE algorithm with sort strategy (S-QUATE) is proposed in [16], which improves the performance of the original QUATRE. In [17], a C-QUATRE algorithm is proposed, which uses a pairwise competition mechanism to improve the performance of QUATRE algorithm. In [18], a new Multi-group QUATRE adapted to global optimization is proposed. By randomly dividing the population into three groups, the population diversity enhancement for multi-group QUATRE is realized. Each group uses a different mutation strategy to improve the efficiency of the algorithm. In the search process, different strategies and appropriate parameters are used to update the mutated scale factor F adaptively to better balance the exploration and development capabilities.

With the gradual expansion of the application of wireless sensor network (WSN), people's demand for WSN is also increasing [19–21]. In life, industry, electric power, military, agriculture and other fields of application has become more extensive [22–24]. However, in the study of WSN, the problem of network coverage has always been concerned by researchers [25]. It is a hot research issue to improve the network coverage rate by some technical means. At the same time, evolutionary algorithm as an effective tool to solve optimization problems, has been applied to the field by researchers. In the past decades, there have been many studies on the application of evolutionary algorithms to WSN coverage optimization [26–28].

In [29], the differential genetic algorithm and penalty function are introduced on the basis of the Artificial Fish Swarm Algorithm (AFSF), and the local searchability of the algorithm is enhanced by the differential genetic algorithm, and this improved algorithm is applied to the WSN coverage problem. In [30], an optimization algorithm based on ant colony algorithm (ACB-SA) is proposed to solve the efficient-energy coverage problem. In [31], Artificial Bee Colony (ABC) is proposed to optimize the coverage of WSN to improve the coverage rate. Reference [9] proposed an improved coverage method of wireless sensor network based on improved PSO algorithm to solve the problem of unreasonable random sensor deployment and distribution. The improved PSO algorithm is applied to the coverage optimization of wireless sensor network, and the maximum coverage is increased. An optimization algorithm of wireless sensor network based on improved binary PSO algorithm is proposed in [32]. In addition, in order to apply GSBPSO to the optimization of wireless sensor networks, a small probability mutation replacement strategy is proposed to replace the individuals that do not meet the coverage requirements in the search process. Based on the original PSO, [33] introduced disturbance factors to make the particles trapped in the local optimal jump out quickly, and make use of the randomness and ergodicity of chaotic motion to carry out local fine search. It is named Adaptive Perturbed Chaotic Particle Spheres Optimization (ADCPSO). Then, the algorithm is optimized with the effective coverage area of WSN as the optimization objective.

To sum up, this paper presents a P-QUATRE, and applies the algorithm to the WSN network layout, to improve the coverage of the network. The work arrangement of other parts is as follows: Firstly, the traditional QUATRE is introduced in detail. Secondly, P-QUATRE is proposed by adding the idea of parallelism to QUATRE. The third part mainly introduces the 0–1 model of WSN. In the fourth part, P-QUATRE is applied to 0–1 model to carry out the simulation experiment, and the experimental results are analyzed in detail. In the end, this is all the research work are summarized.

2 QUasi-Affine TRansformation Evolution

QUATRE was proposed by Meng et al. in 2016. The particle in the algorithm is updated using an affine transformation-like equation in the geometry ($X \mapsto MX + B$), so it is named QUasi-Affine TRansformation Evolution. In this section, we will explain the traditional QUATRE in detail.

In the D-dimensional search space, QUATRE contains many individuals, and each individual is represented by vector x, and the i-th individual is represented by i, as shown in Eq. (1).

$$x_i = (x_{i1}, x_{i2}, ..., x_{iD}) \tag{1}$$

All the individuals are combined to form a matrix X representing the parent population, as shown in Eq. (2). ps is the number of individuals in the population.

$$X = \left[x_1^T, x_2^T, ..., x_{ps}^T\right]^T \tag{2}$$

The population evolves according to Eq. (3), and the evolved population becomes the offspring population, which is represented by X_{next}.

$$\begin{cases} B = X_{gbest} + c \times (X_{r1} - X_{r2}) \\ X_{next} \mapsto M \otimes X + \overline{M} \otimes B \end{cases} \tag{3}$$

where X_{r1} and X_{r2} are the random matrices obtained by randomly arranging the row vectors of the population matrix X. c is the scaling factor, which is an important parameter in the formula. The choice of c value will affect the size of the search range. X_{gbest} is a globally optimal population composed of globally optimal individuals. As shown in Eq. (4), each individual of X_{gbest} is the current globally optimal individual X_{best_i}. M is the selection matrix. A $D \times D$ lower triangular matrix is tiled along the column direction, and then the elements of each row vector are randomly arranged. Finally, on the premise of keeping the row vectors unchanged, the row vectors are randomly arranged in the direction of the column, so as to obtain M. $\overline{M}$ is an anti-selection matrix. If the element is 1 in M, then it is 0 in $\overline{M}$. If it is 0 in M, it is 1 in $\overline{M}$. $\otimes$ stands for multiplying the elements in the corresponding positions in the two matrices.

$$X_{gbest} = \left[X_{best_1}^T, X_{best_2}^T, ..., X_{best_{ps}}^T \right]^T \tag{4}$$

The entire population evolves according to Eq. (3). QUATRE first scrambles the globally optimal population through the difference matrix to obtain the globally optimal population B after scrambles. Then, according to the selection matrix M, the parent population X and the globally optimal population B after scrambles are crossover operations to obtain the child population X_{next}. Individuals in the offspring population X_{next} will be compared with individuals in the parent population X for fitness values, and the winner will be retained in the parent population in the next iteration.

In summary, the steps of QUATRE are as follows: firstly, initialize the population and set parameter c. Then the parent population X is generated, which evolves according to Eq. (3) to generate the daughter population X_{next}. The fitness function values of individuals in X_{next} are compared with those of individuals in X, and individuals with better fitness function values are replaced with individuals with poorer fitness function values. Next, the updated population is taken as the parent population of the new generation to evolve according to Eq. (3), and then a new daughter population is generated. Until the termination condition is satisfied, that is, the solution satisfying the precision is obtained or the maximum number of iterations is reached.

3 The Novel Parallel Quasi-Affine TRansformation Evolution

In this chapter, we will add a parallel strategy to the traditional QUATRE to improve the searchability of the algorithm [34]. This strategy is described in more detail next. We divide the population into n groups, which we define as $G = \{G_1, G_2, ..., G_n\}$. Where n is a positive integer. Communication policies are used when the number of iterations reaches a predetermined value. The strategy is that each population selects the best point in its own group and uses this point to replace the randomly selected values in other populations. This ensures that the population iteration will not converge to an optimal value. Moreover, it can make the particles with poor

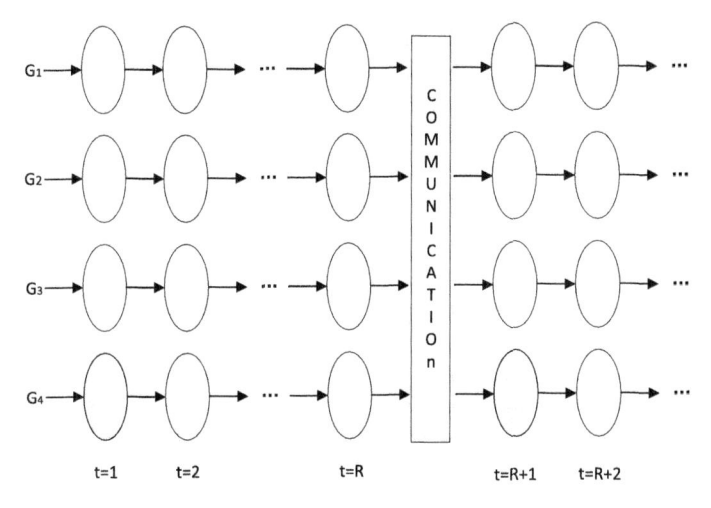

Fig. 1 The novel Parallel QUasi-Affine TRansformation Evolution

fitness function value in the population timely change to a good position to prevent the particles with poor fitness function value from falling into a bad state.

We will give a detailed explanation of the novel Parallel QUasi-Affine TRansformation Evolution (P-QUATRE) below. First of all, the n populations of particles are optimized, respectively, according to the traditional QUATRE. That is, they search for what they think is the best solution in their own territory. Second, when they communicate every R iterations. That is, the winners in their respective groups are randomly substituted for any value in the other groups. This ensures the diversity of the population. Finally, after n times of communication, they get the optimal solution that each group seeks together. P-QUATRE is shown in Fig. 1.

4 The 3D WSN Coverage Model

In the study of WSN, the network coverage is closely related to the perception model of each node and the location deployment of nodes. The perception model can directly affect the quality of service of sensor network. Therefore, it is particularly important to establish a perception model for each node before deploying its location. Common WSN perception models include: 0–1 model, exponential model, statistical model, obstacle model, etc. At the same time, most of the network layout of WSN is studied on a 2D plane. However, in the 3D space of the real world, we need to establish it in the 3D space for research. Therefore, based on the 0–1 model, a 3D space coverage model is established in this paper. Next we will explain the 0–1 model and apply it to 3D space.

0–1 model: In general, assume that in a $m \times n$ region, we think of this region as consisting of $m \times n$ pixels. In addition, it is assumed that the coordinate of sensor node i is (x_i, y_i, z_i), the communication radius of sensor node is r_i, and the spatial coordinate of target pixel j is (x_j, y_j, z_j). Then, the distance between sensor node i and target pixel j is

$$d_{ij} = \sqrt{(x_i - x_j)^2 + (y_i - y_j)^2 + (z_i - z_j)^2} \tag{5}$$

On the premise of barrier-free, the sensing area of the sensor node is a ball with the sensor node i as the center and r_i as the radius. Therefore, if the distance between the target pixel and the sensor node is less than the communication radius of the node, it can be perceived by the node. The result of whether each pixel in this region can be perceived by a sensor is represented by o_{ij}, that is, if it can be perceived, o_{ij} is set to 1. Otherwise, it is set to 0 as shown in the equation

$$o_{ij} = \begin{cases} 1, & if\ d_{ij} < r_i \\ 0, & otherwise \end{cases} \tag{6}$$

In order to make the experiment in this paper closer to the real life, we set up some obstacles on the basis of three-dimensional space. If the straight line between the target pixel j and the sensor node i passes through the obstacle, then $o_{ij} = 0$.

5 Experimental Results

In order to verify the effectiveness of P-QUATRE applied to the 0–1 model, P-QUATRE, PSO, GBMO, and TSA were, respectively, tested on the 0–1 model in this paper, and the experimental results were compared. In this paper, the population number *pop* was set to 30 and the experiment was run 10 times to ensure the fairness of the experiment. The number of iterations $iter_{max}$ is set to 10. Specific parameter Settings are shown in Table 1. In order to verify the performance of P-QUATRE applied to the 0–1 model, we fixed the number of traffic radius and the number of nodes for simulation experiments in the following two sections.

Table 1 Experimental parameter setting

Algorithm	PSO	GBMO	TSA	P-QUATRE
$iter_{max}$	10	10	10	10
run	10	10	10	10
pop	30	30	30	30

Table 2 Coverage of different number of nodes

Node Num	PSO	GBMO	TSA	P-QUATRE
30	0.4608	0.4619	0.4637	0.4712
35	0.5069	0.5097	0.5182	0.5232
40	0.5542	0.5594	0.5629	0.5848
45	0.5962	0.6002	0.6023	0.6189
50	0.6345	0.6408	0.6401	0.6597
55	0.6651	0.6706	0.6727	0.6993

5.1 Experimental Results of Different Number of Nodes

This section mainly carries out simulation experiments on the influence of different number of sensor nodes on the coverage rate. In order to ensure the fairness of the experiment, this experiment is carried out under the condition that the communication radius is constant. In this case, we set the sensor node number ($NodeNum$) to (30, 35,...,55). At this point, the communication radius is set to 5 m. In order to better verify the ability of P-QUATRE to solve the 0–1 model, we applied P-QUATRE, PSO, GBMO, and TSA to the model for comparison. Table 2 and Fig. 2 show the coverage rates of these four algorithms when the number of sensor nodes is different. It can be clearly seen from Fig. 2 that with the increase in the number of nodes, the coverage capacity of PSO, GBMO, TSA, and P-QUATRE is also continuously increasing. It can be seen that the coverage ability of PSO, GBMO, and TSA is not as good as that of P-QUATRE. This proves the effectiveness of P-QUATRE proposed in this paper.

5.2 Simulation Experiment of Different Communication Radius

In order to ensure the fairness of the experiment, this section changes the communication radius of the nodes to carry out the experiment on the premise that the number of sensor nodes remains unchanged. Set the number of sensor nodes to 30. The communication radius was set to (5, 6, .., 10), respectively, for the experiment. In addition, P-QUATRE, PSO, GBMO, and TSA were applied to 0–1 coverage model for comparison. It is obvious from the results in Fig. 3 and Table 3 that as the communication radius continues to increase. The coverage capabilities of PSO, GBMO, TSA, and P-QUATRE are also increasing. Therefore, P-QUATRE is feasible in solving the coverage problem.

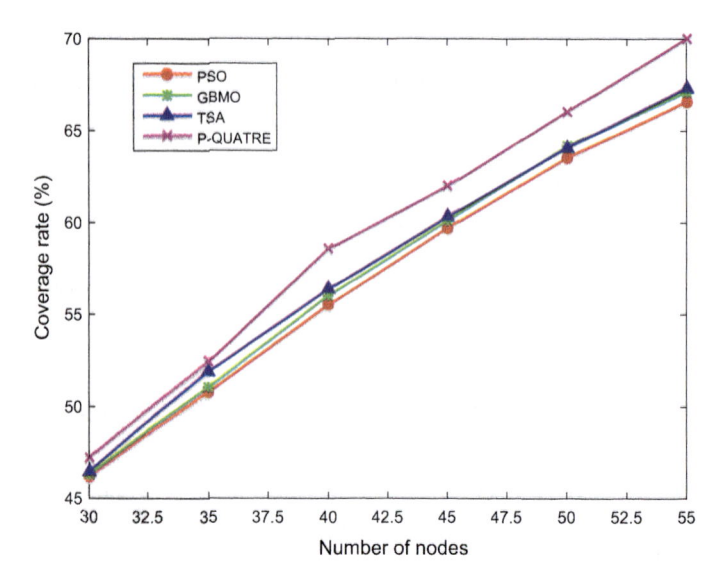

Fig. 2 The experimental results of changing the number of nodes without changing the communication radius

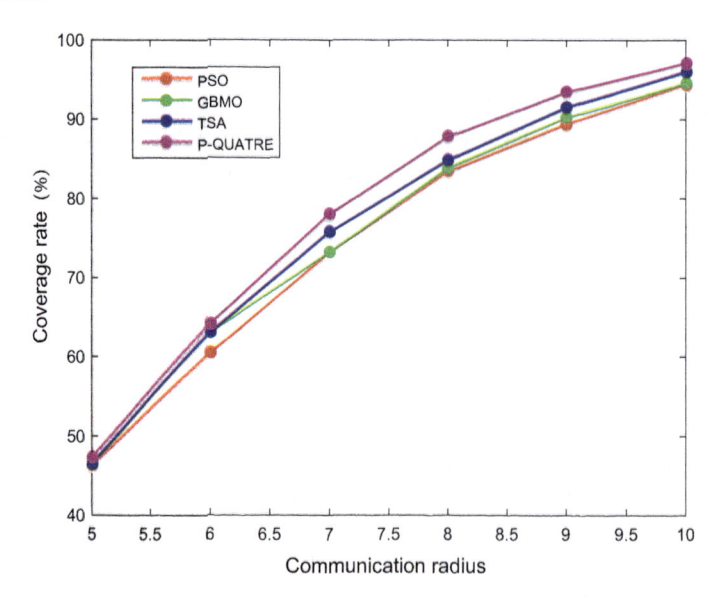

Fig. 3 The experimental results of changing the number of nodes without changing the communication radius

Table 3 Coverage of different communication radius

Radius	PSO	GBMO	TSA	P-QUATRE
5 m	0.4608	0.4619	0.4637	0.4712
6 m	0.6029	0.6297	0.6299	0.6403
7 m	0.7295	0.7299	0.7556	0.7780
8 m	0.8323	0.8367	0.8469	0.8759
9 m	0.8923	0.9000	0.9134	0.9323
10 m	0.9426	0.9439	0.9586	0.9689

6 Conclusions

In the study of WSN, how to improve the network coverage has always been the focus of researchers. This paper proposes a kind of QUATRE with parallel idea to solve the problem of WSN 3D network layout. Through the simulation experiment, the proposed P-QUATRE is compared with other traditional algorithms, and it can be seen that P-QUATRE is more effective in improving the network coverage. However, in the process of research, we found that the algorithm still has problems such as slow running speed. In the following research work, we will further improve its shortcomings.

References

1. Sun, C., Zeng, J., Pan, J., Xue, S., Jin, Y.: A new fitness estimation strategy for particle swarm optimization. Inf. Sci. **221**, 355–370 (2013)
2. Xue, X., Pan, J.-S.: A compact co-evolutionary algorithm for sensor ontology meta-matching. Knowl. Inf. Syst. **56**(2), 335–353 (2018)
3. Chu, S.-C., Huang, H-C., Roddick, J.F., Pan, J.-S.: Overview of algorithms for swarm intelligence. In: International Conference on Computational Collective Intelligence, pp. 28–41. Springer (2011)
4. Pan, J.-S., Meng, Z., Xu, H., Li, X.: A matrix-based implementation of de algorithm: the compensation and deficiency. In International Conference on Industrial, Engineering and Other Applications of Applied Intelligent Systems, pp. 72–81. Springer (2017)
5. Wu, T.-Y., Lin, J.C.-W., Zhang, Y., Chen, C.-H.: A grid-based swarm intelligence algorithm for privacy-preserving data mining. Appl. Sci. **9**(4), 774 (2019)
6. Wu, J.M.-T., Zhan, J., Lin, J.C.-W.: An aco-based approach to mine high-utility itemsets. Knowl.-Based Syst. **116**, 102–113 (2017)
7. Karaboga, D., Basturk, B.: On the performance of artificial bee colony (abc) algorithm. Appl. Soft Comput. **8**(1), 687–697 (2008)
8. Eberhart, R., Kennedy, J.: Particle swarm optimization. In: Proceedings of the IEEE International Conference on Neural Networks, vol. 4, pp. 1942–1948. Citeseer (1995)
9. Kong, H., Yu, B.: An improved method of wsn coverage based on enhanced pso algorithm. In: 2019 IEEE 8th Joint International Information Technology and Artificial Intelligence Conference (ITAIC) (2019)

10. Abdechiri, M., Meybodi, M.R., Bahrami, H.: Gases brownian motion optimization: an algorithm for optimization (gbmo). Appl. Soft Comput. J. **13**(5), 2932–2946 (2013)
11. Kaur, S., Awasthi, L.K., Sangal, A.L., Dhiman, G.: Tunicate swarm algorithm: A new bio-inspired based metaheuristic paradigm for global optimization. Eng. Appl. Artif. Intell. **90**, 103541.1–103541.29 (2020)
12. Meng, Z., Pan, J.-S.: Quasi-affine transformation evolution with external archive (quatre-ear): An enhanced structure for differential evolution. Knowl. Based Syst. **155**(SEP.1), 35–53 (2018)
13. Meng, Z., Pan, J.-S.: Quasi-affine transformation evolutionary (quatre) algorithm: A parameter-reduced differential evolution algorithm for optimization problems. In: 2016 IEEE Congress on Evolutionary Computation (CEC), pp. 4082–4089. IEEE (2016)
14. Zhi-Gang, D., Pan, J.-S., Chu, S.-C., Luo, H.-J., Pei, H.: Quasi-affine transformation evolutionary algorithm with communication schemes for application of rssi in wireless sensor networks. IEEE Access **8**, 8583–8594 (2020)
15. Meng, Z., Chen, Y., Li, X., Yang, C., Zhong, Y.: Enhancing quasi-affine transformation evolution (quatre) with adaptation scheme on numerical optimization. Knowl.-Based Syst. 197
16. Pan, J.S., Meng, Z., Chu, S.C., Roddick, J.F.: Quatre algorithm with sort strategy for global optimization in comparison with de and pso variants (2017)
17. Meng, Z., Pan, J.S.: A competitive quasi-affine transformation evolutionary (c-quatre) algorithm for global optimization. In: 2016 IEEE International Conference on Systems, Man, and Cybernetics (SMC) (2017)
18. Liu, N., Pan, J.S., Wang, J., Nguyen, T.T.: An adaptation multi-group quasi-affine transformation evolutionary algorithm for global optimization and its application in node localization in wireless sensor networks. Sensors **19**(19), 4112 (2019)
19. Pan, J.-S., Dao, T.-K., Pan, T.-S., Nguyen, T.T., Chu, S.C., Roddick, J.F.: An improvement of flower pollination algorithm for node localization optimization in wsn. J. Inf. Hiding Multimedia Signal Process. **8**(2), 486–499 (2017)
20. Pan, J.-S., Nguyen, T.-T., Dao, T.-K., Pan, T.-S., Chu, S.-C.: Clustering formation in wireless sensor networks: a survey. J. Netw. Intell. **2**(4), 287–309 (2017)
21. Pan, J.-S., Meng, Z., Chu, S.-C., Hua-Rong, X.: Monkey king evolution: an enhanced ebb-tide-fish algorithm for global optimization and its application in vehicle navigation under wireless sensor network environment. Telecommun. Syst. **65**(3), 351–364 (2017)
22. Chunming, W., Shang, H.: Qos-aware resource allocation for d2d communications. J. Northeast Electric Power Univ. **40**(2), 89–95 (2020)
23. Chen, C.-M., Lin, Y.-H., Lin, Y.-C., Sun, H.-M.: Rcda: Recoverable concealed data aggregation for data integrity in wireless sensor networks. IEEE Trans. Parallel Distrib. Syst. **23**(4), 727–734 (2011)
24. Renuka, K.M., Kumar, S., Kumari, S., Chen, C.M.: Cryptanalysis and improvement of a privacy-preserving three-factor authentication protocol for wireless sensor networks. Sensors **19**(21), 4625 (2019)
25. Chien-Ming, C., Xiang Bing, W., Tsu-Yang, King-Hang, W.: An anonymous mutual authenticated key agreement scheme for wearable sensors in wireless body area networks. Appl. Sci. **8**(7), 1074 (2018)
26. Chu, S.-C., Xue, X., Pan, J.-S., Xiaojing, W.: Optimizing ontology alignment in vector space. J. Internet Technol. **21**(1), 15–22 (2020)
27. Li, J., Zhang, Q., Zhang, Z., Yin, Y., Zhang, H.: Congestion control and energy optimization routing algorithm for wireless sensor networks. J. Northeast Electric Power Univ. **40**(4), 69–74 (2020)
28. Sun, Z., Yang, D.: D2d radio resource allocation algorithm based on global fairness. J. Northeast Electric Power Univ. **39**(1), 81–87 (2019)
29. Li, S.: Research based on the application of a differential fish swarm algorithm in wsn coverage. Bull. Sci. Technol. (2015)
30. Lee, J.W., Lee, J.J.: Ant-colony-based scheduling algorithm for energy-efficient coverage of wsn. IEEE Sens. J. **12**(10), 3036–3046 (2012)

31. Xing, X., Zhou, G., Chen, T.: Research on coverage optimisation of wireless sensor networks based on an artificial bee colony algorithm. Int. J. Wireless Mobile Comput. **9**(2), 199–204 (2015)
32. Li, K., Feng, Y., Chen, D., Li, S.: A global-to-local searching-based binary particle swarm optimisation algorithm and its applications in wsn coverage optimisation. Int. J. Sens. Netw. **32**(4), 197 (2020)
33. Wang, J.X., Li, X.: Wsn coverage enhancement algorithm based on particle swarm optimization. Adv. Mater. Res. **945–949**, 2386–2393 (2014)
34. Gao, M., Pan, J.-S., Li, J., Zhang, Z., Chai, Q.-W.: 3-d terrains deployment of wireless sensors network by utilizing parallel gases brownian motion optimization. J. Internet Technol. **22**(1), 13–29 (2021)

Parameters Extraction of Solar Cell Using an Improved QUasi-Affine TRansformation Evolution (QUATRE) Algorithm

Jeng-Shyang Pan, Ai-Qing Tian, Tien-Szu Pan, and Shu-Chuan Chu

Abstract Solar simulation model is the mathematics of solar power generation; accurate mathematical model is a very important link in simulation, evaluation, control, and optimization. The model of solar power generation is affected by factors such as environment and light, and its solar model is often nonlinear. In addition, conventional solutions are often unable to accurately determine the exact values of unknown parameters. Meta-heuristic algorithms have received extensive attention in recent years. In this paper, the QUasi-Affine TRansformation Evolution (QUATRE) algorithm is partially improved, and the particle position is rearranged randomly, which to a certain extent prevents the algorithm from falling into the local optimal and more accurately approaching the theoretical optimal value of the algorithm. The improvement of QUARTER applies to the optimization of solar parameters. This article selects a commercial double diode model with a diameter of 57 mm. The result optimized by the improved QUATRE algorithm is promising and superior to other methods.

J.-S. Pan · A.-Q. Tian · S.-C. Chu
College of Computer Science and Engineering, Shandong University of Science and Technology, Qingdao 266590, China
e-mail: jspan@cc.kuas.edu.tw

A.-Q. Tian
e-mail: stones12138@163.com

S.-C. Chu (✉)
College of Science and Engineering, Flinders University, Sturt Rd, Bedford Park, Adelaide, SA 5042, Australia

T.-S. Pan
Department of Electronic Engineering, National Kaohsiung University of Science and Technology, 415 Chien-Kung Road, Kaohsiung 807, Taiwan
e-mail: tpan@nkust.edu.tw

© The Author(s), under exclusive license to Springer Nature Singapore Pte Ltd. 2022
J.-F. Zhang et al. (eds.), *Advances in Intelligent Systems and Computing*, Smart Innovation, Systems and Technologies 268, https://doi.org/10.1007/978-981-16-8048-9_24

1 Introduction

The continuous progress of social civilization has brought about the consumption of traditional energy and the exploration of new energy [1, 2]. The trend of using renewable energy has increased significantly in the past ten years. Among renewable energies, solar cells are used everywhere in the world to generate electricity. This is not only because of the high demand for electricity in the company, but also because of the advantages of simple installation, low maintenance, absence of pollution, and excessive noise.

Solar Power Europe's assessment report shows that the global installed capacity of solar energy systems increases significantly each year. In a solar power generation system, solar panels are connected by a series of solar cells, which can provide offline or network-connected needs. The battery panel is generally composed of batteries connected in series for higher output voltage. A panel with an important output current is realized by increasing the surface of the batteries or by connecting the batteries in parallel.

The solar cell modeling portion is primarily designed to modify the nonlinear volt-ampere curve under the influence of different environments [3–6]. Depending on the mathematical model that has been proposed to date, the behavior of the system in different environments is expressed. They range from simple hypothetical models to advanced models of complex physical variables. The commercial double diode model is widely used. This model contains seven unknown parameters, and the accuracy of these parameters is very important in solar cell simulation, optimization design, performance assessment. The metaheuristic algorithm has the characteristics of a rapid convergence, a clear algorithm logic and a higher precision [7–9]. It has been extensively used in engineering, medical, finance, and other complex linear and nonlinear problems.

2 Related Work

Computational intelligence belongs to the scope of artificial intelligence, and realizes the perception of the evaluation function describing the optimization problem and the solution of the optimization problem through calculation methods and methods. Computational intelligence also comprises more detailed divisions, such as artificial neural networks, fuzzy computing, and evolutionary computing [10–13]. Among them, the main idea of evolutionary computing is to simulate the natural selection process of nature in the calculation process to determine the calculation method, and then realize the solution of complex problems. Intelligent computing mainly includes QUasi-Affine TRansformation Evolution [14–16], Grey Wolf Optimizer(GWO) [17], Pigeon Inspired Optimization (PIO) [18–20], Flower Pollination Algorithm (FPA) [21], Ant Colony Optimization (ACO) [22], and Fish Migration Optimization (FMO) [23–25].

2.1 QUasi-Affine TRansformation Evolution

Meng and Pan first proposed the QUATRE algorithm in 2016 [26]. The evolutionary algorithm is called the affine form of the algorithm, and in this paper, a standard QUATRE is designed [27–30]. The structure of the QUATRE algorithm contains the following Eq. 1:

$$X \Leftarrow \bar{M} \otimes X + M \otimes B, \tag{1}$$

where M is the evolution matrix, X is the individual population matrix, $X = (X_1, X_2, ..., X_i, ..., X_{No})^T$, $i \in [1, No]$, B is the mutant matrix, No is the size of the population, X is the target matrix, $\otimes$ represents a bitwise multiplication operation of matrix elements.

where F is the control factor of the difference matrix with a range of (0, 1] in the Table 1. The outcome of $(X_{r,i} - X_{r,j})$ $(i, j \in [1, 2, ..., 5])$ is deemed to difference matrix. General, $F = 0.7$ is a superior candidate.

In regard to the coevolutionary matrix M, the M matrix in the QUATRE algorithm is transformed from the initial matrix M_{init}. In this article, $No = D$ is taken as an example, D is the dimensionality of the objective function, matrix M_{init} is the lower triangular matrix with element value 1, and equation 2 gives the transformation process from matrix M_{init} to matrix M:

$$\mathbf{M}_{init} = \begin{bmatrix} 1 & & & & & \\ 1 & 1 & & & & \\ 1 & 1 & 1 & & & \\ 1 & 1 & 1 & 1 & & \\ & & \cdots & & & \\ 1 & 1 & 1 & 1 & \cdots & 1 \end{bmatrix} \sim \begin{bmatrix} 1 & 1 & 1 & 1 & & \\ 1 & 1 & 1 & & & \\ & & \cdots & & & \\ & & & 1 & 1 & \\ 1 & & & & & \\ 1 & 1 & 1 & 1 & \cdots & 1 \end{bmatrix} = \mathbf{M}$$

After a two-step operation, the matrix M_{init} is transformed into a matrix M. The transformation is achieved by two consecutive operations: the first step, all the row vectors in the matrix M_{init} all rearranged randomly; the second step, the first step of the matrix row vectors randomly all arranged.

Table 1 The six mutation matrices **B** calculation

Number	QUATRE/B	Equation
1	QUATRE/target/1	$\mathbf{B} = \mathbf{X} + F * (\mathbf{X}_{r1,g} - \mathbf{X}_{r2,g})$
2	QUATRE/rand/1	$\mathbf{B} = \mathbf{X}_{r1,g} + F * (\mathbf{X}_{r2,g} - \mathbf{X}_{r3,g})$
3	QUATRE/best/1	$\mathbf{B} = \mathbf{X}_{gbest,g} + F * (\mathbf{X}_{r2,g} - \mathbf{X}_{r3,g})$
4	QUATRE/target/2	$\mathbf{B} = \mathbf{X} + F * (\mathbf{X}_{r1,g} - \mathbf{X}_{r2,g}) + F * (\mathbf{X}_{r3,g} - \mathbf{X}_{r4,g})$
5	QUATRE/rand/2	$\mathbf{B} = \mathbf{X}_{r1,g} + F * (\mathbf{X}_{r2,g} - \mathbf{X}_{r3,g}) + F * (\mathbf{X}_{r4,g} - \mathbf{X}_{r5,g})$
6	QUATRE/best/2	$\mathbf{B} = \mathbf{X}_{gbest,g} + F * (\mathbf{X}_{r1,g} - \mathbf{X}_{r2,g}) + F * (\mathbf{X}_{r3,g} - \mathbf{X}_{r4,g})$

2.2 Improve QUATRE Algorithm Work

This subsection introduces a QUATRE algorithm with a sorting mechanism, which will re-distribute some particles in the population randomly after each iteration [31–33]. Excellent particles will lead the poor particles to continue iterating. If the redistributed particles encounter a better position in the process of moving, it will guide the entire population to move. Figure 1 shows the idea of the Improve QUATRE (I-QUATRE) algorithm.

Considering the exploration and development of particles, in appropriate settings on the redistribution of the number of particles can prevent local optimization algorithm. If the number of redistributed particles is too large, the algorithm will always be exploration globally, and the local exploitation capability will be weak. If the number of redistributed particles is too small, the improved algorithm will have little effect. The specific number of particles that need to be changed is positioned $k = ceil(No/10)$ in this article, where $ceil$ stands for round-up operation. The number of particle changes is increasing with the number of iterations. The change process is shown in Eq. 2.

$$\omega_\alpha = \frac{k-1}{Max-1}(\alpha - 1) + 1, \qquad (2)$$

where ω_α is the number of particles that need to be redistributed in α iteration, $1 \leqslant \omega \leqslant k$, Max is the maximum number of iterations of the algorithm.

2.3 Double Diode Model of Solar Cell

Both the single diode (SD) model and the double diode (DD) model are model displays of solar cell power generation. The SD model is very simple and clear, but in order to more accurately measure the internal parameters of the solar cell and consider the characteristics of the volt-ampere curve. This article uses DD for simulation. This model is more complicated than monopoles and the solution process is more tortuous, but this method can fully present the internal display of solar cell-related parameters.

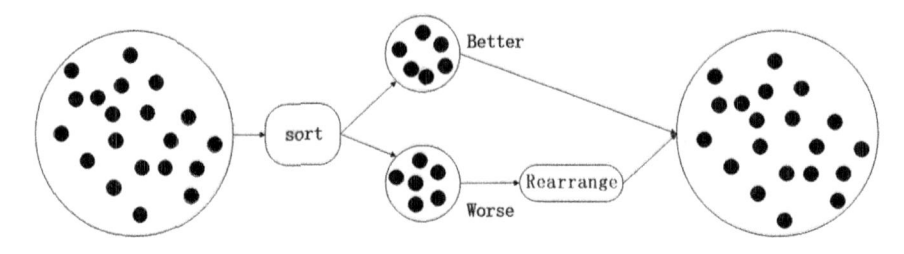

Fig. 1 Illustration of sort scheme in I-QUATRE algorithm

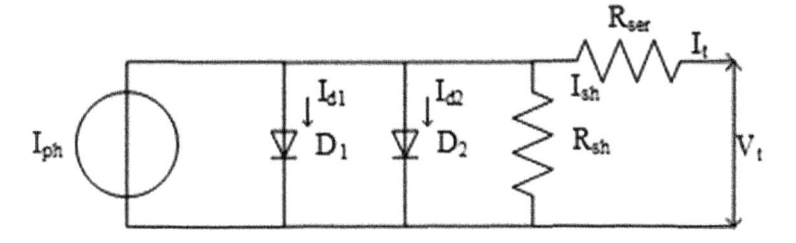

Fig. 2 Double diode model of solar cell

As can be seen from Fig. 2. The DD model has not only series resistance but also parallel resistance. The solar current I_t is generated by light energy. It can be expressed as follows Eq. 3:

$$I_t = I_{ph} - I_{d1} - I_{d2} - I_{sh}, \tag{3}$$

where I_{ph} is the photogenerated current generated by solar light energy; I_t and V_t is the output current and output voltage; I_{d1} and I_{d2} are the first diode current and the second diode current; I_{sh} is the current shunted by parallel resistors R_{sh} and I_{ser} is the current shunted by series resistance R_{ser}. In order to correctly simulate the simulation of solar cells, this paper select Shockley's diode equation. I_{d1}, I_{d2}, and I_{sh} on the output voltage expansion form is the following equation:

$$I_{d1} = I_{sd1} \left[\exp\left(\frac{q\,(V_t + R_{ser} \times I_t)}{n_1 \times k \times T} \right) - 1 \right], \tag{4}$$

$$I_{d2} = I_{sd2} \left[\exp\left(\frac{q\,(V_t + R_{ser} \times I_t)}{n_2 \times k \times T} \right) - 1 \right], \tag{5}$$

$$I_{sh} = \frac{V_t + R_{ser} \times I_t}{R_{sh}}, \tag{6}$$

There, Eq. 3 can be reshow as:

$$\begin{aligned} I_t = I_{ph} &- I_{sd1} \left[\exp\left(\frac{q\,(V_t + R_{ser} \times I_t)}{n_1 \times k \times T} \right) - 1 \right] \\ &- I_{sd2} \left[\exp\left(\frac{q\,(V_t + R_{ser} \times I_t)}{n_2 \times k \times T} \right) - 1 \right] - \frac{V_t + R_{ser} \times I_t}{R_{sh}}, \end{aligned} \tag{7}$$

where in the above equations, q is the electronic quantity of electricity, and its values is 1.602×10^{-19} (*coulombs*); The value of Boltzmann constant k is 1.380×10^{-23} (*J/K*); then, T is the thermodynamic temperature of cell.

There are seven parameters that need to be optimized in the model optimized in this article:

(1) The light generation current I_{ph};
(2) The first diode current I_{sd1};
(3) The second diode current I_{sd2};
(4) Ideality factor of the first diode n_1;
(5) Ideality factor of the second diode n_2;
(6) Series resistance R_{ser};
(7) Shunt resistance R_{sh};

The optimization process is performed by minimizing the objective function designed by the root mean square error (RMSE). $f(V_e, I_e, x)$ is the error function, which indicates between the experimental data and the calculated values. Therefore, x represents the parameter that needs to be optimized and $x = (I_{ph}, I_{sd1}, I_{sd2}, n_1, n_2, R_s$ and $R_{sh})$.

$$f(V_e, I_e, x) = I_e - \left\{ I_{ph} - I_{sd1} \left[\exp\left(\frac{q(V_e + I_e \cdot R_{ser})}{n_1 \cdot k \cdot T}\right) - 1 \right] \right. \\ \left. - I_{sd2} \left[\exp\left(\frac{q(V_e + I_e \cdot R_{ser})}{n_2 \cdot k \cdot T}\right) - 1 \right] - \frac{V_e + I_e \cdot R_{ser}}{R_{sh}} \right\}, \quad (8)$$

$$RMSE = \sqrt{\frac{1}{N} \sum_{i=1}^{N} \left(f(V_e, I_e, x)^2\right)}, \quad (9)$$

Equations 8 and 9 show how to calculate the error of solar cell parameters.

3 Experiment Analysis

3.1 Experiments on Benchmark Functions

In order to evaluate the performance of the I-QUATRE algorithm, this paper selects 23 test functions for evaluation [17], and compares the QUATER, PSO and DE algorithms. The number of iterations of the four algorithms is set to 5000, the population size is set to 60, and each test function is run 30 times and the average value is taken. The experimental results are shown in Table 2.

It can be seen from the algorithm simulation results that I-QUATRE performs well in all test functions, especially the performance in fixed-dimensional functions, which can reach its theoretical optimal value.

Table 2 Comparison between I-QUATRE algorithm and the QUATRE, PSO, DE algorithm under test functions

Test function	QUATRE		I-QUATRE		PSO		DE	
	Mean	Std	Mean	Std	Mean	Std	Mean	Std
F1	1.33×10^{-17}	7.39×10^{-18}	2.17×10^{-35}	2.03×10^{-35}	6.11	1.05	7.30×10^{-12}	6.20×10^{-12}
F2	7.88×10^{-6}	4.03×10^{-9}	1.03×10^{-8}	0.12	10.7443	0.96	0.0927	2.9×10^{-6}
F3	1.69×10^{3}	470.11	5.80×10^{-3}	8.00×10^{-4}	9.17	2.14	5.03	2.25
F4	0.0133	3.5×10^{-3}	1.3×10^{-3}	9.2×10^{-3}	0.82	0.07	0.053	0.12
F5	42.93	22.57	18.71	28.99	712.18	158.26	29.42	0.74
F6	1.46×10^{-17}	8.42×10^{-18}	1.02×10^{-34}	5.62×10^{-34}	7.72	1.43	1.03×10^{-11}	5.83×10^{-12}
F7	0.012	3.91×10^{-3}	4.8×10^{-3}	1.71×10^{-3}	28.47	10.24	0.21	0.05
F8	-1.68	16.24	-6.68	8.50	-3.07	371.81	-5.44	217.19
F9	24.30	2.31	2.47	6.75	194.01	13.46	189.39	14.31
F10	20.01	2.1×10^{-3}	3.21×10^{-10}	8.43×10^{-4}	3.37	0.18	9.37×10^{-7}	4.41×10^{-7}
F11	0	0	1.5×10^{-15}	4.70×10^{-4}	0.85	0.27	4.9×10^{-11}	3.79×10^{-8}
F12	6.9×10^{-3}	7.05×10^{-18}	1.13×10^{-17}	2.54×10^{-5}	0.29	0.07	3.48×10^{-12}	2.76×10^{-12}
F13	3.66×10^{-4}	3.51×10^{-17}	5.96×10^{-17}	2.10×10^{-3}	1.85	0.31	1.23×10^{-11}	1.00×10^{-11}
F14	1	0	1	5.55×10^{-14}	1.862	0.92	1.76	0
F15	1.27×10^{-3}	4.34×10^{-6}	1.69×10^{-3}	2.67×10^{-5}	0.039	2.81×10^{-2}	3.07×10^{-4}	1.78×10^{-19}
F16	-1.03	6.77×10^{-16}	-1.03	1.04×10^{-5}	-0.039	1.11×10^{-2}	-1.03	6.77×10^{-16}
F17	0.39	1.01×10^{-3}	0.39	3.50×10^{-3}	0.51	0.21	0.49	3.21×10^{-2}
F18	47.10	60.53	3.67	21.20	39.42	0.64	10.25	15.46
F19	-3.86	0.19	-3.51	2.71×10^{-15}	-3.85	5.65×10^{-3}	-3.84	0.53
F20	-0.32	1.66	-2.87	0.23	-1.74	0.98	-1.93	0.56
F21	-4.49	2.36	-10.15	9.96×10^{-14}	-8.89	0.95	-9.31	6.63×10^{-6}
F22	-9.51	2.30	-10.40	5.71×10^{-16}	-3.83	1.42	-4.35	9.03×10^{-6}
F23	-9.01	8.48×10^{-4}	-10.53	6.98×10^{-10}	-4.13	7.38	-9.05	8.04×10^{-6}

3.2 Parameters Estimation of Double Diode Model

The output current I_t in the DD model can be obtained by solving Eq. 3. Table 3 shows the limit range of parameter adjustment.

In this paper, the temperature of solar cell is set to be 33 °C. The number of particles were be set to 50, and the iteration of algorithms was 200.

In addition, Table 4 shows the optimal results and RMSE values of the last optimization of the seven parameters of the four types of algorithms. Table 5 lists the comparisons of net currents between calculated and experimental data. In the simulation results, the improved QUATRE algorithm is more consistent with the empirical data.

4 Conclusions

In this article, an improved QUATRE with an efficient redistribution mechanism is proposed, which can precisely and effectively extract the parameters from the solar cell model. The redistribution mechanism ensures that the algorithm falls into

Table 3 The search range for the seven parameters

Number	Parameter	Range
1	I_{ph} (A)	[0, 1]
2	I_{sd1} (μA)	[0, 1]
3	I_{sd2} (μA)	[0, 1]
4	n_1	[1, 2]
5	n_2	[1, 2]
6	R_{ser} (Ω)	[0, 0.5]
7	R_{sh} (Ω)	[0, 100]

Table 4 The result of optimal parameters obtained using I-QUATRE

Number	Parameter	Result
1	I_{ph} (A)	0.7605
2	I_{sd1} (μA)	0.6539
3	I_{sd2} (μA)	0.4000
4	n_1	1.5633
5	n_2	1.8510
6	R_{ser} (Ω)	0.0325
7	R_{sh} (Ω)	80
8	Fitness Value	2.10×10^{-3}

Table 5 Comparisons of net currents for computation and experiment

Experimental data		IQUATRE
$V_t(V)$	$I_t(A)$	Computed $I_t(A)$
−0.2057	0.7640	0.7624
−0.1291	0.7620	0.7616
−0.0588	0.7605	0.7609
0.0057	0.7605	0.7602
0.0646	0.7600	0.7596
0.1185	0.7590	0.7590
0.1678	0.7570	0.7585
0.2132	0.7570	0.7579
0.2545	0.7555	0.7570
0.2924	0.7540	0.7556
0.3269	0.7505	0.7532
0.3585	0.7465	0.7486
0.3873	0.7385	0.7405
0.4137	0.7280	0.7267
0.4373	0.7065	0.7051
0.4590	0.6755	0.6725
0.4784	0.6320	0.6278
0.4960	0.5730	0.5693
0.5119	0.4990	0.4977
0.5265	0.4130	0.4130
0.5398	0.3165	0.3180
0.5521	0.2120	0.2137
0.5633	0.1035	0.1045
0.5736	−0.010	−0.0080
0.5833	−0.123	−0.1244
0.5900	−0.210	−0.2108

the defect of local optimization. Greatly ensure that the algorithm can effectively converge to near the optimal value. Moreover, the experimental results show that the calculated data from I-QUATRE agree fundamentally with the experimental data. Therefore, the I-QUATRE proposed in this paper is an effective method to extract internal parameters from solar components.

References

1. Xue, Y., Sun, W., Quansen, W.: The influence of magmatic rock thickness on fracture and instability law of mining surrounding rock. Geomechan. Eng. **20**(6), 547–556 (2020)
2. Sun, W., Zhou, F., Liu, J., Shao, J.: Experimental study on portland cement/calcium sulfoaluminate binder of paste filling. Eur. J. Environ. Civil Eng. 1–16 (2020)
3. Jervase, J.A., Bourdoucen, H., Al-Lawati, A.: Solar cell parameter extraction using genetic algorithms. Measur. Sci. Technol. **12**(11), 1922 (2001)
4. Zhang, C., Zhang, J., Hao, Y., Lin, Z., Zhu, C.: A simple and efficient solar cell parameter extraction method from a single current-voltage curve. J. Appl. Phys. **110**(6), 064,504 (2011)
5. Gao, X., Cui, Y., Jianjun, H., Guangyin, X., Wang, Z., Jianhua, Q., Wang, H.: Parameter extraction of solar cell models using improved shuffled complex evolution algorithm. Energy Convers. Manage. **157**, 460–479 (2018)
6. Lin, P., Cheng, S., Yeh, W., Chen, Z., Lijun, W.: Parameters extraction of solar cell models using a modified simplified swarm optimization algorithm. Sol. Energy **144**, 594–603 (2017)
7. Kim, W., Choi, W.: A novel parameter extraction method for the one-diode solar cell model. Sol. Energy **84**(6), 1008–1019 (2010)
8. Chan, D.S.H., Phang, J.C.H.: Analytical methods for the extraction of solar-cell single-and double-diode model parameters from IV characteristics. IEEE Trans. Electron Devices **34**(2), 286–293 (1987)
9. Chellaswamy, C., Ramesh, R.: Parameter extraction of solar cell models based on adaptive differential evolution algorithm. Renew. Energy **97**, 823–837 (2016)
10. Xue, X.: A compact firefly algorithm for matching biomedical ontologies. Knowl. Inf. Syst. 1–17 (2020)
11. Xue, X., Lu, J.: A compact brain storm algorithm for matching ontologies. IEEE Access **8**:43,898–43,907 (2020)
12. Xue, X., Wang, Y.: Using memetic algorithm for instance coreference resolution. IEEE Trans. Knowl. Data Eng. **28**(2), 580–591 (2015)
13. Zhang, F., Wu, T.-Y., Wang, Y., Xiong, R., Ding, G., Mei, P., Liu, L.: Application of quantum genetic optimization of lvq neural network in smart city traffic network prediction. IEEE Access **8**, 104,555–104,564 (2020)
14. Liu, N., Pan, J.-S., Liao, X., Chen, G.: A multi-population quasi-affine transformation evolution algorithm for global optimization. In: International Conference on Genetic and Evolutionary Computing, pp 19–28. Springer (2018)
15. Chu, S.-C., Chen, Y., Meng, F., Yang, C, Pan, J.-S., Meng, Z.: Internal search of the evolution matrix in quasi-affine transformation evolution (quatre) algorithm. J. Intell. Fuzzy Syst. (Preprint), 1–12 (2020)
16. Meng, Z., Chen, Y., Li, X., Yang, C., Zhong, Y.: Enhancing quasi-affine transformation evolution (quatre) with adaptation scheme on numerical optimization. Knowl.-Based Syst. 105908 (2020)
17. Pei, H., Pan, J.-S., Chu, S.-C., QingWei, C., Tao, L., ZhongCui, L.: New hybrid algorithms for prediction of daily load of power network. Appl. Sci. **9**(21), 4514 (2019)
18. Duan, H., Qiao, P.: Pigeon-inspired optimization: a new swarm intelligence optimizer for air robot path planning. Int. J. Intell. Comput. Cybern. **7**(1), 24–37 (2014)
19. Tian, A.-Q., Chu, S.-C., Pan, J.-S., Cui, H., Weimin, Z.: A compact pigeon-inspired optimization for maximum short-term generation mode in cascade hydroelectric power station. Sustainability **12**(3), 767 (2020)
20. Tian, A.-Q., Chu, S.-C., Pan, J.-S., Yongquan, L.: A novel pigeon-inspired optimization based mppt technique for pv systems. Processes **8**(3), 356 (2020)
21. Pan, J.-S., Dao, T.-K., Pan, T.-S., Nguyen, T.-T., Chu, S.-C., Roddick, J.F.: An improvement of flower pollination algorithm for node localization optimization in wsn. J. Inf. Hiding Multimedia Signal Process. **8**(2), 486–499 (2017)

22. Chu, S.-C., Roddick, J.F., Su, C.-J., Pan, J.-S.: Constrained ant colony optimization for data clustering. In: Pacific Rim International Conference on Artificial Intelligence, pp 534–543. Springer (2004)
23. Pan, J.-S., Pan, J.-S., Tsai, P.-W., Liao, Y.-B.: Fish migration optimization based on the fishy biology. In: 2010 Fourth International Conference on Genetic and Evolutionary Computing, Shenzhen, pp. 783–786 (2010). https://doi.org/10.1109/ICGEC.2010.198
24. Song, P.-C., Chu, S.-C., Pan, J.-S., Yang, H.: Phasmatodea population evolution algorithm and its application in length-changeable incremental extreme learning machine. In: 2020 2nd International Conference on Industrial Artificial Intelligence (IAI). https://doi.org/10.1109/IAI50351.2020.9262236
25. Chai, Q.-W., Chu, S.-C., Pan, J.-S., Zheng, W.-M.: Applying adaptive and self assessment fish migration optimization on localization of wireless sensor network on 3-D terrain. J. Inf. Hiding Multimedia Signal Process. **11**(2), 90–102 (2020)
26. Meng, Z., Pan, J.-S., Huarong, X.: Quasi-affine transformation evolutionary (quatre) algorithm: a cooperative swarm based algorithm for global optimization. Knowl.-Based Syst. **109**, 104–121 (2016)
27. Meng, Z., Pan, J.-S.: Quasi-affine transformation evolution with external archive (quatre-ear): an enhanced structure for differential evolution. Knowl.-Based Syst. **155**, 35–53 (2018)
28. Liu, N., Pan, J.-S., Jason Yang Xue (2019) An orthogonal quasi-affine transformation evolution (o-quatre). In: Advances in Intelligent Information Hiding and Multimedia Signal Processing: Proceedings of the 15th International Conference on IIH-MSP in Conjunction with the 12th International Conference on FITAT, July 18–20, Jilin, China, vols. 2 and 157, pp. 57–66. Springer
29. Pan, J.-S., Meng, Z., Huarong, X., Li, X.: Quasi-affine transformation evolution (quatre) algorithm: A new simple and accurate structure for global optimization. In: International Conference on Industrial Engineering and Other Applications of Applied Intelligent Systems, pp. 657–667. Springer (2016)
30. Liu, N., Pan, J.-S., Wang, J., Nguyen, T.-T.: An adaptation multi-group quasi-affine transformation evolutionary algorithm for global optimization and its application in node localization in wireless sensor networks. Sensors **19**(19), 4112 (2019)
31. Kumari, S., Chaudhary, P., Chen, C.M., Khan, M.K.: Questioning key compromise attack on Ostad-Sharif et al.'s authentication and session key generation scheme for healthcare applications. IEEE Access **7**, 39717–39720 (2020)
32. Wu, T.Y., Lee, Z., Obaidat, M.S., Kumari, S., Kumar, S., Chen, C.M.: An authenticated key exchange protocol for multi-server architecture in 5G networks. IEEE Access **8**, 28096–28108 (2020)
33. Sun, H.M., Wang, K.H., Chen, C.M.: On the security of an efficient time-bound hierarchical key management scheme. IEEE Trans. Dependable Secure Comput. **6**(2), 159–160 (2009)

QUATRE Algorithm for 5G Heterogeneous Network Downlink Power Allocation Problem

Pei-Cheng Song, Shu-Chuan Chu, Anhui Liang, and Jeng-Shyang Pan

Abstract As a new evolutionary algorithm, QUasi-Affine TRansformation Evolutionary (QUATRE) algorithm has excellent performance and scalability, and many different specific high-performance variants have been developed, but there are few applications in the actual field. The 5G heterogeneous network can effectively improve the existing communication network, provide the possibility for higher throughput and stronger coverage, and at the same time have stronger flexibility, but a more complex network structure will bring cross-layer interference and resource allocation problem. This article mainly tests the effects of the QUATRE algorithm and its two variants on the energy efficiency optimization problem of the downlink of 5G heterogeneous networks. The final results show that the QUATRE and its two variants can achieve good results in this practical problem.

1 Introduction

For actual problems in the real world, there are many different methods that can be used, but basically we still have to choose a more suitable and effective method for the specific problem. When solving optimization problems, it is necessary to establish an optimization model and determine the optimization goal [23]. For the optimization goal of a specific model, if mathematical methods cannot be used to obtain accurate results, then evolutionary computing optimization algorithms are often used for

---

P.-C. Song · S.-C. Chu · J.-S. Pan (✉)
College of Computer Science and Engineering, Shandong University of Science and Technology, Qingdao 266590, China
e-mail: jspan@ieee.org

P.-C. Song
e-mail: spacewe@outlook.com

A. Liang
College of Electronic and Information Engineering, Shandong University of Science and Technology, Qingdao 266590, China
e-mail: liangah8@sdust.edu.cn

processing [2]. Since the genetic algorithm (GA) [8] and particle swarm algorithm (PSO) [12] were proposed, a variety of algorithms have been applied to different types of practical problems, such as differential evolution (DE) [4, 17, 22], ant colony optimization (ACO) [5], fish migration optimization (FMO) [3], grey wolf optimizer (GWO) [9, 19], butterfly optimization algorithm (BOA) [1], charged system search algorithm (CSS) [11], and central force optimization (CFO) [6]. Different algorithms have not completely similar performance, and they are good at solving different types of problems [21, 26, 28], but PSO and DE algorithms are more adaptable and can achieve good results in a variety of different application problems.

The algorithm used in this paper is the QUATRE algorithm, which improves on the problems of the PSO and DE algorithms, and proposes a co-evolutionary architecture based on quasi-affine transformation [18]. The QUATRE algorithm improves the problem of slow convergence in PSO. Compared with the DE, it has fewer parameters and can achieve a more reasonable search. Since then, a variety of variants have been developed based on the QUATRE algorithm, including competitive quasi-affine transformation evolutionary (C-QUATRE) algorithm [15], QUATRE algorithm with sort strategy (S-QUATRE) [20], orthogonal quasi-affine transformation evolution (O-QUATRE) algorithm [13], quasi-affine transformation evolution with external archive (QUATRE-EAR) [16], etc. But related algorithms are not widely applied to practical problems, so this article tries to test the effect of QUATRE algorithm and its variants on simple practical problems.

There are many issues in the 5G network that need to be optimized to meet a variety of complex requirements, such as security, performance, and energy consumption [7, 24, 25, 27]. The 5G network has high throughput and strong coverage requirements, and there are many different types of base stations in the network, such as macro-cell base stations (MBS), Microcell Base Stations, Picocell Base Stations, femtocell base stations (FBS), and Relay Base Stations [10]. Heterogeneous networks, as a network convergence model that can meet users' high-speed and multi-services, can be used to deal with the complex needs of the 5G network. Macro/Femtocell, as a heterogeneous network, can be used to improve network throughput and coverage, but after the deployment of heterogeneous networks, the resource allocation of the entire network needs to be optimized in order to obtain higher network performance and energy efficiency [14]. This paper uses the QUATRE algorithm and its two variants to deal with the 5G heterogeneous network downlink energy efficiency optimization problem, tests the performance of the QUATRE algorithm and its variants on the actual problem, and discusses the impact of different improvement methods on the final result.

The content of this article is arranged as follows. Sect. 2 introduces the QUATRE and its variants. The third section introduces the downlink optimization problem of 5G heterogeneous network. The fourth section uses the QUATRE algorithm and its variants to deal with the proposed 5G heterogeneous network downlink optimization problem, and shows the relevant results. The fifth section summarizes.

2 QUATRE Algorithm and Its Variants

This section mainly introduces the principle of the original QUATRE algorithm and two variants C-QUATRE and S-QUATRE. C-QUATRE is an improved QUATRE algorithm using a paired competition mechanism, and S-QUATRE is an improved QUATRE algorithm using a sorting technique.

2.1 QUasi-Affine TRansformation Evolutionary Algorithm

In 2016, the QUATRE algorithm was proposed. The evolution formula used is similar to the affine transformation in geometry. The detailed evolution formula is

$$\widehat{Z} = \bar{M} \otimes \hat{Z} + M \otimes B \quad (1)$$

$\widehat{Z}$ is the population matrix composed of all individuals, $\widehat{Z} = (Z_1, Z_2, \ldots Z_i, \ldots, Z_{Np})$, $i \in [1, Np]$, B is the matrix used to guide evolution, the symbol $\otimes$ represents the multiplication of the elements at the corresponding positions of the two matrices, M is the co-evolution matrix, and $\bar{M}$ represents the incidence matrix of matrix M.

M needs to be generated by the initial matrix M_{init}, which is a lower triangular matrix with a dimension of D. When $Np = D$, the generation method is as shown in the formula

$$M_{init} = \begin{bmatrix} 1 & & & \\ 1 & 1 & & \\ & \cdots & & \\ 1 & 1 & \cdots & 1 \end{bmatrix} \sim \begin{bmatrix} 1 & 1 & & \\ & \cdots & & \\ 1 & 1 & \cdots & 1 \\ & & 1 & \end{bmatrix} = M \quad (2)$$

According to Eq. 2, for the matrix M_{init}, first arrange the elements of each row randomly, and then randomly arrange all the row vectors, then the matrix M can be generated. When Np is not equal to D, different methods need to be used to generate M_{init} matrix. When $Np > D$, the M_{init} matrix can be composed of s D-dimensional lower triangular matrices, ensuring that $Np = s * D + k$, where k represents the first k rows of the D-dimensional lower triangular matrix.

Next, the $\bar{M}$ matrix is obtained by inverting each element of the matrix M. In addition, the generation of B also requires a certain method. Refer to [18], the formula for generating B in the QUATRE algorithm used in this article is

$$B = Z_{gbest} + F \times (Z_{r1} - Z_{r2}) \quad (3)$$

Z_{gbest} is the optimal individual in the matrix $\widehat{Z}$, F represents the parameter, in this article $F = 0.7$, Z_{r1} and Z_{r2} are generated by randomly permuting the row vectors of matrix $\widehat{Z}$.

2.2 C-QUATRE Algorithm

The C-QUATRE is an improved QUATRE algorithm that uses pairwise comparison and competition techniques. It is also a simple improvement based on the QUATRE algorithm. Assuming that the population size is Np and Np is an even number, then the entire population is equally divided into two groups, the first $1/2$ and the second $1/2$ group have the same number of individuals.

Each individual in the first group is sequentially compared with the corresponding individual in the second group. If the current individual is better, keep it, and the other individual uses Formula 1 to adjust. The basic logic is that among the two individuals to be compared, the good individuals are retained, and the poor individuals are adjusted. This process is performed $Np/2$ times, and then the above process is repeated. According to [15], the improved C-QUATRE is better than the original QUATRE under test conditions, and the overall performance of the QUATRE has been greatly improved.

2.3 S-QUATRE Algorithm

S-QUATRE algorithm is an improved QUATRE algorithm using sorting technology. Based on the basic principles of the QUATRE algorithm, the S-QUATRE algorithm sorts the $\widehat{Z}$ according to the fitness value, and then divides it into two groups evenly.

It is necessary to ensure that the population number Np is an even number, $Np/2$ individuals with better fitness values form the first group, and the remaining $Np/2$ individuals form the second group. Then use Eq. 1 to evolve the $Np/2$ individuals in the second group, and the other outstanding individuals are retained in the next generation.

It should be noted that after the evolution of the second group of individuals, it needs to be compared with the second group of individuals before the evolution, and keep the better individuals. Then combine the second group with the first group before sorting, and repeat the above process until the algorithm stop condition is met. According to [20], the introduction of sorting technology enhances the stability of the algorithm and slightly improves the performance of the original QUATRE.

3 5G Heterogeneous Network Downlink Power Allocation

Energy efficiency optimization of 5G heterogeneous networks is an important and novel practical problem, and there have been many related studies [7, 14]. The huge throughput demand and higher coverage in the 5G network will consume more energy and require reasonable allocation. This section mainly introduces the practical problem of 5G heterogeneous network downlink energy efficiency optimization.

The first is the basic structure of 5G heterogeneous networks. Since there are multiple types of base stations in 5G networks, this article only introduces the Macro/Femtocell heterogeneous network, which consists of two types of base stations, Macrocell and Femtocell. The Macrocell base station has a large coverage area, consumes a lot of energy, and is more complicated to deploy. The Femtocell base station is the opposite. Its coverage area is small, the transmission power is lower, and the smaller size is easier to deploy. The combination of the two base stations can provide more effective network services for hotspots and coverage holes. However, because the two base stations usually share spectrum, it will cause cross-layer interference, and reasonable resource allocation is required to avoid resource waste and improve energy efficiency.

This paper uses a joint resource allocation method and uses an evolutionary algorithm to simultaneously allocate channels and power to try to optimize the energy efficiency of the entire 5G heterogeneous network downlink. Next, a mathematical model of 5G heterogeneous network downlink needs to be established. The evaluation standard used in this article is network energy efficiency (E), and the calculation formula is

$$E = \frac{C}{P} \qquad (4)$$

For Eq. 4, C represents the throughput of the entire network, which is the sum of the throughput of all users served by FBS and MBS. P represents the power consumption of the entire network.

For a certain user i in the network, its throughput calculation formula is

$$C_i = \frac{B}{K} \times (1 + log_2(\frac{|g_i \sqrt{p_i l_i}|^2}{\sum_w |g_i^w \sqrt{p_i^w l_i^w}|^2 + P_0})) \qquad (5)$$

For Eq. 5, B represents bandwidth, K is the number of sub-channels. g_i represents the channel gain between the base station and current user i, p_i represents the power allocated by the base station to the user i, l_i means the path loss between the base station and the user i. g_i^w, p_i^w, l_i^w, respectively, represent the channel gain, power allocation and path loss between the w-th interfering base station and user i. w represents the number of all other interfering base stations, and P_0 represents additive white Gaussian noise.

This article does not consider the interference when different users served by the same base station are in different channels, only the downlink is considered.

In the constructed two-layer heterogeneous network, the number of Macrocell base stations is 1, providing services for M macrocell users (MUE), and the number of FBSs is N, providing services for N_f femtocell users (FUE). Since the QUATRE algorithm used in this article can be used to minimize optimization problems, the objective function constructed based on the above description is

$$\text{Consider } \mathbf{x} = [\mathbf{c}_{mue}, \mathbf{c}_{fue}, \mathbf{p}_{mue}, \mathbf{p}_{fue}];$$
$$\mathbf{c}_{mue} = [cm_1, ..., cm_i, ..., cm_M],$$
$$\mathbf{c}_{fue} = [cf_1, ..., cf_i, ..., cf_{N*N_f}],$$
$$\mathbf{p}_{mue} = [pm_1, ..., pm_i, ..., pm_M],$$
$$\mathbf{p}_{fue} = [pf_1, ..., pf_i, ..., pf_{N*N_f}]$$
$$\min \ f(\mathbf{x}) = \frac{1}{E}$$
$$= \frac{\sum \mathbf{p}_{mue} + \sum \mathbf{p}_{fue}}{\sum_1^M C_{mue}^i + \sum_1^{N*N_f} C_{fue}^j}$$
$$st. \ y_1(\mathbf{x}) = \sum \mathbf{p}_{mue} - P_{MBS} + P_{cl} \leq 0,$$
$$y_2(\mathbf{x}) = \sum_1^{N_f} pf_i - P_{FBS} + P_{cl} \leq 0,$$
$$\text{Variable range } 0.5 < cm_i < K + 0.5$$
$$0.5 < cf_i < K + 0.5$$

As shown above, P_{MBS} means the total power of a macrocell base station, and P_{FBS} means the total power of a femtocell base station. P_{cl} means circuit loss. cm_i means the channel assigned to the i-th MUE, cf_i means the channel assigned to the i-th FUE. pm_i means the power allocated to the i-th MUE, pf_i means the power allocated to the i-th FUE.

4 Algorithm Experiment Results

In this section, the parameters are first set up, based on the model of the 5G heterogeneous network downlink energy efficiency optimization problem described in the third section, and then the QUATRE algorithm and its two variants are used for optimization, and the results of the PSO algorithm are compared.

The parameter settings required to generate the model in this article are as follows (Table 1):

Table 1 Experimental settings for parameters

Symbol	Parameter	Symbol	Parameter
MBS coverage radius	600 m	B	100 MHz
P_{MBS}	6000 mW	K	40
FBS coverage radius	60 m	P_{cl}	0.5 mW
P_{FBS}	500 mW		

Table 2 Test results of different algorithms

M	N	N_f	f_{PSO}	f_Q	f_{C-Q}	f_{S-Q}	E_{PSO}	E_Q	E_{C-Q}	E_{S-Q}
5	5	5	5.48E+01	5.01E+01	5.01E+01	5.01E+01	1.79E-02	1.99E-02	2.00E-02	2.00E-02
10	5	5	5.28E+01	4.76E+01	4.76E+01	4.76E+01	1.84E-02	2.09E-02	2.10E-02	2.10E-02
15	5	5	1.31E+02	1.19E+02	1.19E+02	1.19E+02	7.47E-03	8.37E-03	8.41E-03	8.41E-03
20	5	5	7.91E+01	7.14E+01	7.14E+01	7.14E+01	1.24E-02	1.39E-02	1.40E-02	1.40E-02
10	5	5	6.47E+01	5.98E+01	5.98E+01	5.98E+01	1.53E-02	1.67E-02	1.67E-02	1.67E-02
10	5	10	5.17E+01	4.60E+01	4.60E+01	4.60E+01	1.90E-02	2.17E-02	2.18E-02	2.18E-02
10	5	15	3.16E+08	4.87E+01	9.42E+01	4.87E+01	1.74E-02	2.05E-02	2.05E-02	2.05E-02
10	5	20	3.64E+09	1.13E+09	1.13E+09	1.13E+09	1.61E-02	1.83E-02	1.87E-02	1.87E-02

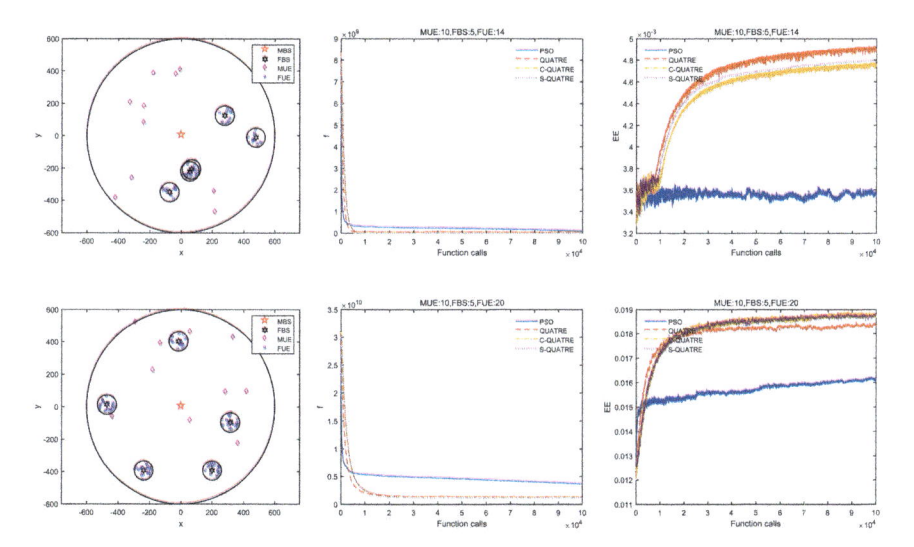

Fig. 1 Comparison of different algorithms for different network models

Based on the above, this article tested the performance of QUATRE, C-QUATRE, S-QUATRE, and PSO in different numbers of MUE, FBS, and FUE. The number of evaluations of each algorithm on each model is 100,000, and the average value is recorded after 11 tests. The number of particles for the four algorithms is 20.

The results are shown in Table 2.

In order to show the effects of different algorithms more intuitively, this article selects two test results for display, as shown in Fig. 1.

As shown in Fig. 1, the first column represents the two generated network models, the second column represents the changes in the fitness values of the four algorithms, and the third column represents the changes in the energy efficiency of the four algorithms.

It can be seen from Fig. 1 that the QUATRE algorithm and its two variants have little difference in fitness value as the number of evaluations increase, and the overall effect is slightly better than that of the PSO algorithm. But for energy efficiency, the

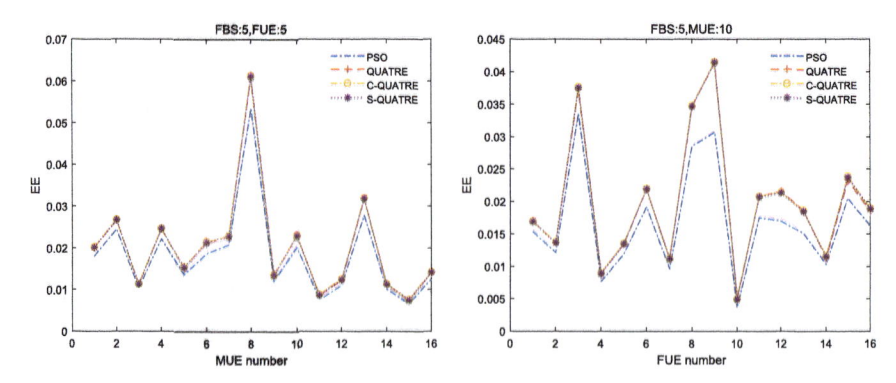

Fig. 2 Comparison of the results of different algorithms when the number of FUE or MUE changes

energy efficiency change trends of the three QUATRE algorithms are different, and the performance of the three QUATRE algorithms fluctuates in different situations, which needs further discussion.

Finally, the performance of the four algorithms when FUE and MUE changes are shown, as shown in Fig. 2.

It can be seen from the figure that the QUATRE algorithm and its two variants do not show much difference in performance when the number of FUE or MUE changes.

5 Conclusion

The main purpose of this paper is to test the performance of the QUATRE algorithm and its variants in practical application problems. First, the energy efficiency optimization requirements in 5G heterogeneous networks are selected as the actual problem of algorithm optimization. Next, this article selected the QUATRE algorithm and its two variants, tested the performance of the three algorithms in the 5G heterogeneous network downlink energy efficiency optimization problem, and compared them with the PSO. Finally, the test results of different algorithms are displayed. The results show that the overall effect of the three QUATRE algorithms on this problem is not much different, but the performance will change when the scale of the problem is larger. More detailed analysis and discussion on the adaptation of algorithm improvement methods to different application models require more experiments.

References

1. Arora, S., Singh, S.: Butterfly optimization algorithm: a novel approach for global optimization. Soft. Comput. **23**(3), 715–734 (2019). https://doi.org/10.1007/s00500-018-3102-4
2. Boussaïd, I., Lepagnot, J., Siarry, P.: A survey on optimization metaheuristics. Inf. Sci. **237**, 82–117 (2013)
3. Chai, Q.W., Chu, S.C., Pan, J.S., Zheng, W.M.: Applying adaptive and self assessment fish migration optimization on localization of wireless sensor network on 3-D terrain. J. Inf. Hiding Multimedia Signal Process. **11**(2), 90–102 (2020)
4. Das, S., Suganthan, P.N.: Differential evolution: A survey of the state-of-the-art. IEEE Trans. Evol. Comput. **15**(1), 4–31 (2010)
5. Dorigo, M., Maniezzo, V., Colorni, A.: Ant system: optimization by a colony of cooperating agents. IEEE Trans. Syst., Man, Cybern., Part B (Cybern.) **26**(1), 29–41 (1996)
6. Formato, R.A.: Central force optimization. Prog. Electromagn. Res. **77**, 425–491 (2007)
7. Ge, X., Yang, J., Gharavi, H., Sun, Y.: Energy efficiency challenges of 5G small cell networks. IEEE Commun. Mag. **55**(5), 184–191 (2017)
8. Holland, J.H.: Genetic algorithms. Sci. Am. **267**(1), 66–73 (1992)
9. Hu, P., Pan, J.S., Chu, S.C.: Improved binary grey wolf optimizer and its application for feature selection. Knowl.-Based Syst. **195**, 105,746 (2020). https://doi.org/10.1016/j.knosys.2020.105746
10. Kaimaletu, S., Krishnan, R., Kalyani, S., Akhtar, N., Ramamurthi, B.: Cognitive interference management in heterogeneous femto-macro cell networks. In: 2011 IEEE International Conference on Communications (ICC), pp. 1–6 (2011)
11. Kaveh, A., Talatahari, S.: A novel heuristic optimization method: charged system search. Acta Mech. **213**(3–4), 267–289 (2010)
12. Kennedy, J., Eberhart, R.: Particle swarm optimization. In: Proceedings of ICNN'95-International Conference on Neural Networks, vol. 4, pp. 1942–1948. IEEE (1995)
13. Liu, N., Pan, J.S., Xue, J.Y.: An orthogonal quasi-affine transformation evolution (O-QUATRE). In: Advances in Intelligent Information Hiding and Multimedia Signal Processing: Proceedings of the 15th International Conference on IIH-MSP in conjunction with the 12th International Conference on FITAT, July 18–20, Jilin, China, Vols 2, 157, pp. 57–66. Springer (2019)
14. López-Pérez, D., Chu, X., Vasilakos, A.V., Claussen, H.: On distributed and coordinated resource allocation for interference mitigation in self-organizing lte networks. IEEE/ACM Trans. Netw. **21**(4), 1145–1158 (2013)
15. Meng, Z., Pan, J.S.: A competitive quasi-affine transformation evolutionary (C-QUATRE) algorithm for global optimization. In: 2016 IEEE International Conference on Systems, Man, and Cybernetics (SMC), pp. 001,644–001,649. IEEE (2016)
16. Meng, Z., Pan, J.S.: Quasi-affine transformation evolution with external archive (QUATRE-EAR): an enhanced structure for differential evolution. Knowl.-Based Syst. **155**, 35–53 (2018)
17. Meng, Z., Pan, J.S., Kong, L.: Parameters with adaptive learning mechanism (PALM) for the enhancement of differential evolution. Knowl.-Based Syst. **141**, 92–112 (2018)
18. Meng, Z., Pan, J.S., Xu, H.: Quasi-affine transformation evolutionary (QUATRE) algorithm: a cooperative swarm based algorithm for global optimization. Knowl.-Based Syst. **109**, 104–121 (2016)
19. Mirjalili, S., Mirjalili, S.M., Lewis, A.: Grey wolf optimizer. Adv. Eng. Softw. **69**, 46–61 (2014). https://doi.org/10.1016/j.advengsoft.2013.12.007
20. Pan, J.S., Meng, Z., Chu, S.C., Roddick, J.F.: QUATRE algorithm with sort strategy for global optimization in comparison with DE and PSO variants. In: The Euro-China Conference on Intelligent Data Analysis and Applications, pp. 314–323. Springer (2017)
21. Pan, J.S., Meng, Z., Chu, S.C., Xu, H.R.: Monkey king evolution: an enhanced ebb-tide-fish algorithm for global optimization and its application in vehicle navigation under wireless sensor network environment. Telecommun. Syst. **65**(3), 351–364 (2017)

22. Pan, J.S., Meng, Z., Xu, H., Li, X.: A matrix-based implementation of de algorithm: the compensation and deficiency. In: International Conference on Industrial, Engineering and Other Applications of Applied Intelligent Systems, pp. 72–81. Springer (2017)
23. Rao, S.S.: Engineering Optimization: Theory and Practice. Wiley, New York (2019)
24. Wang, E.K., Liu, X., Chen, C.M., Kumari, S., Shojafar, M., Hossain, M.S.: Voice-transfer attacking on industrial voice control systems in 5G-aided IIoT domain. IEEE Trans. Ind. Inf. 1 (2020). https://doi.org/10.1109/TII.2020.3023677
25. Wang, P., Chen, C.M., Kumari, S., Shojafar, M., Tafazolli, R., Liu, Y.N.: HDMA: hybrid D2D message authentication scheme for 5G-enabled VANETs. IEEE Trans. Intell. Transp. Syst. 1–10 (2020). https://doi.org/10.1109/TITS.2020.3013928
26. Wolpert, D.H., Macready, W.G.: No free lunch theorems for optimization. IEEE Trans. Evol. Comput. **1**(1), 67–82 (1997)
27. Wu, T.Y., Lee, Z., Obaidat, M.S., Kumari, S., Kumar, S., Chen, C.M.: An authenticated key exchange protocol for multi-server architecture in 5G networks. IEEE Access **8**, 28096–28108 (2020)
28. Xue, X., Lu, J.: A compact brain storm algorithm for matching ontologies. IEEE Access **8**, 43898–43907 (2020)

Reversible Image Watermarking Based on Deep Learning

Jianchuan He, Linlin Tang, Jiawei Chen, Tao Qian, Shuhan Qi, Yang Liu, and Jiajia Zhang

Abstract Reversible image watermarking refers to technology that can restore an image to its original state after extracting the watermark. The scheme based on prediction error expansion (PEE) can achieve greater embedding capacity and less image distortion than other methods, so it has been widely researched in recent years. Prediction results of the predictor used by PEE are still not accurate enough, which limits the development of PEE. In this paper, a reversible watermarking predictor based on a deep neural network is proposed. Compared with other predictors, the prediction error histogram generated by our proposed predictor distributes more sharply. At the same time, because of the better prediction results, the watermarked image is closer to the original image. Experimental results show that the proposed method is effective and superior to the existing methods.

J. He · L. Tang (✉) · J. Chen · T. Qian · S. Qi · Y. Liu · J. Zhang
Harbin Institute of Technology, Shenzhen, China
e-mail: hittang@126.com

J. He
e-mail: 863646236@qq.com

J. Chen
e-mail: 17628039029@163.com

T. Qian
e-mail: qt41@qq.com

S. Qi
e-mail: shuhanqi@cs.hitsz.edu.cn

Y. Liu
e-mail: liu.yang@hit.edu.cn

J. Zhang
e-mail: zhangjiajia@hit.edu.cn

© The Author(s), under exclusive license to Springer Nature Singapore Pte Ltd. 2022
J.-F. Zhang et al. (eds.), *Advances in Intelligent Systems and Computing*, Smart Innovation, Systems and Technologies 268, https://doi.org/10.1007/978-981-16-8048-9_26

1 Introduction

Reversible image watermarking is a special digital watermarking technology. After embedding the watermark into the image, the watermark and the original image bearing the watermark can be extracted without loss. Reversible watermarking is mainly used for image authentication and copyright protection. For example, in military and medical fields, images need to be embedded with watermarks as image authentication, but permanent image distortion is strictly prohibited, because the slightest deviation will lead to completely different results in these fields. Besides, in the field of Cloud storage, the Cloud manager can insert additional data into the media file for content authentication, but the manager cannot permanently damage the image, so the insertion operation should be reversible.

A reversible image watermarking scheme should have high embedding capacity and low distortion to the original image. Early reversible watermarking schemes were based on lossless compression [1], However, the compression ratio is relatively low, which cannot provide enough payload, and noise generated by compression reduces the quality of the image. Then, techniques like Histogram Shifting (HS) [2, 3] and Deference Expansion (DE) [3, 4] appeared, which provide higher embedding capacity and lower distortion. HS technique proposed by Ni et al. [2] is to embed additional data into peak point after moving the value between the peak value and zero value point of the gray histogram. DE technique proposed by Tian et al. [4] has made important progress in reversible watermarking by extending the difference between adjacent pixels to produce an empty least significant bit (LSB). Thodi et al. [3, 5] promoted the technology of HS and DE to form a prediction error expansion (PEE) scheme. Distribution of prediction error histogram (PEH) is sharper than gray histogram, which can effectively improve the embedding capacity. As an extension of HS and DE, PEE is the most effective and widely used reversible watermarking technology presently. PEE uses the difference between pixel value and its predicted value, namely prediction error (PE), to extend and embed, instead of enlarging the difference between adjacent pixels, which greatly reduces distortion to the original image caused by the watermark.

PEE is mainly divided into two steps. The first step is to estimate prediction error (PE) and generate prediction error histogram (PEH). In order to generate PEH more sharply, a lot of research on predictors has been carried out. Thodi et al. [5] proposed to use adjacent 2×2 pixels for prediction, embed them from left to right and from top to bottom, and then extract them in the reverse. Sachnev et al. [6] proposed a rhombus predictor. As shown in Fig. 1, the image is divided into a non-adjacent point set and cross set as in a chessboard, and then pixels in the point set are used to predict the pixels in the cross set. Their solution is to get the predicted value by simply averaging the top, bottom, left, and right adjacent pixels (the rhombus) of the point-set pixels. Later, Jia et al. [7] proposed a weighted average rhombus predictor based on Sachnev's method, which adaptively adjusted weight according to the difference between the surrounding pixels and the mean value, and achieved better results. In addition, Tang et al. [8] also improved the rhombus predictor, mainly by calculating

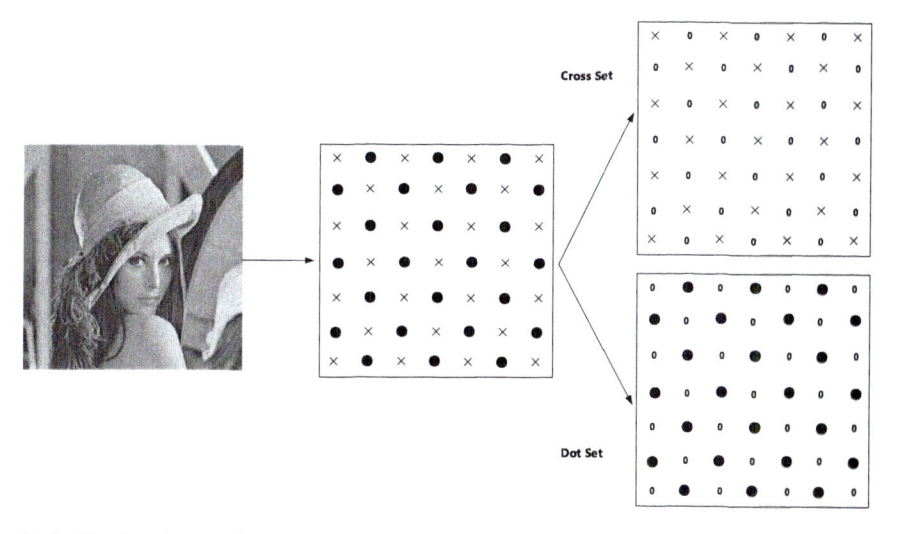

Fig.1 The rhombus predictor

fluctuation around pixels and removing one or two of the four adjacent pixels, so as to achieve higher prediction accuracy. Lin et al. [9] proposed a globally optimal least squares predictor; Dragoi [10] proposed a locally optimal least squares predictor with higher prediction accuracy, but it needed to embed a large number of additional parameters. Hwang et al. [11] also improved the least squares predictor. Naskar et al. [12] proposed a median edge predictor based on the JPEG compression standard. Yang et al. [13] proposed a gradient predictor based on CALIC coding, which performed better than the median edge predictor. The best performance predictor of PEE is based on the checkerboard-based rhombus predictor as it can take pixels in all four directions into account and is relatively easy to implement.

The second step of PEE is to determine the best embedding scheme according to prediction error distribution and embedding capacity, so as to minimize the distortion. The scheme proposed by Thodi et al. [3] is to directly shift difference beyond peak value of prediction error, and embed watermark at the peak value by modifying prediction error, as shown in Fig. 2. Sachnev et al. [6] proposed to calculate the local

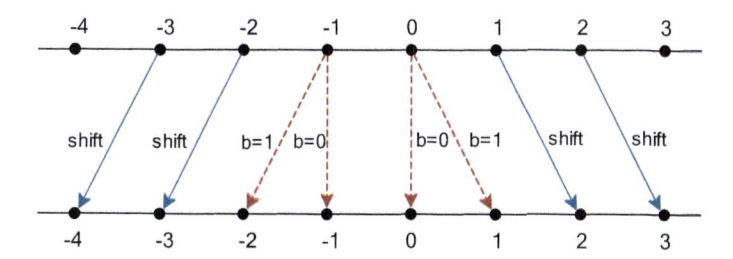

Fig. 2 Watermark embedding of PEE

complexity of each pixel; the watermark is first embedded into the pixel with lower local complexity, which achieved less distortion. Ou et al. [14] proposed 2D-PEH, which generated 2D-PEH by combining two adjacent prediction errors, and then designed a more effective embedding strategy based on 2D-PEH, that is, pairwise PEE, which further reduced embedding distortion. Li et al. [15] proposed a multi-histogram embedding scheme, which further increased the embedding capacity.

This paper proposes a new and effective predictor based on deep learning [17–19]. The picture is first divided into a point set and a cross set according to a chessboard, and then one set is used to predict another set. The data set used in the experiment is 90000 pictures selected from the test set of the ImageNet data set. Experiments show that compared with traditional predictors, the predictor in this paper will greatly improve the prediction accuracy, which directly leads to a watermark image closer to the original image under the same watermark embedding scheme. As far as we know, this job is the first detailed report on applying deep learning to reversible image watermarking.

The rest of this paper is organized as follows. We describe the framework of PEE-based reversible watermarking in Sect. 2. Then the proposed method is introduced in detail in Sect. 3, followed by experiments in Sect. 4. Finally, we conclude this paper in Sect. 5.

2 PEE-Based Reversible Watermarking

This part mainly introduces the overall steps of the PEE-based reversible watermarking scheme.

Firstly, divide picture into a chessboard as shown in Fig. 1; point-set pixels are used to modify and embed the watermark, and cross-set pixels are unchanged, and are used to predict the value of the point-set pixels. Scan all pixels of the point set in a certain order to obtain a sequence of pixel values $(x_1, x_2, \ldots, x_N)$. Use cross set to predict point set to get the predicted pixel value sequence $(\hat{x}_1, \hat{x}_2, \ldots, \hat{x}_N)$. For pixel x_i of point set, prediction value is $\hat{x}_i$, so prediction error of x_i is

$$e_i = x_i - \hat{x}_i \tag{1}$$

Then we get the sequence of prediction error $(e_1, e_2, \ldots, e_N)$; they will be fixed one by one in traditional PEE. For prediction error e_i, according to the thresholds T_1, T_2 that control the embedding capacity (the most commonly used thresholds 0 and -1 are taken as examples here), the prediction error will be extended or shifted as

$$e_i' = \begin{cases} e_i + b, & \text{if } e_i = 0 \\ e_i - b, & \text{if } e_i = -1 \\ e_i + 1, & \text{if } e_i > 0 \\ e_i - 1, & \text{if } e_i < -1 \end{cases} \tag{2}$$

where b = 0,1 is the embedded watermark information. After that, pixels of point set x_i are modified as follows:

$$x_i' = \hat{x}_i + e_i' \tag{3}$$

Finally, after receiving the image with embedded information, the receiver makes the same division and uses cross set to predict point set to get the same prediction sequence $(\hat{x}_1, \hat{x}_2, \ldots, \hat{x}_N)$; modified prediction error can then be extracted:

$$e_i' = x_i' - \hat{x}_i \tag{4}$$

The true prediction error e_i can be obtained by e_i' and the threshold $0, -1$ as follows:

$$e_i = \begin{cases} 0, & if\ e_i' = 0\ or\ e_i' = 1 \\ -1, & if\ e_i' = -1\ or\ e_i' = -2 \\ e_i' + 1, & if\ e_i' < -1 \\ e_i' - 1, & if\ e_i' > 0 \end{cases} \tag{5}$$

At the same time, the original watermark information can be obtained:

$$b = \begin{cases} 0\ if\ e_i' = 0\ or\ e_i' = -1 \\ 1\ if\ e_i' = 1\ or\ e_i' = -2 \end{cases} \tag{6}$$

The pixel value of the original image can be recovered according to the real prediction error and the predicted value of the pixel:

$$x_i = e_i + \hat{x}_i \tag{7}$$

3 Proposed Method

3.1 Training of Predictor

Based on vigorous development of deep learning in the direction of image denoising, our prediction task and image denoising task have a great correlation, so the model we used is a model SGN originally applied in the field of image denoising [16]; the overall training method of the model is shown in Fig. 3.

A training set of 90,000 images is selected from ImageNet's test set. Firstly, the image is randomly scaled and then clipped to 512 × 512. Clipped image is taken as the label image: img. M_1 and M_2 are matrices sized 512 × 512 which satisfy the following conditions:

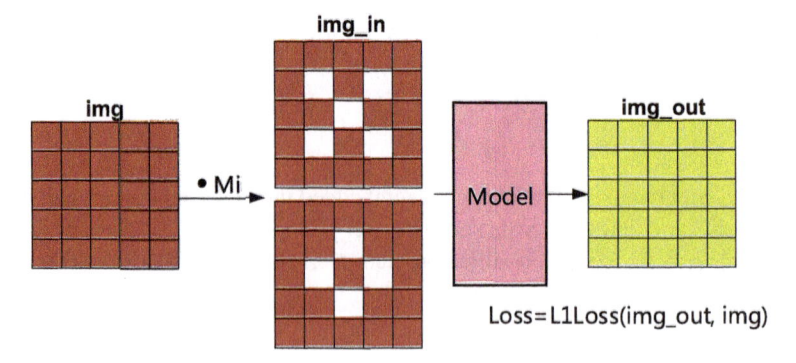

Fig. 3 The training of SGN model

$$M_1[i][j] = \begin{cases} 1, if i, j = 0 \, or \, 511 \\ (i+j) mod 2, else \end{cases} \quad (8)$$

$$M_2[i][j] = \begin{cases} 1, if i, j = 0 \, or \, 511 \\ (i+j+1) mod \, 2, else \end{cases} \quad (9)$$

Then dot multiply img by a random $M_i(i = 1, 2)$(equivalent to adding noise to the image), which obtains the input of model img_in. The loss function used in the network is L1Loss.

$$L^{l_1} = \frac{1}{N} \sum_{p \in P} |img_{out}(p) - img(p)| \quad (10)$$

where P set represents the set of positions (i, j) which satisfy $M_i(i, j) = 0$.

3.2 Watermark Embedding and Extracting

Overall steps of watermark embedding are shown in Fig. 4a. The first one is to preprocess the image. LSB replacement method is used to embed necessary parameters in the first line to extract the watermark, such as the number of embedded layers, the bitmap size composed of overflow bits (0 to 255; may overflow after modification), the threshold parameters selected for watermark embedding, etc. And these LSB bits and the compressed bitmap are added to the watermark as a part of the watermark.

Then preprocessed image img is to dot multiply M_1 to obtain the input of SGN model: $img\_in = img \times M_1$. img_in is input into the trained SGN model, and the output picture img_out is obtained. The next step is to fix the model output. Use the pixels in img_in which have not been modified compared to img to replace the corresponding position pixel in img_out. Thus, the prediction image is obtained.

(a) watermark embed (b) watermark extract

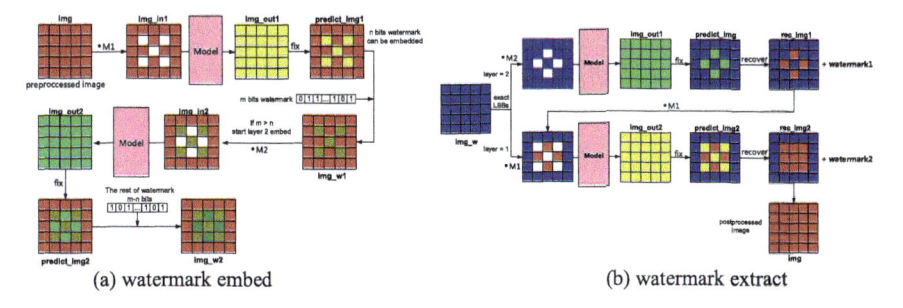

Fig. 4 Watermark embedding and extracting

Then, embed the watermark bits into *img* according to prediction error. If the position provided by the first layer which can embed the watermark is not enough to embed all of the watermark, all the positions in layer 1 will be embedded. And then embed layer 2. The whole process of layer 2 embedding is similar to that of layer 1.

The step of watermark extraction is completely the reverse process of watermark embedding, as shown in Fig. 4b.

4 Experiment

The effectiveness of the proposed scheme was compared with other traditional predictors; methods compared with are the methods of Sachnev [6], Jia [7], and Tang [8]. The images used in the experiment are from Standard Test Images, as shown in Fig. 5.

The mean square error (MSE) was used to reflect the difference between the predicted image and the original image. At the same time, in order to be fair, the same scheme was used when embedding the watermark. After sorting pixel-based local complexity, the embedding position was selected from low to high in the order of the complexity.

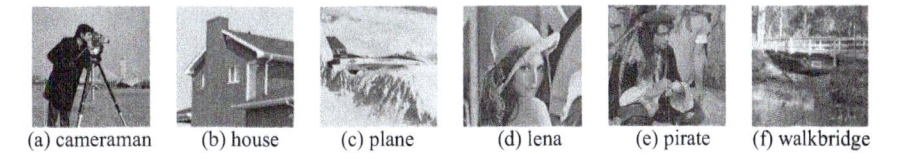

(a) cameraman (b) house (c) plane (d) lena (e) pirate (f) walkbridge

Fig. 5 Standard test images

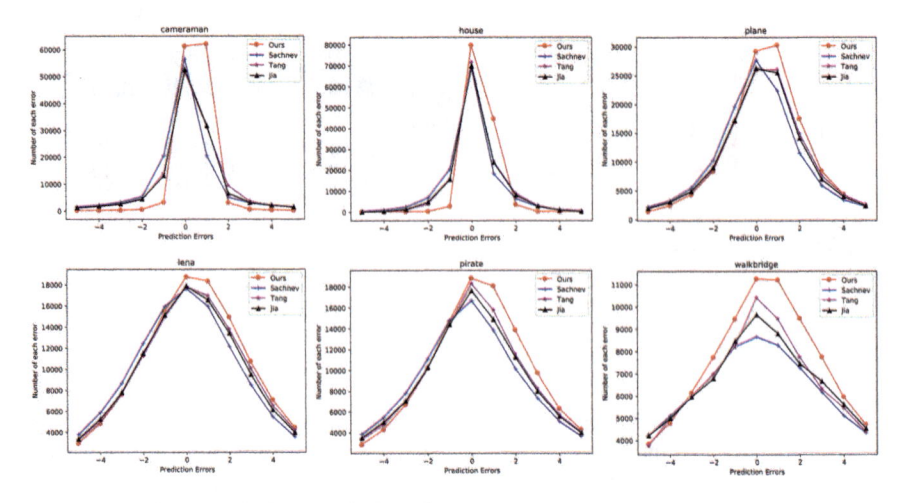

Fig. 6 PEH of each method on standard test images

Table 1 MSE of each method on standard test images

	Cameraman	House	Plane	Lena	Pirate	Walkbridge
Ours	**0.335**	**0.235**	**3.229**	**6.838**	**14.691**	**44.995**
Sachnev	5.544	0.907	10.665	12.157	24.448	59.729
Tang	6.294	0.762	10.412	11.842	24.231	68.930
Jia	4.801	0.771	9.469	11.524	22.482	57.145

4.1 Prediction Accuracy

Predicition accuracy was measured by MSE of the predicted image and the original image. Figure 6 shows the prediction error of each image when using different predictors. We can see that the histogram predicted by the proposed method in this paper is more sharply distributed.

Then calculate the MSE of each image in 4 methods; the results are shown in Table 1. As we can see, the proposed method achieved a lower mean square error, so our proposed deep learning-based predictor is obviously better than the traditional rhombus-based predictor in generating prediction errors.

4.2 Watermark Embedding Performance

For a better comparison, the same embedding method is used for different predictors. Embedding capacity is from 5000 to 150,000 bits. Results are shown in Fig. 7. From this figure, we can see that the PSNR of our method is significantly higher than

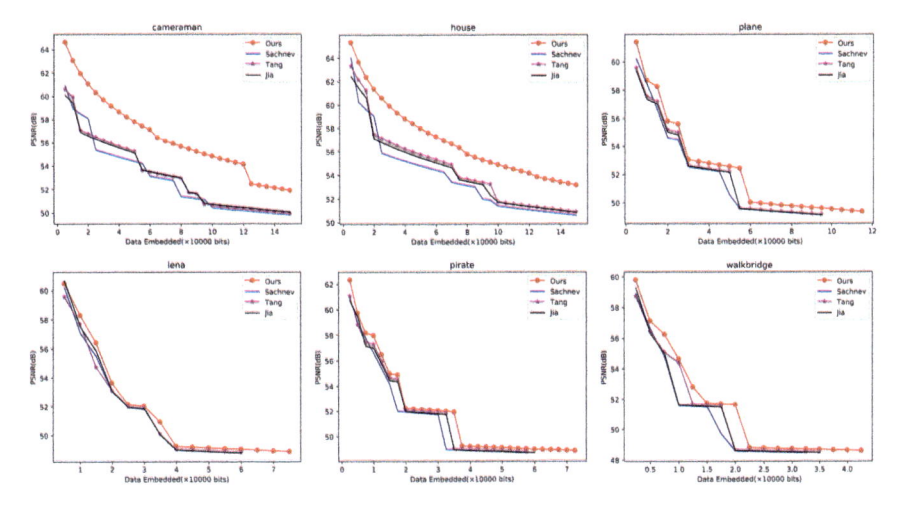

Fig.7 PSNR of each method on standard test images

that of other methods when embedding the same length watermark, especially on simple pictures like *cameraman* and *house*. The maximum embedding capacity of our method on the same image is obviously higher than that of other methods.

5 Conclusions

In our paper, combining deep learning and PEE, a new predictor based on a deep neural network is proposed, which can achieve high accuracy in the pixel prediction step of PEE, that is the PEH is more sharp than other predictors. The proposed predictor can be trained well by using the test set of the ImageNet dataset. The experiment shows that our proposed predictor performs well which achieves higher embedding capacity and less distortion compared with traditional predictors.

Acknowledgements This work was supported by Shenzhen Science and Technology Plan Fundamental Research Funding JCYJ20180306171938767 and Shenzhen Foundational Research Funding JCYJ20180507183527919.

References

1. Celik, M.U., Sharma, G., Tekalp, A.M., et al.: Lossless generalized-LSB data embedding. IEEE Trans. Image Process. **14**(2), 253–266 (2005)
2. Li, X., Li, B., Yang, B., et al.: General framework to histogram-shifting-based reversible data hiding. IEEE Trans. Image Process. **22**(6), 2181–2191 (2013)

3. Thodi, D.M., Rodríguez, J.J.: Expansion embedding techniques for reversible watermarking. IEEE Trans. Image Process. **16**(3), 721–730 (2007)
4. Tian, J.: Reversible data embedding using a difference expansion. IEEE Trans. Circ. Syst. Video Technol. **13**(8), 890–896 (2003)
5. Thodi, D.M., Rodriguez, J.J.: Prediction-error based reversible watermarking. In: 2004 International Conference on Image Processing, 2004. ICIP'04., pp. 1549–1552 (2004)
6. Sachnev, V., Kim, H.J., Nam, J., et al.: Reversible watermarking algorithm using sorting and prediction. IEEE Trans. Circ. Syst. Video Technol. **19**(7), 989–999 (2009)
7. Jia, Y., Yin, Z., Zhang, X., et al.: Reversible data hiding based on reducing invalid shifting of pixels in histogram shifting. Signal Process. **163**, 238–246 (2019)
8. Tang, X., Zhou, L., Liu, D., et al.: Reversible data hiding based on improved rhombus predictor and prediction error expansion. In: 2020 IEEE 19th International Conference on Trust, Security and Privacy in Computing and Communications (TrustCom), pp. 13–21 (2020)
9. Lin, S.-L., Huang, C.-F., Liou, M.-H., et al.: Improving histogram-based reversible information hiding by an optimal weight-based prediction scheme. J. Inf. Hiding Multimed. Signal Process. **4**(1), 19–33 (2013)
10. Dragoi, I.-C., Coltuc, D.: Local-prediction-based difference expansion reversible watermarking. IEEE Trans. Image Process. **23**(4), 1779–1790 (2014)
11. Hwang, H.J., Kim, S., Kim, H.J.: Reversible data hiding using least square predictor via the LASSO. EURASIP J. Image Video Process. **2016**(1), 1–12 (2016)
12. Naskar, R., Chakraborty, R.: Reversible watermarking utilising weighted median-based prediction[J]. IET Image Proc. **6**(5), 507–520 (2012)
13. Yang, W.-J., Chung, K.-L., Liao, H.-Y.M., et al.: Efficient reversible data hiding algorithm based on gradient-based edge direction prediction. J. Syst. Softw. **86**(2), 567–580 (2013)
14. Ou, B., Li, X., Zhao, Y., et al.: Pairwise prediction-error expansion for efficient reversible data hiding. IEEE Trans. Image Process. **22**(12), 5010–5021 (2013)
15. Li, X., Zhang, W., Gui, X., et al.: Efficient reversible data hiding based on multiple histograms modification. IEEE Trans. Inf. Forensics Secur. **10**(9), 2016–2027 (2015)
16. Gu, S., Li, Y., Gool, L.V., et al.: Self-guided network for fast image denoising. In: Proceedings of the IEEE/CVF International Conference on Computer Vision, pp. 2511–2520 (2019)
17. Wang, E.K., Chen, C.M., Hassan, M.M., Almogren, A.: A deep learning based medical image segmentation technique in Internet-of-Medical-Things domain. Futur. Gener. Comput. Syst. **108**, 135–144 (2020)
18. Wang, K., Chen, C.M., Hossain, M.S., Muhammad, G., Kumar, S., Kumari, S.: Transfer reinforcement learning-based road object detection in next generation IoT domain. Comput. Netw. 108078 (2021)
19. Tseng, K.K., Zhang, R., Chen, C.M., Hassan, M.M.: DNetUnet: a semi-supervised CNN of medical image segmentation for super-computing AI service. J. Supercomput. **77**(4), 3594–3615 (2021)

Image Encryption with Logistic Chaotic Model Using C-QUATRE Algorithm

Xiao-Xue Sun, Jeng-Shyang Pan, Tsu-Yang Wu, Lingping Kong, and Shu-Chuan Chu

Abstract With the continuous update and progress of the network, a large number of public and private images and other multimedia information are transmitted. When lots of images are transmitted, security is an important aspect. Image encryption is the principal problem to be solved. Among the image encryption methods, the chaotic logistic function is one of the simplest and common methods. In a general encryption system, the initial secret key directly generated by the one value and parameter of the chaotic map is easily cracked by the exhaustive attack. Before generating the initial secret key, this paper applies Competitive QUasi-Affine TRansformation Evolution (C-QUATRE) algorithm to optimize the initial key based on global optimization capability. This novel image chaotic encryption model dramatically improves the effect.

1 Introduction

Nowadays, information security on the Internet has aroused extensive concern for people. The primary means of protecting data on the network is encrypting data, including text, images, video, audio, and so on. Image is one of the most widely circulated data information on the Internet. Therefore, image security is a significant issue. Different from the text information, image data has its characteristics. In general, images contain a large amount of data, and there are some requirements, such as real-time encryption and decryption, fidelity, and so on. To accomplish this goal,

---

X.-X. Sun · J.-S. Pan · T.-Y. Wu · S.-C. Chu (✉)
College of Computer Science and Engineering, Shandong University of Science and Technology, Qingdao 266590, China
e-mail: xues1123@163.com

J.-S. Pan
e-mail: jspan@cc.kuas.edu.tw

L. Kong
Faculty of Electrical Engineering and Computer Science, VŠB-Technical University of Ostrava, 708 00 Ostrava, Czechia
e-mail: konglingping2007@163.com

many image encryption methods have been proposed [22–24]. Image encryption can be classified according to different ways, including the space where the encryption operation is located, the position of the pixel, and whether the gray value is changed. Another category of image encryption is called chaotic encryption.

Matthews firstly puts forward the application of chaos theory in the encryption system [10]. Fridrich applies some reversible chaotic maps to create a new symmetric encryption technology [3]. In 1998, Fridrich proposed to encrypt images with a reversible two-dimensional chaotic map and proposed a generalized method of discretizing the chaotic map into integer points [4]. In 2011, an image encryption method using the differential evolution (DE) algorithm was proposed [5]. In 2017, a chaotic encryption method based on genetic algorithm (GA) appeared [25].

Swarm intelligence is one of the crucial fields of artificial intelligence research. Swarm intelligence optimization algorithm includes many examples, such as the classic particle swarm optimization (PSO) algorithm and cat swarm optimization (CSO) algorithm [2, 21]. In recent years, some new algorithms have emerged including cuckoo search (CS) [19] algorithm, QUATRE algorithm [6, 7, 14–16], and phasmatodea population evolution (PPE) algorithm [18]. Most swarm intelligence algorithms are simple and effective that can be applied to various optimization problems. QUATRE is a simple algorithm with a few parameters and excellent performance [1, 8, 9, 11, 17, 20]. It can select the optimal solution for certain performance in a limited domain. C-QUATRE is a pairwise comparison mechanism algorithm [12]. The overall performance of the C-QUATRE algorithm is better than the QUATRE algorithm. In this paper, C-QUATRE is applied to optimize the logistic chaotic scrambling method, which encrypts the image. Only with the corresponding key can you decrypt the obtained encrypted image.

The remainder of this paper is organized as follows. Section 2 will introduce some existing related work. In Sect. 3, the encryption scheme based on C-QUATRE is shown. In Sect. 4, experiments are conducted to verify the effectiveness of this method and discuss the results through experiments. Finally, the conclusion and outlook are made in Sect. 5.

2 Related Work

In this chapter, the preliminaries are introduced including the basic logistic map function and C-QUATRE algorithm.

2.1 One-Dimensional Logistic Map

Chaos refers to the seemingly random and irregular movement that occurs in a deterministic system. It originates from nonlinear dynamic systems. The original kinetic equation can be expressed by the following equation.

Image Encryption with Logistic Chaotic Model ...

$$x_{n+1} = f(x_n), n = 0, 1, \ldots \tag{1}$$

Here, the function $f(\cdot)$ can be a nonlinear function.

The logistic map is one of the most commonly used chaotic maps with simple expression and good performance. It is defined as in Eq. 2.

$$f(x) = \mu x(1-x), x \in [0, 1], \tag{2}$$

where the system control parameter is set as μ, which is a constant with the range of $(0, 4]$. The initial value of the system is set as x_0. The output of the chaotic sequence is defined as x_n where x_n belongs to $[0, 1]$, so μ should satisfy $0 < \mu \leq 4$. When $3.569945672 \cdots < \mu \leq 4$ and x_n belongs to $[0, 1]$, the logistic map sequence x_n the characteristics of aperiodic and sensitive dependence on initial conditions. This system is in a state of chaos. These two characteristics are required for keys and keystreams in cryptography. The logistic map is simple without losing the complex characteristics of the chaos, so it is usually used in image encryption.

2.2 C-QUATRE Algorithm

In 2016, Meng et al. propose a new QUATRE algorithm [13]. The evolution formula adopted by this algorithm is similar to the affine transformation of geometry. In geometry, the affine transformation is from the affine space X to Y ($f : X \to Y$) with equation as $X \mapsto MX + B$. The evolution way is adopted in the QUATRE algorithm as in Eq. 3.

$$\widehat{\mathbf{X}} = \overline{\mathbf{M}} \otimes \widehat{\mathbf{X}} + \mathbf{M} \otimes \mathbf{B}, \tag{3}$$

where $\widehat{\mathbf{X}}$ is the population matrix. $\widehat{\mathbf{X}} = \{X_1, X_2, \ldots, X_i, \ldots, X_p\}^T$ and $i \in [1, p]$. p is the population number; $\mathbf{B}$ is the evolutionary guidance matrix. $\mathbf{M}$ is the co-evolution matrix, and $\overline{\mathbf{M}}$ is the correlation matrix of $\mathbf{M}$. The sign $\otimes$ refers to the bitwise multiplication of elements in a matrix. The matrix $\mathbf{M}$ is transformed from the initial matrix $\mathbf{M}_{int}$, which is a unitary lower triangular matrix.

For the population matrix $\mathbf{X}$, there is an evolutionary guidance matrix $\mathbf{B}$ opposite to it. $\mathbf{B}$ has eight different generation methods. This paper uses the most commonly used one way: $'QUATRE/best/1$ and the equation is Eq. 4.

$$\mathbf{B}_{r,G} = \mathbf{X}_{g,G} + F \otimes (\mathbf{X}_{r_1,G} - \mathbf{X}_{r_2,G}), \tag{4}$$

where r_1, r_2 denotes the index of the population and G represents the current number of iterations.

C-QUATRE used in this paper is an improved QUATRE algorithm with a pairwise comparison mechanism. The main idea of C-QUATRE is dividing the individuals of the population into a better group and a worse group by pairwise comparison. The process of comparison is measured by the fitness function. The winning individual directly enters the next-generation population, while the losing individual uses the C-QUATRE algorithm to evolve. Finally, the next-generation population matrix is obtained by randomly arranging all the particles in the population. The overall performance of the C-QUATRE algorithm is better than the QUATRE algorithm. By introducing the mechanism of pairwise comparison, the overall optimization effect of the algorithm is greatly improved.

3 Image Encryption Scheme Optimized by C-QUATRE

The section mainly employs the C-QUATRE algorithm to optimize the encryption method for the simple of the chaotic logistic model, that is, select the optimal solution to achieve the best performance of encryption.

C-QUATRE is used to optimize the encryption method, and the specific encryption process is shown in Fig. 1.

The encryption method proposed in this paper divides the original into four sub-blocks and then encrypts the sub-graphs, respectively. The procedures described below are taken as an example in a sub-image. As shown in Fig. 1, two secret keys are selected. One part is directly used to encrypt the original image, and the other part is optimized through the C-QUATRE algorithm to obtain the second key. The key generation is based on the original sub-image.

3.1 Image Encryption

The specific encryption model consists of the following steps. First, the original image O with the size $M \times N$ is divided into four equal sub-images O_k, ($k = 1, 2, 3, 4$). The logistic chaotic map parameters μ_1, μ_2 are defined. The encryption operation is divided into two parts as below. The steps in the first part include the following:

Step 1: Initialize the key from the original image. Each sub-image O_k is regarded as a pixel matrix. The key C is directly selected from the first eight pixels at the first-row vector. The initial value of the logistic map is obtained by $C = [C1, C2, \ldots, C7, C8]$ where C_i represents the number of pixel values. C is converted into ASCII code by Eq. 5. Therefore, the length of the string after conversion becomes 64 bits.

$$A = [C_{1,1}, \ldots, C_{1,8}, C_{2,1}, \ldots, C_{2,8}, \ldots, C_{8,8}]. \tag{5}$$

Step 2: Calculate the initial value of the logistic chaotic map by Eq. 6.

```
                    ┌─────────────────┐
                    │ Original image  │
                    └────────┬────────┘
                             ▼
                    ┌─────────────────┐
                    │    Sub-image    │
                    └────────┬────────┘
                    ┌────────┴────────┐
                    ▼                 ▼
        ┌───────────────────┐ ┌───────────────────┐
        │ The first eight   │ │ Initial population│
        │ pixel values of   │ └─────────┬─────────┘
        │ the sub-image     │           │
        └─────────┬─────────┘           ▼
           Logistic chaotic   ┌───────────────────┐
             function         │ C_QUATRE algorithm│
                  ▼           │   optimization    │
        ┌───────────────────┐ └─────────┬─────────┘
        │   Secret key-C    │   Logistic chaotic
        └─────────┬─────────┘     function
                  ▼                     ▼
        ┌───────────────────┐ ┌───────────────────┐
        │ The first encrypted│ │   Secret key-S    │
        │     sub-image     │ └───────────────────┘
        └─────────┬─────────┘
                  ▼
            ┌───────────┐
            │ Diffusion │
            └─────┬─────┘
                  ▼
          ┌─────────────┐
          │Final encrypted│
          │  sub-image  │
          └─────────────┘
```

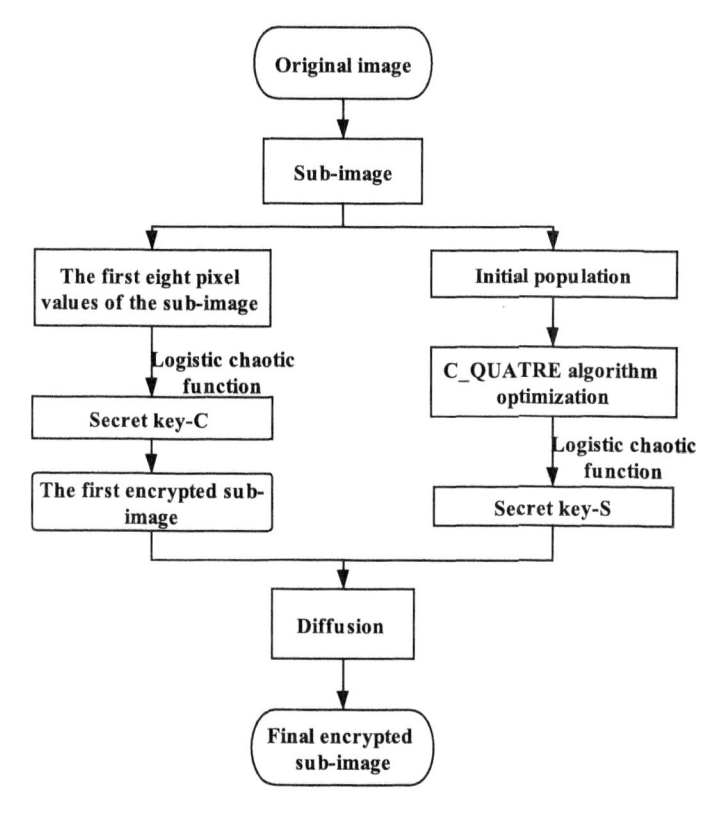

Fig. 1 This is the process displaying the whole encryption model by C-QUATRE

$$U_0 = \frac{C_{1,1} \times 2^{63} + C_{1,2} \times 2^{62} + \cdots + C_{8,7} \times 2^1 + C_{8,8} \times 2^0}{2^{64}}. \tag{6}$$

Step 3: The chaotic value is obtained after 100 iterations to eliminate the transient effect of the chaotic map, that is, from an aperiodic steady state to a periodic steady state). The chaotic value is taken as the initial value into the logistic one-dimensional quadratic map to iterate $M \times N - 1$. The required sequence is transformed into a matrix with $M \times N$, which is recorded as $\mathbf{L}_k^1$.

Step 4: Perform the first encryption operation on the image $\mathbf{O}_k$, which is calculated by Eq. 7.

$$\mathbf{O}_k' = round(255 \times \mathbf{L}_k^1) \otimes \mathbf{O}_k. \tag{7}$$

Step 5: Splice four encrypted sub-images $\mathbf{O}_k'$ to obtain the first encrypted image $\mathbf{O}'$ after performing the above steps for the four sub-images, respectively.

The specific steps in the second part are as follows.

Step 1: Randomly select 64 members from the matrix $\mathbf{O}_k$ where each row vector represents a member, and then select the first eight pixels from each row as the individual X_i of the initial population.

Step 2: Determine the difference based on the average value of the pixel changes between the images on the initial population $\mathbf{X}$ and apply the C-QUATRE to select the optimal value $\mathbf{X}_g$ which is the second key $\mathbf{S}$.

Step 3: Calculate the initial value of the chaotic logistic map according to Eqs. 5 and 6.

Step 4: This step is the same as Step 3 of the first part, and the final matrix got is recorded as $\mathbf{L}_k^2$.

Step 5: The encrypted sub-image is obtained by Eq. 8.

$$\mathbf{O}_k^{''} = round(255 \times \mathbf{L}_k^2) \otimes \mathbf{O}_k^{'}. \qquad (8)$$

Step 6: Splice four encrypted sub-images $\mathbf{O}_k^{''}$ to gain the final encrypted image $\mathbf{O}^{''}$ after performing the above steps for the four sub-images, respectively. The specific steps in the second part are as follows.

Step 1: Randomly select 64 members from the matrix $\mathbf{O}_k$ where each row vector represents a member, and then select the first eight pixels from each row as the individual X_i of the initial population.

Step 2: Determine the difference based on the average value of the pixel changes between the images on the initial population $\mathbf{X}$ and apply the C-QUATRE to select the optimal value $\mathbf{X}_g$ which is the second key $\mathbf{S}$.

Step 3: Calculate the initial value of the chaotic logistic map according to Eqs. 5 and 6.

Step 4: This step is the same as Step 3 of the first part, and the final matrix got is recorded as $\mathbf{L}_k^2$.

Step 5: The encrypted sub-image is obtained by $\mathbf{O}_k^{''} = round(255 \times \mathbf{L}_k^2) \otimes \mathbf{O}_k^{'}$.

Step 6: Splice four encrypted sub-images $\mathbf{O}_k^{''}$ to gain the final encrypted image $\mathbf{O}^{''}$ after performing the above steps for the four sub-images, respectively.

3.2 Image Decryption

Because the process of the cryptographic stream generated by the chaotic system and the C-QUATRE algorithm belongs to the symmetric key algorithm, the decryption process is the reverse operation of encryption. The specific steps are as follows.

Step 1: The key C_k and the key S_k are assigned to two logistic maps as initial values and parameters for iteration. After 100 iterations, the initial value of the logistic one-dimensional quadratic map is obtained. The periodic steady-state chaotic sequence is obtained after enjoying the initial value for $M/timeN - 1$. The matrices $\mathbf{L}_k^{1'}$ and $\mathbf{L}_k^{2'}$ are acquired by chaotic sequence.

Step 2: The encrypted image $\mathbf{O}_k^{''}$ is decrypted by Eq. 9.

$$\mathbf{I}_k^{'} = round(255 \times \mathbf{L}_k^{2'}) \otimes \mathbf{O}_k^{''}. \tag{9}$$

Step 3: The plain-text image $\mathbf{I}_k$ is restored as $\mathbf{I}_k = round(255 \times \mathbf{L}_k^{1'}) \otimes \mathbf{O}_k^{'}$.

4 Experiment and Result Analysis

The experiment takes image "Pepper" with 512×512 as an example, the encryption and decryption simulation operate on $MATLAB\_R2019b$. The original test image and four sub-images after division are shown in Fig. 2a and b.

The original image contains a lot of information, so the pixel values of the histogram about the original image are not evenly distributed and have large fluctuations. As can be seen from Fig. 2, the original image has obvious peak and valley distribution compared with the histogram of the encrypted image. In contrast, the histogram of the encrypted image is relatively stable. In addition, the encryption algorithm is tested for the correlation of adjacent pixels, that is, the ability to resist statistical attacks is tested. An outstanding image encryption algorithm should reduce the correlation between adjacent pixels. Table 1 shows the test results from three directions.

(a) Original Pepper   (b) Four sub-images   (c) Histogram of the original image

(d) Encrypted Pepper   (e) Decrypted Pepper   (f) Histogram of the encryption image

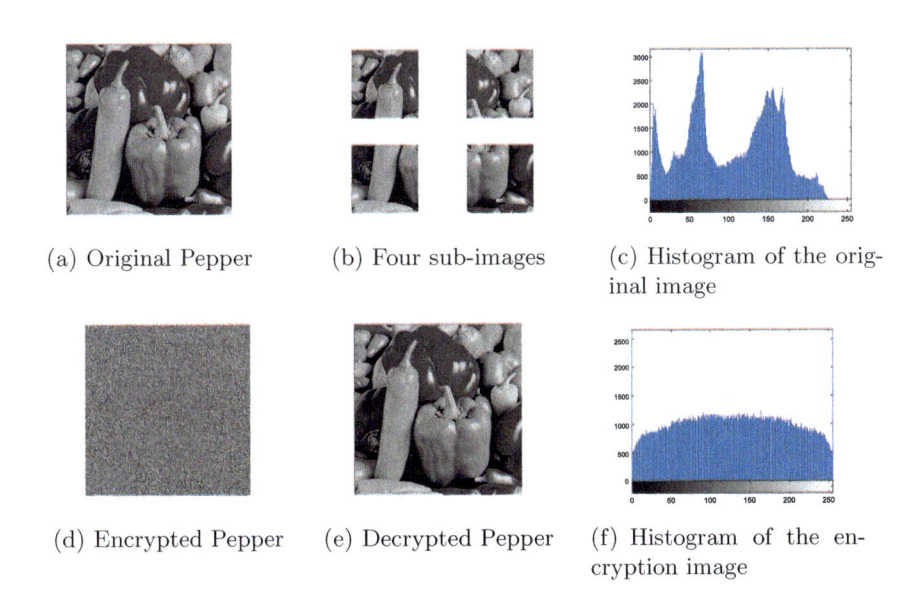

Fig. 2 The original test image, histograms, encrypted image, and decrypted image

Table 1 Correlation coefficient

	Horizontal pixels	Vertical pixels	Diagonal pixels
Original image	0.9729	0.9821	0.9587
Encrypted image with C-QUATRE	0.0374	**0.0121**	**0.0042**
Encrypted image with GA [25]	**0.0186**	0.0246	0.0092

It can be seen from the data that the encryption scheme using C-QUATRE has a better effect, and the correlation between the encrypted image and the original image is completely separate.

5 Conclusion

As a global search optimization algorithm, C-QUATRE has the advantages of simplicity, good robustness, and fast convergence. The process of generating the key by combining this algorithm with the chaotic logistic system is still a symmetric key. According to the experimental results of image encryption, the new key optimized by the C-QUATRE algorithm is more complex and random than the password generated by the ordinary pseudo-random sequence generator. Besides, the encryption scheme proposed in this paper ensures the security of the secret key. In future work, we can combine a variety of encryption technologies and perform the encryption in the frequency domain.

References

1. Chu, S.C., Chen, Y., Meng, F., Yang, C., Pan, J.S., Meng, Z.: Internal search of the evolution matrix in quasi-affine transformation evolution (QUATRE) algorithm. J. Intell. Fuzzy Syst. 1–12 (2020)
2. Chu, S.C., Tsai, P.W., Pan, J.S.: Cat swarm optimization. In: Pacific Rim International Conference on Artificial Intelligence, pp. 854–858. Springer (2006)
3. Fridrich, J.: Image encryption based on chaotic maps. In: 1997 IEEE International Conference on Systems, Man, and Cybernetics. Computational Cybernetics and Simulation, vol. 2, pp. 1105–1110. IEEE (1997)
4. Fridrich, J.: Symmetric ciphers based on two-dimensional chaotic maps. Int. J. Bifurc. Chaos **8**(06), 1259–1284 (1998)
5. Hassan, M.A.S., Abuhaiba, I.S.I.: Image encryption using differential evolution approach in frequency domain. arXiv preprint arXiv:1103.5783 (2011)
6. Jiang, B.Q., Pan, J.S.: A parallel quasi-affine transformation evolution algorithm for global optimization. J. Netw. Intell. **4**(2), 30–46 (2019)

7. Liu, N., Pan, J.S., Liao, X., Chen, G.: A multi-population quasi-affine transformation evolution algorithm for global optimization. In: International Conference on Genetic and Evolutionary Computing, pp. 19–28. Springer (2018)
8. Liu, N., Pan, J.S., Wang, J., et al.: An adaptation multi-group quasi-affine transformation evolutionary algorithm for global optimization and its application in node localization in wireless sensor networks. Sensors **19**(19), 4112 (2019)
9. Liu, N., Pan, J.S., et al.: A bi-population quasi-affine transformation evolution algorithm for global optimization and its application to dynamic deployment in wireless sensor networks. EURASIP J. Wirel. Commun. Netw. **2019**(1), 175 (2019)
10. Matthews, R.: On the derivation of a "chaotic" encryption algorithm. Cryptologia **13**(1), 29–42 (1989)
11. Meng, Z., Chen, Y., Li, X., Yang, C., Zhong, Y.: Enhancing quasi-affine transformation evolution (QUATRE) with adaptation scheme on numerical optimization. Knowl.-Based Syst. 105908 (2020)
12. Meng, Z., Pan, J.S.: A competitive quasi-affine transformation evolutionary (C-QUATRE) algorithm for global optimization. In: 2016 IEEE International Conference on Systems, Man, and Cybernetics (SMC), pp. 001,644–001,649. IEEE (2016)
13. Meng, Z., Pan, J.S.: Quasi-affine transformation evolutionary (QUATRE) algorithm: a parameter-reduced differential evolution algorithm for optimization problems. In: 2016 IEEE Congress on Evolutionary Computation (CEC), pp. 4082–4089. IEEE (2016)
14. Meng, Z., Pan, J.S.: Quasi-affine transformation evolution with external archive (QUATRE-EAR): an enhanced structure for differential evolution. Knowl.-Based Syst. **155**, 35–53 (2018)
15. Meng, Z., Pan, J.S., Xu, H.: Quasi-affine transformation evolutionary (quatre) algorithm: A cooperative swarm based algorithm for global optimization. Knowl.-Based Syst. **109**, 104–121 (2016)
16. Pan, J.S., Meng, Z., Xu, H., Li, X.: Quasi-affine transformation evolution (QUATRE) algorithm: a new simple and accurate structure for global optimization. In: International Conference on Industrial, Engineering and Other Applications of Applied Intelligent Systems, pp. 657–667. Springer (2016)
17. Pan, J.S., Sun, X.X., Chu, S.C., Abraham, A., Yan, B.: Digital watermarking with improved SMS applied for QR code. Eng. Appl. Artif. Intell. **97**, 104,049
18. Song, P.C., Chu, S.C., Pan, J.S., Yang, H.: Phasmatodea population evolution algorithm and its application in length-changeable incremental extreme learning machine. In: 2020 2nd International Conference on Industrial Artificial Intelligence (IAI), pp. 1–5. IEEE (2020)
19. Song, P.C., Pan, J.S., Chu, S.C.: A parallel compact cuckoo search algorithm for three-dimensional path planning. Appl. Soft Comput. **94**, 106,443 (2020)
20. Sun, X.X., Pan, J.S., Chu, S.C., Hu, P., Tian, A.Q.: A novel pigeon-inspired optimization with QUasi-Affine TRansformation evolutionary algorithm for DV-Hop in wireless sensor networks. Int. J. Distrib. Sens. Netw. **16**(6), 1550147720932,749 (2020)
21. Tsai, P.W., Pan, J.S., Chen, S.M., Liao, B.Y., Hao, S.P.: Parallel cat swarm optimization. In: 2008 International Conference on Machine Learning and Cybernetics, vol. 6, pp. 3328–3333. IEEE (2008)
22. Tseng, K.K., Zhang, R., Chen, C.M., Hassan, M.M.: Dnetunet: a semi-supervised cnn of medical image segmentation for super-computing ai service. J. Supercomput. 1–22 (2020)
23. Wang, E.K., Chen, C.M., Hassan, M.M., Almogren, A.: A deep learning based medical image segmentation technique in internet-of-medical-things domain. Futur. Gener. Comput. Syst. **108**, 135–144 (2020)
24. Wu, T.Y., Fan, X., Wang, K.H., Pan, J.S., Chen, C.M.: Security analysis and improvement on an image encryption algorithm using chebyshev generator. J. Internet Technol. **20**(1), 13–23 (2019)
25. Zhang, Y., Zhang, Q., Liao, H., Wu, W., Li, X., Niu, H.: A fast image encryption scheme based on public image and chaos. In: 2017 International Conference on Computing Intelligence and Information System (CIIS), pp. 270–276. IEEE (2017)

Visualization of Population Convergence Results by Sammon Mapping in Multi-objective Optimization

Václav Snášel, Lingping Kong, and Jeng-Shyang Pan

Abstract In many (large)-objective optimization problems, high-dimensional data are involved. Hence, the results of population convergence and population distance in decision space are high-dimensional geometrical objects which are difficult to analyze and interpret. A popular method Parallel Coordinates Plot scales well to high-dimensional data. However, the parallel coordinates plot is not as straightforward as the classic scatter plot to illustrate the information contained in a solution set. Metrics indicators measure the population which tries to solve this but they reduce the population optimization information into a single value. In this paper, we present a way of visualization of the distance evolution on population in multi-objective optimization of both fitness values and the decision space. We take advantage and modify Sammon projection, which is an algorithm that maps a high-dimensional space to a space of lower dimensionality, to project the population into a 2D space for observing the convergence of algorithms and the relation among populations and objectives.

1 Introduction

Due to the population-based nature, evolutionary algorithms (EAs) can approximate the desired solution of a multi-objective optimization problem (MOP) in a single run. Researchers have developed EAs to deal with MOPs under the tag of multi-objective evolutionary algorithms (MOEAs). Recent developments [26] on MOEAs

---

V. Snášel · L. Kong (✉)
Faculty of Electrical Engineering and Computer Science, VŠB -Technical University of Ostrava, Ostrava, Czech Republic
e-mail: konglingping2007@163.com

V. Snášel
e-mail: vaclav.snasel@vsb.cz

J.-S. Pan
College of Computer Science and Engineering, Shandong University of Science and Technology, Qingdao 266590, China

are mainly focused on decomposition-based method (such as MOEA/D [25] and PPS-MOEA/D [10]), indicator-based method (like AR-MOEA [21]), preference-based method (such as SPEA2 [29] and PESA-II [3]) and Hybrid MOEAs (like MSEA [22]). Preference-based MOEAs use the rank of the members of a population, which are determined by both the Pareto dominance and the preference information from the decision-maker. For example, AGE-MOEA [18](Adaptive Geometry Estimation-based MOEA), C-TAEA [15] (Two-Archive EAs) and CA-MOEA [12] (Clustering-based adaptive MOEA), these three very recent approaches share the same procedure of non-dominated sorting from NSGA-II [6] or NSGAIII [13].

Real-world optimization problems usually involve several conflicting objectives for which a trade-off must be found [8]. For such a MOP, the multiple conflicting objectives imply that no single solution is globally optimal unless a weight vector (encoding the priorities of different objectives) can be assigned to these objectives [1]. A basic MOP function can be expressed as in Eq. 1, where $F(\mathbf{x})$ is the vector of object functions values with $F(\mathbf{x}) = \{f_1(\mathbf{x}), \ldots, f_M(\mathbf{x})\}$ and the k-dimensional decision variable vector is $\mathbf{x} = \{x_1, \ldots, x_D\}$ in the decision space. In (1), $C(\mathbf{x}) \leq 0$ expresses the easiest constraints, e.g., the bounds on the decision variables. Specially, if an MOP has more than three objectives (i.e., $M > 3$), it is often known as a many-objective optimization problem (MaOP) nowadays.

$$\begin{aligned} optimize\ F(\mathbf{x}) &= min\{f_1(\mathbf{x}), \ldots, f_M(\mathbf{x})\} \\ subject\ to:\ \mathbf{x} &\in R^D\ |C(\mathbf{x}) \leq 0 \end{aligned} \qquad (1)$$

Although many proposed MOEAs have shown promising performance on MOPs with two or three objectives, their performance deteriorates rapidly as the number of objectives increases on MaOPs [17], mainly due to the phenomenon known as dominance resistance [19, 21]. To address this issue, many MaOPs have been proposed, such as AGE-MOEA [18], EIMEGO [24] and two_arch2 [23] on many-objective problems and WOF [27, 28] and LSMOF [11] for large objective problems. Two-dimensional and three-dimensional data can be easily displayed. In multi-objective optimization, a population with multi-objective values is not easy to use a classic scatter plot as a basic tool for viewing solution vectors. Parallel Coordinates Plot though scales well to high-dimensional data, however, the parallel coordinates plot is not as straightforward as the classic scatter plot to illustrate the information contained in a solution set. It could be difficult for people to comprehend the plot in a higher dimensional space. It is a need to design a tool that allows us to observe/perceive the quality of a population convergence, the shape and distribution of generational population, the relation between populations and objectives. In this paper, we present a way of visualization of the distance evolution on the generational population.

Overall, this paper is divided into four sections. Following the introduction, Sect. 2 gives the background of Sammon mapping. The details of instruction of plotting population are provided in Sect. 3 and with its results. Section 4 concludes the paper.

2 Background

The rapid development of population-based algorithms in handling many (large)-objective optimization problems requires viable methods of visualizing a many(high)-dimensional solution set. Metrics indicators measure the population which tries to solve this but they reduce the population involving information into a single value [14]. Parallel coordinates plot is popular in high-dimensional data presentation. While the population convergence and the quality information may not be easy to observe by parallel coordinates plot [2, 16]. As the low-dimensional graphical representation of the populations could be much more informative than just a single value, this paper presents a way to view the population convergence results. The modified Sammon mapping maps the generational population such that the distances between each iterative population will be preserved and observed.

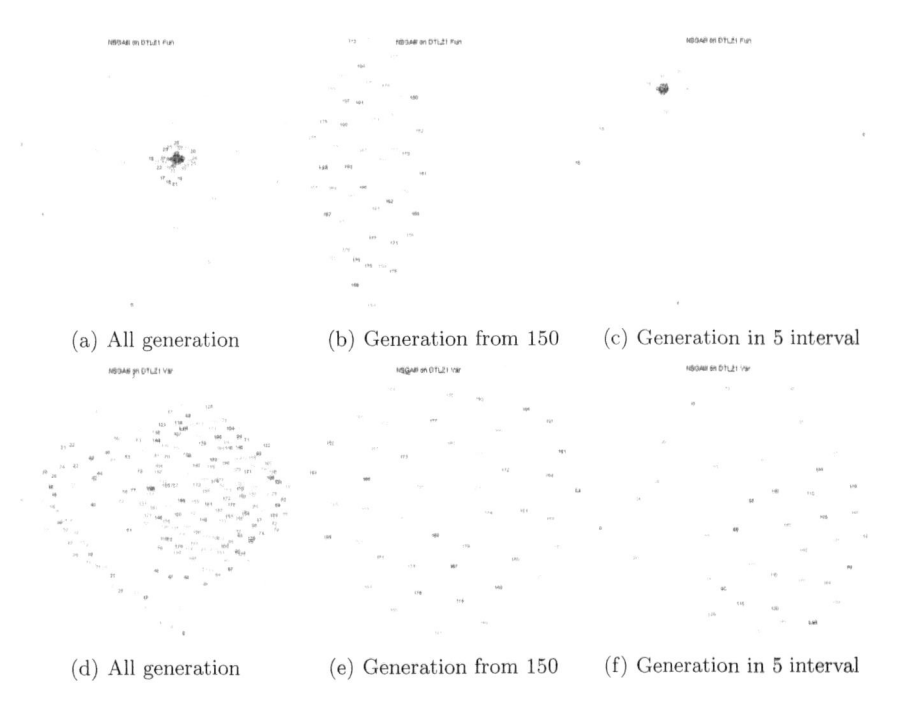

(a) All generation   (b) Generation from 150   (c) Generation in 5 interval

(d) All generation   (e) Generation from 150   (f) Generation in 5 interval

Fig. 1 NSGAIII on DTLZ1 for $M = 4$; **a, b, c** depict the population convergence on objective values (Fun), where each point is with $N \times M$ dimensions; **d, e, f** depict the population convergence on variable values (Var), where each point is with $N \times D$ dimensions. The number on each point is the generation number of a population. The first generation starts with 0 in RED, and the last generation is labeled as 'Last' in BLACK

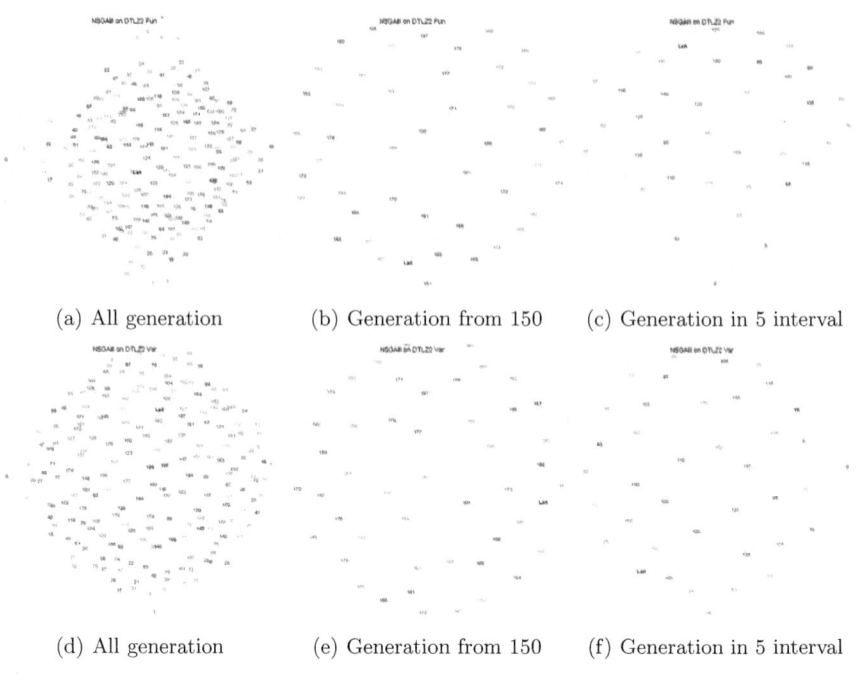

(a) All generation   (b) Generation from 150   (c) Generation in 5 interval

(d) All generation   (e) Generation from 150   (f) Generation in 5 interval

Fig. 2 NSGAIII on DTLZ2 for M = 4; **a, b, c** depict the population convergence on objective values (Fun), where each point is with $N \times M$ dimensions; **d, e, f** depict the population convergence on variable values (Var), where each point is with $N \times D$ dimensions. The number on each point is the generation number of a population. The first generation starts with 0 in RED, and the last generation is labeled as 'Last' in BLACK

2.1 Sammon Mapping

Sammon projection [20] is a feature extraction algorithm that maps a high-dimensional space to a space of lower dimensionality for pattern recognition and exploratory data analysis [4]. This tool is a simple nonlinear projection method by trying to preserve the structure of inter-point distances in (n)-dimensional space in the lower dimension projection.

Sammon mapping tries to preserve inter-pattern distances of N points in (n)-dimensional by minimizing an error criterion which following error function, termed as Sammon's error as Eq. 2.

$$E = \frac{1}{\sum_{i<j} d_{ij}^*} \sum_{i<j} \frac{(d_{ij}^* - d_{ij})^2}{d_{ij}^*} \qquad (2)$$

where the i and j present the two objects(points), the distance d_{ij}^* is the distance between them in the original space and d_{ij} denotes the projected distance in projection dimension. There are many ways to minimize the error, such as gradient descent,

(a) All generation   (b) Generation from 150   (c) Generation in 5 interval

(d) All generation   (e) Generation from 150   (f) Generation in 5 interval

Fig. 3 NSGAIII on DTLZ3 for M = 4; **a, b, c** depict the population convergence on objective values (Fun), where each point is with $N \times M$ dimensions; **d, e, f** depict the population convergence on variable values (Var), where each point is with $N \times D$ dimensions. The number on each point is the generation number of a population. The first generation starts with 0 in RED, and the last generation is labeled as 'Last' in BLACK

Genetic algorithms, simulated annealing or other heuristic approaches, which usually are iterative methods.

3 Population Projection by Sammon Mapping

Suppose the following $\mathcal{P} = \{\mathcal{P}_1, \mathcal{P}_2, \ldots, \mathcal{P}_{t_{max}}\}$ represents the t_{max} generations of population, and $\mathcal{P}_i = \{p_1, p_2, \ldots, p_N\}$ with N solutions where $i \in \{1, 2 \ldots, t_{max}\}$. The solution p_j is expressed in D-dimensional decision variable space as $p_j(\mathbf{x}) = \{x_1, x_2, \ldots, x_D\}$ in which $F(\mathbf{x}) = \{f_1(\mathbf{x}), f_2(\mathbf{x}), \ldots, f_M(\mathbf{x})\}$ where $j \in \{1, 2, \ldots, N\}$ and M is the objective number of a solution. Two population projections are displayed by Sammon mapping, the population with objective value and population with decision variables with two figures.

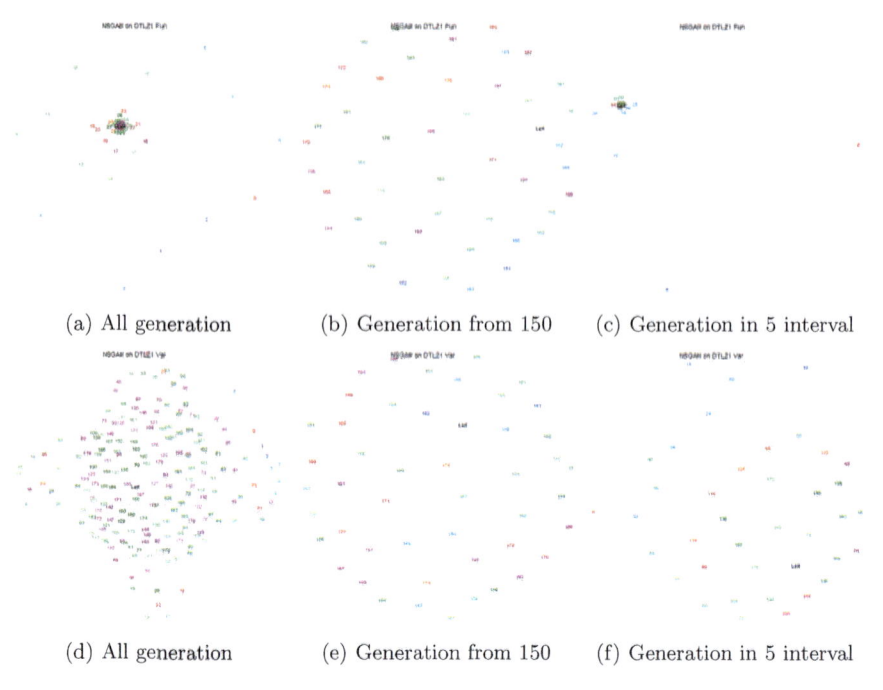

(a) All generation   (b) Generation from 150   (c) Generation in 5 interval

(d) All generation   (e) Generation from 150   (f) Generation in 5 interval

Fig. 4 NSGAIII on DTLZ1 for M = 5; **a, b, c** depict the population convergence on objective values (Fun), where each point is with $N \times M$ dimensions; **d, e, f** depict the population convergence on variable values (Var), where each point is with $N \times D$ dimensions. The number on each point is the generation number of a population. The first generation starts with 0 in RED, and the last generation is labeled as 'Last' in BLACK

3.1 Mapping Setting

A population with M objective value is displayed as a point in projection, each point is a $N \times M$-dimensional data in original space. There are t_{max} points in space which are t_{max} generations in the multi-objective optimization process. The second display is for the decision space of a population, in which a population is considered as a point, each point is a $N \times D$-dimensional data in original space. There are also t_{max} points in space.

The Sammon projection includes two steps. The first computes the projection and the second creates the image for the display. We modify the code from https://www.codeproject.com/Articles/43123/Sammon-Projection to show our population results. The projected space is $2D$ space, and the iterative number is set to 200 for all experiments. The dataset is generated by jMetal [9]. We choose NSGAIII [5] as the algorithm and three multi-objective problems DTLZ1, DTLZ2 and DTLZ3 [7] on four and five objective data. The algorithm runs once on each problem with 199 maximum iterations, which will be 199 points total in each projection. The population size is set 35, as in the NSGAIII algorithm, and it uses reference points to preserve

(a) All generation	(b) Generation from 150	(c) Generation in 5 interval
(d) All generation	(e) Generation from 150	(f) Generation in 5 interval

Fig. 5 NSGAIII on DTLZ2 for M = 5; **a, b, c** depict the population convergence on objective values (Fun), where each point is with $N \times M$ dimensions; **d, e, f** depict the population convergence on variable values (Var), where each point is with $N \times D$ dimensions. The number on each point is the generation number of a population. The first generation starts with 0 in RED, and the last generation is labeled as 'Last' in BLACK

the diversity of the solution set. Hence, we set the space division number small with $H1 = 4$ for $M = 4$ and $H1 = 3$ for $M = 5$, where $H1$ is the simplex-lattice design reference vectors on the outer boundaries [21] and M is the objective number.

3.2 Figure Illustration

The population with objective values as a $N \times M$-dimensional point is labeled with the name *NSGAIII on DTLZ Fun*, and the population with decision variable as a $N \times D$-dimensional point is with a name as *NSGAIII on DTLZ Var*. Each data from a problem is shown in three figures, (a) all generation, which depicts all t_{max} generation population with t_{max} points in figures; (b) generation from 150, which means the starting point (population) is from 150 generation to the end generation, and total points in the figure will be $t_{max} - 150$; (c) generation in 5 interval, which shows the points with every 5 generation population, and the total points in the figure will be $t_{max}/5$. For simply the observation, the first generation population

(a) All generation (b) Generation from 150 (c) Generation in 5 interval

(d) All generation (e) Generation from 150 (f) Generation in 5 interval

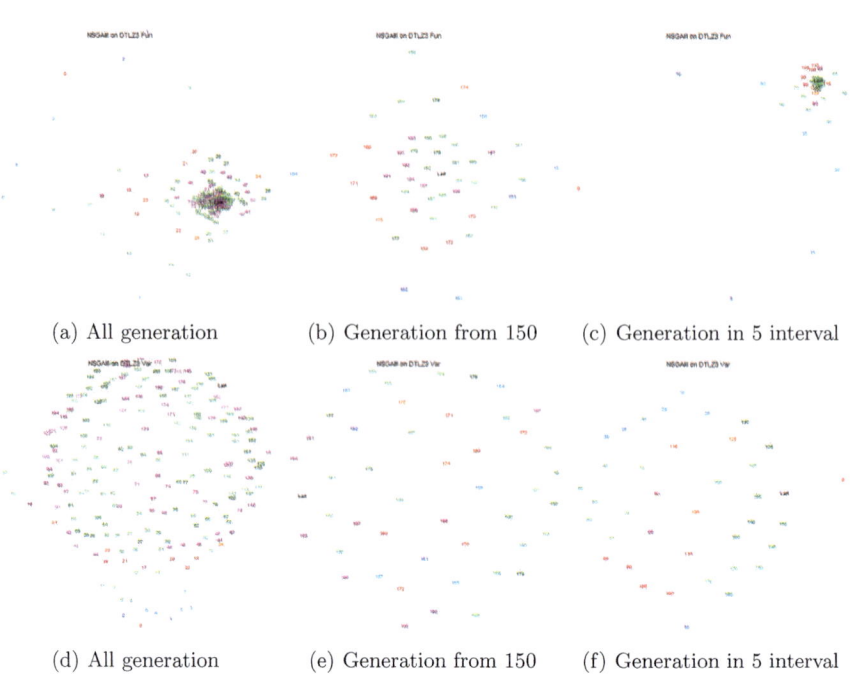

Fig. 6 NSGAIII on DTLZ3 for M = 5; **a, b, c** depict the population convergence on objective values (Fun), where each point is with $N \times M$ dimensions; **d, e, f** depict the population convergence on variable values (Var), where each point is with $N \times D$ dimensions. The number on each point is the generation number of a population. The first generation starts with 0 in RED, and the last generation is labeled as 'Last' in BLACK

point is depicted in red with 0 symbol. The last generation population is depicted in black with 'Last' symbol. Other population is labeled with its generation number $\{1, 2, \ldots, t_{max} - 1\}$,

From Figs. 1a, 2a, 3a, 4a, 5a, 6a, all population illustration, we can see that for the problem DTLZ1, DTLZ3, during the last stage of generations, the population are getting closer to a fixed point on objective values. The fast convergence can be partly shown as the early generations are far away from the last generations, and the post-stage generations are compact to a small cluster. The decision variables represented the decision space convergence partly which shows population moving as in Figs. 1d, 2d, 3d, 4d, 5d, 6d. As in the early stage, the unstable of population optimization shows that some points are scattered in large distances. While we still could see that some adjacent populations construct a convergence curve toward the final points, as in Fig. 4f, 10 ∼ 50 generations are mostly in the outer boundary, the 50 ∼ 100 generations mostly construct a second rounded circle inside of outer boundary. In the end, the 170 ∼ 190 generations are around with the 'Last' generations.

4 Conclusion and Future Work

The use for the population projection is visualization, which is useful for preliminary analysis in population convergence and population exploration tendency in decision space. Though a rough visualization of the population distribution can be partly obtained by Sammon projection, it could be used as the first step in convergence analysis. So it is possible to cross-verify the convergence speed and convergence trapping results. Especially, the exploration tendency in decision space can be an assistant measurement that predicts the next step location of the population and reduces the evaluation times. Of course, it cannot entirely replace quality metrics and prediction in assessing multi-objective population evolution. The clarity of some features by Sammon mapping can be affected by limited datasets and experiments; in this regard, a straightforward thought is to continue this exploration as our future work, to find a better visualization tool to display the generational population in the conflicting multi-objective problems so that people could see some meaningful patterns.

Acknowledgements This work was supported by the ESF in "Science without borders" project, $reg.nr.CZ.02.2.69/0.0/0.0/16\_027/0008463$ within the Operational Programme Research, Development and Education, and by the Ministry of Education, Youth and Sports of the Czech Republic in project "Metaheuristics Framework for Multi-objective Combinatorial Optimization Problems (META MO-COP)", reg.no.LTAIN19176.

References

1. Azzouz, R., Bechikh, S., Said, L.B.: Dynamic multi-objective optimization using evolutionary algorithms: a survey. In: Recent Advances in Evolutionary Multi-objective Optimization, pp. 31–70. Springer (2017)
2. Chen, C.M., Chen, Y.H., Lin, Y.H., Sun, H.M.: Eliminating rouge femtocells based on distance bounding protocol and geographic information. Expert Syst. Appl. **41**(2), 426–433 (2014)
3. Corne, D.W., Jerram, N.R., Knowles, J.D., Oates, M.J.: Pesa-ii: Region-based selection in evolutionary multiobjective optimization. In: Proceedings of the 3rd Annual Conference on Genetic and Evolutionary Computation. pp. 283–290. Morgan Kaufmann Publishers Inc (2001)
4. De Ridder, D., Duin, R.P.: Sammon's mapping using neural networks: a comparison. Pattern Recogn. Lett. **18**(11–13), 1307–1316 (1997)
5. Deb, K., Jain, H.: An evolutionary many-objective optimization algorithm using reference-point-based nondominated sorting approach, part i: solving problems with box constraints. IEEE Trans. Evol. Comput. **18**(4), 577–601 (2013)
6. Deb, K., Pratap, A., Agarwal, S., Meyarivan, T.: A fast and elitist multiobjective genetic algorithm: Nsga-ii. IEEE Trans. Evol. Comput. **6**(2), 182–197 (2002)
7. Deb, K., Thiele, L., Laumanns, M., Zitzler, E.: Scalable test problems for evolutionary multi-objective optimization. In: Evolutionary Multiobjective Optimization, pp. 105–145. Springer (2005)
8. Derrac, J., García, S., Molina, D., Herrera, F.: A practical tutorial on the use of nonparametric statistical tests as a methodology for comparing evolutionary and swarm intelligence algorithms. Swarm Evol. Comput. **1**(1), 3–18 (2011)

9. Durillo, J.J., Nebro, A.J., Alba, E.: The jmetal framework for multi-objective optimization: design and architecture. In: IEEE Congress on Evolutionary Computation, pp. 1–8. IEEE (2010)
10. Fan, Z., Li, W., Cai, X., Li, H., Wei, C., Zhang, Q., Deb, K., Goodman, E.: Push and pull search for solving constrained multi-objective optimization problems. Swarm Evol. Comput. **44**, 665–679 (2019)
11. He, C., Li, L., Tian, Y., Zhang, X., Cheng, R., Jin, Y., Yao, X.: Accelerating large-scale multi-objective optimization via problem reformulation. IEEE Trans. Evol. Comput. **23**(6), 949–961 (2019)
12. Hua, Y., Jin, Y., Hao, K.: A clustering-based adaptive evolutionary algorithm for multiobjective optimization with irregular pareto fronts. IEEE Ttrans. Cybern. **49**(7), 2758–2770 (2018)
13. Jain, H., Deb, K.: An evolutionary many-objective optimization algorithm using reference-point based nondominated sorting approach, part ii: handling constraints and extending to an adaptive approach. IEEE Trans. Evol. Comput. **18**(4), 602–622 (2013)
14. Kovács, A., Abonyi, J.: Vizualization of fuzzy clustering results by modified Sammon mapping. In: Proceedings of the 3rd International Symposium of Hungarian Researchers on Computational Intelligence, pp. 177–188 (2002)
15. Li, K., Chen, R., Fu, G., Yao, X.: Two-archive evolutionary algorithm for constrained multi-objective optimization. IEEE Trans. Evol. Comput. **23**(2), 303–315 (2018)
16. Li, M., Zhen, L., Yao, X.: How to read many-objective solution sets in parallel coordinates [educational forum]. IEEE Comput. Intell. Mag. **12**(4), 88–100 (2017)
17. Pan, J.S., Kong, L., Sung, T.W., Tsai, P.W., Snášel, V.: α-fraction first strategy for hierarchical model in wireless sensor networks. J. Internet Technol. **19**(6), 1717–1726 (2018)
18. Panichella, A.: An adaptive evolutionary algorithm based on non-Euclidean geometry for many-objective optimization. In: Proceedings of the Genetic and Evolutionary Computation Conference. pp. 595–603 (2019)
19. Purshouse, R.C., Fleming, P.J.: On the evolutionary optimization of many conflicting objectives. IEEE Trans. Evol. Comput. **11**(6), 770–784 (2007)
20. Sammon, J.W.: A nonlinear mapping for data structure analysis. IEEE Trans. Comput. **100**(5), 401–409 (1969)
21. Tian, Y., Cheng, R., Zhang, X., Cheng, F., Jin, Y.: An indicator-based multiobjective evolutionary algorithm with reference point adaptation for better versatility. IEEE Trans. Evol. Comput. **22**(4), 609–622 (2017)
22. Tian, Y., He, C., Cheng, R., Zhang, X.: A multistage evolutionary algorithm for better diversity preservation in multiobjective optimization. In: IEEE Transactions on Systems, Man, and Cybernetics: Systems (2019)
23. Wang, H., Jiao, L., Yao, X.: Two_arch2: An improved two-archive algorithm for many-objective optimization. IEEE Trans. Evol. Comput. **19**(4), 524–541 (2014)
24. Zhan, D., Cheng, Y., Liu, J.: Expected improvement matrix-based infill criteria for expensive multiobjective optimization. IEEE Trans. Evol. Comput. **21**(6), 956–975 (2017)
25. Zhang, Q., Li, H.: Moea/d: A multiobjective evolutionary algorithm based on decomposition. IEEE Trans. Evol. Comput. **11**(6), 712–731 (2007)
26. Zhou, A., Qu, B.Y., Li, H., Zhao, S.Z., Suganthan, P.N., Zhang, Q.: Multiobjective evolutionary algorithms: a survey of the state of the art. Swarm Evol. Comput. **1**(1), 32–49 (2011)
27. Zille, H.: Large-scale multi-objective optimisation: new approaches and a classification of the state-of-the-art. PhD Thesis (2019). http://dx.doi.org/10.25673/32063
28. Zille, H., Ishibuchi, H., Mostaghim, S., Nojima, Y.: A framework for large-scale multiobjective optimization based on problem transformation. IEEE Trans. Evol. Comput. **22**(2), 260–275 (2017)
29. Zitzler, E., Laumanns, M., Thiele, L.: Spea2: improving the strength pareto evolutionary algorithm. TIK-Rep. **103** (2001)

A Novel Binary QUasi-Affine TRansformation Evolution (QUATRE) Algorithm and Its Application for Feature Selection

Fei-Fei Liu, Shu-Chuan Chu, Xiaopeng Wang, and Jeng-Shyang Pan

Abstract QUasi-Affine TRansformation Evolution (QUATRE) algorithm is a new intelligent computing algorithm based on matrix iteration behavior. Binary QUATRE (BQUATRE) is a binary version that can be used to solve binary application problems. From continuous to binary arithmetic is a crucial part of the binary version. In order to convert the continuous type to the binary type, this paper proposes a simple and effective conversion method. After the benchmark function test, it proves that the improved binary QUATRE method has strong competitiveness. Finally, the feature selection problem can be successfully solved in the UCI data set, and a higher classification accuracy can be obtained with a smaller number of features.

1 Introduction

As a new type of meta-heuristic algorithm, the swarm intelligence algorithm can effectively solve most parameter optimization problems. However, the research in this area is still in its infancy compared with other mature algorithms, and there are still many problems that need further research. The research lacks theoretical analysis and basic support of mathematics, and does not have universal significance.

F.-F. Liu · S.-C. Chu · X. Wang · J.-S. Pan (✉)
College of Computer Science and Engineering, Shandong University of Science and Technology, Qingdao 266590, China
e-mail: jspan@cc.kuas.edu.tw

F.-F. Liu
e-mail: 13210318436@163.com

S.-C. Chu
e-mail: scchu0803@gmail.com

X. Wang
e-mail: 931087291@qq.com

S.-C. Chu
College of Science and Engineering, Flinders University, Sturt Rd, Bedford Park, Adelaide, SA 5042, Australia

With the continuous development of swarm intelligence algorithms, researchers in this area continue to try to apply swarm intelligence algorithms to various fields to solve various problems. So far, the results obtained are satisfactory. According to different research results, the swarm intelligence algorithm performs well in discrete and continuous solution spaces, and they are all objects worthy of study. In addition, they also have outstanding performance in combinatorial optimization problems. The development of intelligent computing today has mainly formed three major parts, namely Fuzzy, Neural Network, and Evolutionary Computation. Cluster intelligent algorithms include Particle Swarm Optimization (PSO), QUasi-Affine TRansformation Evolution [1–3], Gray Wolf Optimizer (GWO) [4], and Pigeon Inspired Optimization (PIO) [5–7].

Since entering the information age, all walks of life have gradually begun to popularize informatization, which has led to an explosion in data volume, resulting in more and more high-latitude data sets, which makes classification very difficult. However, not all features are suitable for classification. In addition to related features, the data set also includes irrelevant and redundant features, which even leads to low classification accuracy and a dimensionality disaster. In the classification process, more and more dimensions will bring other noises to the entire data set, which will damage the classification ability of the data set. Therefore, it is very important to select a more suitable feature subset from the original features to obtain all features similar or even better classification results.

Feature selection is an important task in data mining and machine learning, which can reduce the dimensionality of data and improve the performance of learning algorithms [8–12]. Feature selection is the process of selecting some representative features from a set of initial features to reduce the dimensionality of the feature space. It is one of the key issues in data mining and machine learning. For data mining and machine learning, a good learning sample is key to training a classifier. Whether the sample contains irrelevant or redundant features directly affects the performance of the classifier. The purpose of feature selection is to find the minimum feature subset necessary to solve the problem. By removing irrelevant and redundant features from the original features to reduce the dimensionality of the data, accelerate the learning process, simplify the learning model, and improve the performance of the algorithm. An effective feature selection method is key to finding an optimal feature subset.

2 Related Work

QUasi-Affine TRansformation Evolution algorithm creatively combines PSO and DE algorithms to create a new type of algorithm, which has higher convergence accuracy and faster speed.

2.1 QUasi-Affine TRansformation Evolution

Meng and Pan first proposed the QUATRE algorithm in 2016 [13]. The evolutionary algorithm is called the affine form of the algorithm, and in this paper, a standard QUATRE is designed [14–17]. The structure of the QUATRE algorithm contains the following Eq. 1:

$$X \Leftarrow \bar{M} \otimes X + M \otimes B, \qquad (1)$$

where M is the evolution matrix, X is the individual population matrix, $X = (X_1, X_2, ..., X_i, ..., X_{No})^T$, $i \in [1, No]$, B is the mutant matrix, No is the size of the population, X is the target matrix, and $\otimes$ represents a bitwise multiplication operation of matrix elements.

F is the control factor of the different matrix with a range of (0, 1] in Table 1. The outcome of $(\mathbf{X}_{r,i} - \mathbf{X}_{r,j})$ ($i, j \in [1, 2, ..., No]$) is deemed to a different matrix. Generally, $F = 0.7$ is superior candidate.

In regards to the coevolutionary matrix M, the M matrix in the QUATRE algorithm is transformed from the initial matrix M_{init}. In this article, $No = D$ is taken as an example, D is the dimensionality of the objective function, matrix M_{init} is the lower triangular matrix with element value 1, and the below equation gives the transformation process from matrix M_{init} to matrix M:

$$\mathbf{M}_{init} = \begin{bmatrix} 1 & & & & & \\ 1 & 1 & & & & \\ 1 & 1 & 1 & & & \\ 1 & 1 & 1 & 1 & & \\ & & \cdots & & & \\ 1 & 1 & 1 & 1 & \cdots & 1 \end{bmatrix} \sim \begin{bmatrix} 1 & 1 & 1 & 1 & & \\ 1 & 1 & 1 & & & \\ & & \cdots & & & \\ & & & 1 & 1 & \\ 1 & & & & & \\ 1 & 1 & 1 & 1 & \cdots & 1 \end{bmatrix} = \mathbf{M}.$$

Table 1 The six mutation matrix **B** calculation

Number	QUATRE/B	Equation
1	QUATRE/target/1	$\mathbf{B} = \mathbf{X} + F * (\mathbf{X}_{r1,g} - \mathbf{X}_{r2,g})$
2	QUATRE/rand/1	$\mathbf{B} = \mathbf{X}_{r1,g} + F * (\mathbf{X}_{r2,g} - \mathbf{X}_{r3,g})$
3	QUATRE/best/1	$\mathbf{B} = \mathbf{X}_{gbest,g} + F * (\mathbf{X}_{r2,g} - \mathbf{X}_{r3,g})$
4	QUATRE/target/2	$\mathbf{B} = \mathbf{X} + F * (\mathbf{X}_{r1,g} - \mathbf{X}_{r2,g}) + F * (\mathbf{X}_{r3,g} - \mathbf{X}_{r4,g})$
5	QUATRE/rand/2	$\mathbf{B} = \mathbf{X}_{r1,g} + F * (\mathbf{X}_{r2,g} - \mathbf{X}_{r3,g}) + F * (\mathbf{X}_{r4,g} - \mathbf{X}_{r5,g})$
6	QUATRE/best/2	$\mathbf{B} = \mathbf{X}_{gbest,g} + F * (\mathbf{X}_{r1,g} - \mathbf{X}_{r2,g}) + F * (\mathbf{X}_{r3,g} - \mathbf{X}_{r4,g})$

After a two-step operation, the matrix M_{init} is transformed into a matrix M. The transformation is achieved by two consecutive operations: 1. all the row vectors in the matrix M_{init} are all rearranged randomly; 2. the first step of the matrix row vectors randomly all arranged.

2.2 Binary QUATRE Algorithm Work

The binary version of the QUATRE algorithm is a very effective method to solve the feature selection problem [18–23]. In the continuous QUATRE algorithm, since the value in the matrix changes continuously, the change of the matrix is linear. But in the binary version, since the values in the matrix can only be 0 and 1, we thought of a simple and effective way to solve this problem. Generally speaking, a transfer function is used to solve the problem of converting continuous values into numerical methods with only 0 and 1.

Firstly, the numerical change of our initialization matrix is limited to a continuous range. Secondly, when passing the evaluation criteria, the matrix must be converted into a binary version before the evaluation can be done. Therefore, this paper designs a simple conversion criterion. For the matrix to be judged, first normalize the matrix, and then use the random number generator. If any value in the matrix is greater than a random number, then take 1 otherwise take 0. Then, use the changed matrix fitness value evaluation. Then update the global optimal value and the mark of the global optimal position. Finally, return to the unchanged matrix and continue to iterate according to the continuous algorithm. The method is used as shown in Algorithm 1.

3 Feature Selection

In the process of classification, too high dimensions will cause the problem of dimensionality disaster. In the process of dimensionality reduction, high dimensions may retain some unrelated dimensions, and low dimensions may also remove some of the more important features. Therefore, feature selection has become a more important issue. In the judgment process of the classifier, feature selection will have a very important impact on accuracy and complexity. The classification accuracy of new instances after feature selection is an important evaluation criterion for feature subsets. Feature selection includes three methods: filter, wrap, and embed.

Algorithm 1 Binary version conversion method

Input:
1: Matrix M;
2: Historical best value $gBestvalue$;
3: Evaluation function $F()$;
4: Search range;

Output:
5: The label index of the matrix $index$;
6: Matrix fitness value $value$;
7: //Enter method;
8: Normalize matrix M to get matrix $M1$;
9: **for** $i = 1$ to $size(M1, 1)$ **do**
10:    **for** $j = 1$ to $size(M1, 2)$ **do**
11:       $r1 = random\ number$;
12:       **if then**$M1(i, j) > r1$
13:          $M2 = 1$;
14:       **else**
15:          $M2 = 0$;
16:       **end if**
17:    **end for**
18: **end for**
19: Evaluation function $F()$ to evaluate matrix $M2$;
20: Update the optimal value by $gBestvalue$ and select the label of the optimal location $label$;

3.1 Feature Selection Data Set

The source of these data sets is UCI machine learning [24]. Each category has different instances and the size of the database. The specific information can be seen in Table 2.

3.2 KNN and K-Fold Validation

The K-Nearest Neighbor (KNN) classification algorithm is one of the simplest methods in data mining classification technology where the meaning of k-nearest neighbors means that each sample can be represented by the k-nearest neighbors. KNN is a classification algorithm, which inputs instance-based learning. If most of the k-most similar samples in the feature space (i.e., the nearest neighbor in the feature space) of a sample belong to a certain category, the sample also belongs to this category, where k is usually an integer not greater than 20. In the KNN algorithm, the selected neighbors are all objects that have been correctly classified. This method determines the category to which the samples to be classified belong based on the category of the nearest sample or samples in the decision of classification.

Table 2 The details of the simulation data sets

Data Sets	Instances	Number of classes (k)	Number of features (d)	Size of classes
Wine	178	3	13	59,71,48
WDBC	569	2	30	357212
Glass	214	6	9	29,76,70,17,13,9
WBC	683	2	9	444239
Vowel	871	6	3	72,89,172, 151,207,180
Vehicle	846	3	18	199217218212
Thyroid	215	3	5	150,35,30
Sonar	208	2	60	97111
Seeds	210	3	7	70,70,70
Robotnavigation	5456	4	25	82620972205329
Jain	373	2	2	276,97
Ionosphere	351	2	34	126225
Heartstatlog	270	2	13	150120
Haberman	306	2	3	225,81
Ecoli	336	8	8	143,77,2,2, 259,20,5,52
Diabetes	768	2	8	268500
CMC	1473	3	9	629333511
Cancer	683	2	9	444239
Bupa	345	2	6	145200
Blood	748	2	4	570178
Balancescale	625	3	4	49288288

$$dis(x, y) = \left(\sum_{k=1}^{n} |x_k - y_k|^t \right)^{\frac{1}{t}}. \qquad (2)$$

In KNN, the distance between objects is calculated as a non-similarity index between each object to avoid the matching problem between objects. Here, the distance generally uses Euclidean distance or Manhattan distance as shown in Eq. 2.

Where x_k, y_k are two instances in the collection, and t is a variable constant. When $t = 1$, dis represents the Manhattan distance, and when $t = 2$, dis refers to the Euclidean distance.

Cross-Validation is sometimes called circular estimation; it is a practical method for statistically cutting data samples into smaller subsets. Therefore, the analysis can be performed on a subset first, and the other subsets are used for subsequent confirmation and verification of the analysis. The initial subset is called the training set. The other subsets are called validation sets or test sets. As an evolution of Holdout

Method, K-fold Cross Validation (K-CV) divides the original data into k groups (usually divided equally), and each subset of data is used as a validation set, and the remaining k-1 group of subset data is used as a training set k models, and the average number of classification accuracy of the final verification sets of these k models is used as the performance index of the classifier under this K-CV. k is generally greater than or equal to 2, and it is generally taken from 3 in actual operation. If the amount of data in the original data set is small, k will take the value 2. The K-CV experiment needs to build a total of k models and calculate the average recognition rate of k test sets. In practice, k must be large enough to allow enough training samples in each round. Generally speaking, $k = 10$ (as an empirical parameter) is quite sufficient.

3.3 Evaluation Standard Fitness Value Function

$$Fitness\ Function = k_1 \times Error(KNN) + k_2 \times \frac{|SE|}{|AL|}. \qquad (3)$$

In general classification prediction, the accuracy or error rate of the classification will be selected as the most important evaluation standard. In the process of feature selection, what should be considered is the accuracy of the classification and the number of subsets after feature selection. Equation 3 is the evaluation standard used in this article.

$$Error(KNN) = \frac{Wrong_{num}}{Correct_{num} + Wrong_{num}} \qquad (4)$$

where k_1 usually takes 0.99, and k_2 takes 0.01 as known in [25]. SE is the subset feature after feature selection. AL is the characteristic number of the data set. $Error_{(KNN)}$ represents the classification error after K-fold cross-validation test, which is expressed in Eq. 4. The correct and wrong classifications after each classification are, respectively, indicated by Correct$_{num}$ and Wrong$_{num}$.

4 Experiment Analysis

In order to evaluate the performance of the BQUATRE algorithm, this paper selects 23 test functions for evaluation [4], and compares the BGWO algorithm. The number of iterations of the two algorithms is set to 500, the population size is set to 20, and each test function is run 30 times and the average value is taken. The experimental results are shown in Table 3.

It can be seen from the algorithm simulation results that BQUATRE performs well in some test functions, especially the performance in unimodal and multi-modal

Table 3 Comparison between BQUATRE algorithm and BGWO algorithm under test functions

Test Function	BQUATRE			BGWO		
	Mean	Std	Min	Mean	Std	Min
F1	0	0	0	3.72	1.4852	1
F2	0	0	0	4.14	1.6037	1
F3	0	0	0	146.64	103.3889	2
F4	1	0	1	1	0	1
F5	24.82	53.316	0	0	0	0
F6	7.5	0	7.5	15.18	3.4133	7.5
F7	6.0×10^{-4}	7.0×10^{-4}	0	41.4402	22.2694	0.0003
F8	−25.2441	0	−25.2441	−25.2441	0	−25.2441
F9	0	0	0	3.64	1.3056	0
F10	0	0	0	1.3351	0.2498	0.7171
F11	0	0	0	0.1807	0.0762	0.0477
F12	1.669	0	1.669	2.269	0.2724	1.669
F13	0	0	0	0	0	0
F14	12.6705	0	12.6705	12.6705	0	12.6705
F15	0.1484	0	0.1484	0.1484	0	0.1484
F16	0	0	0	0	0	0
F17	27.7029	0	27.7029	27.7029	0	27.7029
F18	600	0	600	600	0	600
F19	−0.3348	0	−0.3348	−0.3341	0.0049	−0.3348
F20	−0.1657	0	−0.1657	−0.1453	0.0429	−0.1657
F21	−5.0552	0	−5.0552	−5.0552	0	−5.0552
F22	−5.0877	0	−5.0877	−5.0877	0	−5.0877
F23	−5.1285	0	−5.1285	−5.1285	0	−5.1285

functions, which can reach its theoretical optimal value. It can be seen in Table 4. The classification accuracy of the binary method proposed in this paper is worse than that of BGWO in only one database of Jain.

In addition, from the analysis of fitness function value, it can be seen that the BQUATRE algorithm proposed in this paper has strong competitiveness.

It can be seen from the algorithm simulation results that BQUATRE performs well in all test functions, especially the performance in fixed-dimensional functions, which can reach its theoretical optimal value.

Table 4 Comparison between BQUATRE algorithm and BGWO algorithm under data sets

Data Sets	BQUATRE			BGWO		
	Mean	Std	Min	Mean	Std	Min
	Accuracy	Number	Fitness value	Accuracy	Number	Fitness value
Wine	0.9844	6.6	0.0242	0.9560	4.6	0.0402
WDBC	0.9800	13	0.0327	0.9749	10.2	0.0407
WBC	0.4431	3.2	0.5525	0.3694	4.2	0.5468
Vowel	0.9762	3.6	0.0357	0.9753	3.2	0.0426
Vehicle	0.8759	3	0.1673	0.8644	3	0.1670
Thyroid	0.7598	10.6	0.2432	0.7365	8.2	0.2576
Sonar	0.9533	2.2	0.0631	0.9333	2	0.0663
Seeds	0.8888	30.4	0.1201	0.8310	27.2	0.1301
Robotnavigation	0.9490	3.6	0.0645	0.8616	2	0.0741
Jain	0.9114	8.4	0.0620	0.9354	4.4	0.1017
Ionosphere	1.0000	2	0.0500	1.0000	2	0.0500
Heartstatlog	0.9873	1	0.0211	0.9905	1	0.0246
Haberman	1.0000	12	0.0049	1.0000	4.6	0.0171
Glass	0.8828	7	0.1247	0.8528	5.2	0.1384
Ecoli	0.7966	1	0.2087	0.7605	1.2	0.2099
Diabetes	0.8648	5.8	0.1637	0.8161	4.6	0.1699
CMC	0.7856	5.2	0.2267	0.7691	5.2	0.2362
Cancer	0.5630	3.4	0.4307	0.5395	3.6	0.4341
Bupa	0.9712	4.2	0.0417	0.9631	3.8	0.0507
Blood	0.6861	3.8	0.3122	0.6427	3.2	0.3299
Balancescale	0.7709	0.8	0.2248	0.7593	0.2	0.2248

5 Conclusions

This paper proposes a method to change the continuous QUATRE algorithm into the binary version algorithm. The method of this change is simple to implement and performs better in 23 test functions compared with the BGWO algorithm. Subsequently, the improved BQUATRE algorithm in this paper is applied to feature selection. From the results of simulation experiments, the classification accuracy of the BQUATRE algorithm is higher, and the number of selected features is more in line with needs.

References

1. Liu, N., Pan, J.-S., Liao, X., Chen, G.: A multi-population quasi-affine transformation evolution algorithm for global optimization. In: International Conference on Genetic and Evolutionary Computing, pp 19–28. Springer (2018)
2. Chu, S.-C., Chen, Y., Meng, F., Yang, C., Pan, J.-S., Meng, Z.: Internal search of the evolution matrix in quasi-affine transformation evolution (quatre) algorithm. J. Intell. Fuzzy Syst. (Preprint), 1–12 (2020)
3. Meng, Z., Chen, Y., Li, X., Yang, C., Zhong, Y.: Enhancing quasi-affine transformation evolution (quatre) with adaptation scheme on numerical optimization. Knowl.-Based Syst. 105908 (2020)
4. Pei, H., Pan, J.-S., Chu, S.-C., QingWei, C., Tao, L., ZhongCui, L.: New hybrid algorithms for prediction of daily load of power network. Appl. Sci. **9**(21), 4514 (2019)
5. Duan, H., Qiao, P.: Pigeon-inspired optimization: a new swarm intelligence optimizer for air robot path planning. Int. J. Intell. Comput. Cybern. **7**(1), 24–37 (2014)
6. Tian, A.-Q., Chu, S.-C., Pan, J.-S., Cui, H., Weimin, Z.: A compact pigeon-inspired optimization for maximum short-term generation mode in cascade hydroelectric power station. Sustainability **12**(3), 767 (2020)
7. Tian, A.-Q., Chu, S.-C., Pan, J.-S., Yongquan, L.: A novel pigeon-inspired optimization based mppt technique for pv systems. Processes **8**(3), 356 (2020)
8. Mao, Y., Zhou, X.B., Xia, Z., Yin, Z., Sun, Y.X.: Survey for study of feature selection algorithms. Moshi Shibie yu Rengong Zhineng/Pattern Recognit. Artif. Intell. **20**(2), 211–218 (2007)
9. Aksu, D., Üstebay, S., Aydin, M.A., Atmaca, T.: Intrusion detection with comparative analysis of supervised learning techniques and fisher score feature selection algorithm. In: International Symposium on Computer and Information Sciences, pp. 141–149. Springer, Cham (2018)
10. Zhang, Y., Gong, D.W., Sun, X.Y., Guo, Y.N.: A PSO-based multi-objective multi-label feature selection method in classification. Sci. Rep. **7**(1), 1–12 (2017)
11. Wang, X., Chen, R.-C., Yan, F.: High-dimensional data clustering using K-means subspace feature selection. J. Netw. Intell. **4**(3), 80–87 (2019). August
12. Xiao, L.: Clustering research based on feature selection in the behavior analysis of MOOC users. J. Inf. Hiding Multimedia Signal Process. **10**(1), 147–155 (2019). January
13. Meng, Z., Pan, J.-S., Huarong, X.: Quasi-affine transformation evolutionary (quatre) algorithm: a cooperative swarm based algorithm for global optimization. Knowl.-Based Syst. **109**, 104–121 (2016)
14. Meng, Z., Pan, J.-S.: Quasi-affine transformation evolution with external archive (quatre-ear): an enhanced structure for differential evolution. Knowl.-Based Syst. **155**, 35–53 (2018)
15. Liu, N., Pan, J.-S., Xue, J.Y.: An orthogonal quasi-affine transformation evolution (o-quatre). In: Advances in Intelligent Information Hiding and Multimedia Signal Processing: Proceedings of the 15th International Conference on IIH-MSP in Conjunction with the 12th International Conference on FITAT, July 18–20, Jilin, China, Vols. 2, 157, pp 57–66. Springer (2019)
16. Pan, J.-S., Meng, Z., Huarong, X., Li, X.: Quasi-affine transformation evolution (quatre) algorithm: a new simple and accurate structure for global optimization. In: International Conference on Industrial Engineering and Other Applications of Applied Intelligent Systems, pp. 657–667. Springer, Berlin (2016)
17. Liu, N., Pan, J.-S., Wang, J., Nguyen, T.-T.: An adaptation multi-group quasi-affine transformation evolutionary algorithm for global optimization and its application in node localization in wireless sensor networks. Sensors **19**(19), 4112 (2019)
18. Kou, X., Feng, J.: Matching ontologies through compact monarch butterfly algorithm. J. Netw. Intell. **5**(4), 191–197 (2020). November
19. Chu, S.-C., Huang, H.-C., Roddick, J.F., Pan, J.-S.: Overview of algorithms for swarm intelligence. ICCCI **1**, 28–41 (2011)
20. Pan, J.-S., Wang, X., Chu, S.-C., Nguyen, T.-T.: A multi-group grasshopper optimisation algorithm for application in capacitated vehicle routing problem. Data Sci. Pattern Recognit. **4**(1), 41–56 (2020)

21. Xue, X., Yang, H., Zhang, J.: Using population-based incremental learning algorithm for matching class diagrams. Data Sci. Pattern Recognit. **3**(1), 1–8 (2019)
22. Cai, D.: A new evolutionary algorithm based on uniform and contraction for many-objective optimization. J. Netw. Intell. **2**(1), 171–185 (2017). Feb
23. Kuang, F.-J., Zhang, S.-Y.: A novel network intrusion detection based on support vector machine and tent chaos artificial bee colony algorithm. J. Netw. Intell. **2**(2), 195–204 (2017). May
24. Dua, D., Graff, C.: UCI Machine Learning Repository. University of California, Irvine, School of Information and Computer Sciences (2017). http://archive.ics.uci.edu/ml
25. Emary, E., Zawbaa, H.M., Hassanien, A.E.: Binary grey wolf optimization approaches for feature selection. Neurocomputing **172**, 371–381 (2016)

Networks and Security

On the Security of a Lightweight Three-Factor-Based User Authentication Protocol for Wireless Sensor Networks

Shuangshuang Liu, Zhiyuan Lee, Lili Chen, Tsu-Yang Wu, and Chien-Ming Chen

Abstract In 2020, Yu et al. proposed a secure, lightweight three-factors user authentication protocol named SLUA-WSA for wireless sensor networks. SLUA-WSA can effectively prevent security threats, ensure anonymity, untraceability, and mutual authentication. However, in this paper, we demonstrate that Yu et al.'s protocol is still vulnerable to known session-specific temporary information attacks.

1 Introduction

Internet of things (IoT) [9, 12] has received increasing attention recently. With the help of wireless sensor networks (WSN), RFID technologies [13, 18], laser scanners, GPS, etc., IoT can real-time acquire any object that needs to be connected or monitored. Each object in an IoT environment can be sensed and interconnected via the Internet within a dynamic network [8, 23, 24].

WSN [6, 19] are an important part of IoT. A WSN is a network made up of a large number of sensor nodes that are randomly deployed. These sensor nodes work together to perceive environmental information and communicate it to a gateway node via a self-organizing network. WSNs are widely used in environmental monitoring, underwater monitoring, medical care, battlefield situational awareness, etc., due to their self-organization and reliability characteristics.

S. Liu · Z. Lee · L. Chen · T.-Y. Wu · C.-M. Chen (✉)
College of Computer and Engineering, Shandong University of Science and Technology, Qingdao 266590, China
e-mail: chienmingchen@ieee.org

S. Liu
e-mail: ShuangLiu0309@163.com

Z. Lee
e-mail: jlizhiyuan@163.com

WSN can provide us a better life experience. However, all sensitive information transmitted via public channels may be easily eavesdropped on by attackers; for these reasons, secure and efficient authenticated key exchange (AKE) protocols are necessary [3, 5, 7, 17]. AKE protocols are required for establishing secure wireless communications over a public and insecure channel. However, various AKE protocols have been proven insecure [1, 2, 10, 15, 20, 22]. In 2017, Wang et al. [21] proposed an elliptic curves cryptography (ECC)-based three-factor AKE protocol for WSN. Unfortunately, this protocol [21] is vulnerable to internal attacks because the random nonces of legitimate users are stored in the gateway node's database, and internal users can access and modify it, resulting in a user login failure. In 2019, aiming at the defects of Wang et al.'s protocol [21], Lu et al. [14] proposed another three-factor AKE protocol for WSN based on ECC. However, Mo and Chen [15] proved that their protocol [14] still cannot resist a known temporary session key attack; thus, they proposed another ECC-based AKE protocol for WSN. They claimed that their protocol could block any attack and provide user anonymity, untraceability, and authentication. Unfortunately, Yu et al. [26] demonstrated that their protocol [15] cannot resist camouflage, replay, and session key exposure attacks. Yu et al. [26] then proposed a new secure and lightweight three-factor authentication protocol, named SLUA-WSN, for WSN. SLUA-WSN uses fuzzy extractor technologies to improve security. Compared with the existing authentication protocols, SLUA-WSN can prevent various kinds of attacks, including sensor node capture, replay, privilege insider, and camouflage attack, and ensure secure untraceability, user anonymity, and mutual authentication.

In this paper, we further pointed out that Yu et al.'s protocol [26] is still vulnerable to known session-specific temporary information attacks. In a known session-specific temporary information attack, if a random nonce is disclosed to an adversary, this adversary can obtain the parameters needed to calculate the session key according to the random nonce. Finally, the adversary can compute the session key. In fact, various AKE protocols are also vulnerable to known session-specific temporary information attacks. In 2020, Moghadam et al. [16] proposed an efficient authentication and key agreement scheme in WSN. However, Lee et al. [11] discovered that Moghadam et al.'s scheme cannot prevent insider and known session-specific temporary information attacks. Very recently, a certificateless-AKA privacy preserving authentication scheme is proposed [4]. However, Yahuza et al. [25] also proved that if the random number in the protocol is known, the session key can still be calculated.

The remainder of the paper is organized as follows. Section 2 briefly reviews Yu et al.'s Protocol. The detailed steps can refer to their paper [26]. In Sect. 3, we demonstrate that Yu et al.'s protocol is vulnerable to temporary information exposure attacks. Finally, Sect. 4 concludes.

Table 1 Notations used in this paper

Notations	Descriptions
U_i	User i
GWN	Gateway node
S_j	Sensor node j
ID_i	U_i's identity
PW_i	U_i's password
R_u, R_g, R_s	Random numbers of user, gateway and sensors, respectively
SID_j	S_j's identity
K_{GWN}	Master key of GWN
X_{pub}	Public key of GWN
E/E_p	Elliptic curve E defined on the finite field F_p with order p
G	A group for an elliptic curve
P	The generator of G
E_k/D_k	Symmetric key encryption/decryption with key k
SK	Session key
T_i	Timestamp
BIO	Biometric of U_i
$h(.)$	Hash function
$\oplus$	XOR operation
$\parallel$	Concatenation operation

2 Review of Yu et al.'s Protocol

Here, we briefly review Yu et al.'s protocol. This protocol includes the following three phases, the pre-deployment process, the registration phase, and the login and authentication phase. Table 1 listed the notations used in this paper.

2.1 Pre-deployment Phase

1. GWN selects a unique identity SID_j for sensor S_j and computes $X_j = h(SID_j \parallel K_{GWN})$. Then, GWN sends $\{SID_j, X_j\}$ to the S_j through a secure channel.
2. Upon receiving the messages from GWN, S_j stores them in its memory.

2.2 User Registration Phase

Assume that a user U_i desires to register to GWN, the following steps are performed.

1. U_i computes $Gen(BIO) = (R_i, P_i)$ and $MPW_i = h(PW_i||R_i)$ and sends ID_i, MPW_i to GWN over a secure channel.
2. While receiving ID_i, MPW_i from U_i, GWN generates a random nonce r_g and calculates $MID_i = h(ID_i||h(K_{GWN}||r_g))$, $X_i = h(MID_i||r_g||K_{GWN})$, $Q_i = h(MID_i||MPW_i) \oplus X_i$ and $W_i = h(MPW_i||X_i)$ and then stores r_g in its secure database. After that, GWN stores $\{Q_i, W_i, MID_i\}$ in a smart card and issues this smart card to U_i.

2.3 Login and Authentication Phase

Here we show the authentication process of SLUA-WSN, and the detailed steps are as follows.

1. U_i inserts his smart card into the terminal and inputs ID_i and PW_i. Then, the U_i imprints BIO_i and computes R_i =Rep(BIO_i, P_i), $MPW_i = h(PW_i||R_i)$, $X_i = h(MID_i||MPW_i) \oplus Q_i$, and $W_i^* = h(MPW_i||X_i)$, and then checks $W_i^* \stackrel{?}{=} W_i$. If the condition is valid, the U_i generates a random nonce R_u and a timestamp T_1. The U_i computes $M_1 = X_i \oplus R_u$, $CID_i = (ID_i||SID_j) \oplus h(MID_i||R_u||X_i)$, and $M_{UG} = h(ID_i||R_u||X_i||T_1)$, and sends $\{M_1, MID_i, CID_i, M_{UG}, T_1\}$ to the GWN over an insecure channel.
2. GWN checks the validity of T_1 and calculates $X_i = h(MID_i||r_g||K_{GWN})$, $R_u = M_1 \oplus X_i$, $(ID_i||SID_j) = CID_i \oplus h(MID_i||R_u||X_i)$ and $M_{UG}^* = h(ID_i||R_u||X_i||T_1)$ and then, checks $M_{UG}^* \stackrel{?}{=} M_{UG}$. If the condition is correct, GWN calculates $M_2 = (R_u||R_g) \oplus h(SID_j||X_j||T_2)$ and $M_{GS} = h(MID_i||SID_j||R_u||R_g||X_j||T_2)$, and sends $\{M_2, MID_i, M_{GS}, T_2\}$ to S_j.
3. S_j checks the validity of T_2 and computes $(R_u||R_g) = M_2 \oplus h(SID_j||X_j||T_2)$ and $M_{GS}^* = h(MID_i||SID_j||R_u||R_g||X_j||T_2)$ and checks $M_{GS}^* \stackrel{?}{=} M_{GS}$. If it is valid, S_j generates a random nonce R_s and timestamp T_3 and calculates $M_3 = R_s \oplus h(R_u||SID_j||X_j||T_3)$, $M_SG = h(R_s||R_g||SID_j||X_j||T_3)$, SK $= h(R_u||R_s)$, and $M_SU = h(SK||R_s||R_u||SID_j||MID_i)$, and then sends $\{M_3, M_{SG}, M_{SU}, T_3\}$ to GWN over an insecure channel.
4. GWN checks the validity of T_3 and calculates $R_s = M_3 \oplus h(R_u||SID_j||X_j||T_3)$ and $M_{SG}^* = h(R_s||R_g||SID_j||X_j||T_3)$, and checks $M_{SG}^* \stackrel{?}{=} M_{SG}$. If it is valid, GWN generates a timestamp T_4 and computes $MID_i^{new} = h(ID_i||h(K_{GWN}||R_g))$, $X_i^{new} = h(MID_i^{new}||R_g||K_{GWN})$, $M_4 = (MID_i^{new}||X_i^{new}||R_s||R_g) \oplus h(MID_i||X_i||T_4)$, and $M_{GU} = h(R_u||R_g||MID_i||X_i||T_4)$ and sends $\{M_4, M_SU, M_GU, T_4\}$ to U_i.

5. U_i checks the validity of T_4 and computes $(MID_i^{new}||X_i^{new}||R_s||R_g) = M_4 \oplus h(MID_i||X_i||T_4)$ and $M_{GU}^* = h(R_u||R_g||MID_i||X_i||T_4)$, and then checks $M_{GU}^* \stackrel{?}{=} M_{GU}$. If the condition is valid, U_i computes $SK = h(R_u||R_s)$ and $M_{SU}^* = h(SK||R_s||R_u||SID_j||MID_i)$, and checks $M_{SU}^* \stackrel{?}{=} M_{SU}$. If the condition is correct, U_i computes $Q_i^{new} = h(MID_i^{MPW}||MPW_i) \oplus X_i^{new}$, and $W_i^{new} = h(MPW_i||X_i^{new})$ and replaces $\{Q_i, W_i, MID_i\}$ with $Q_i^{new}, W_i^{new}, MID_i^{new}$. Consequently, the U_i, the GWN and S_j are mutually authenticated successfully.

3 Cryptanalysis of Yu et al.'s Scheme

In this section, we demonstrate that Yu et al.'s [26] scheme is vulnerable to known session-specific temporary information attacks.

3.1 The Adversary Model

We assume that the adversary E has the following capacities.

1. E has limited/completed control over the messages that were transmitted over the public channel, such as intercepting, modifying, and deleting the transmitted message.
2. E can obtain session-specific temporary information.

3.2 Known Session-Specific Temporary Information Attacks

In an AKE scheme, a known session-specific temporary information attack means that if a random nonce is disclosed to an adversary, this adversary can obtain the parameters needed to calculate the session key according to the random nonce. This section shows that Yu et al.'s scheme is vulnerable to known session-specific temporary information attacks. Figure 1 illustrates the whole procedures of this attack. The detailed steps are listed as follows:

1. If an adversary E obtains random number R_u, E can compute X_i that is equal to $R_u \oplus M_1$ by utilizing the public message $\{M_1, MID_i, CID_i, M_{UG}, T_1\}$.
2. With the message $\{M_4, M_{SU}, M_{GU}, T_4\}$ and $\{M_1, MID_i, CID_i, M_{UG}, T_1\}$ transmitted over public channel, E can obtain M_4, T_4 and MID_i.

Fig. 1 Our attack

User(U_i)	Gateway node (GWN)	Sensor node(S_j)
Selects ID_i, PW_i		
Imprints biometric BIO_i		
$R_i = \text{Rep}(BIO_i, P_i)$		
$MPW_i = h(PW_i \| R_i)$		
$X_i = h(MID_i \| MPW_i) \oplus Q_i$		
$W_i^* = h(MPW_i \| X_i)$		
checks $W_i^* \stackrel{?}{=} W_i$		
Generates a random nonce		
R_u and a timestamp T_1		
$M_1 = X_i \oplus R_u \;\rightarrow\; X_i = M_1 \oplus R_u$		
$CID_i = (ID_i \| SID_j) \; h \oplus$		
$(MID_i \| R_u \| X_i)$		
$M_{UG} = h(ID_i \| R_u \| X_i \| T_1)$		
$\{M_1, MID_i, CID_i, M_{UG}, T_1\} \longrightarrow$		
	Checks T_1	
	$X_i = h(MID_i \| r_g \| K_{GWN})$	
	$R_u = M_1 \oplus X_i$	
	$(ID_i \| SID_j) = CID_i \oplus h(MID_i \| R_u \| X_i)$	
	$M_{UG}^* = h(ID_i \| R_u \| X_i \| T_1)$	
	Checks $M_{UG}^* \stackrel{?}{=} M_{UG}$	
	Generates random nonce R_g and timestamp T_2	
	Computes	
	$M_2 = (R_u \| R_g) \oplus h(SID_j \| X_j \| T_2)$	
	$M_{GS} = h(MID_i \| SID_j \| R_u \| R_g \| X_j \| T_2)$	
	$\{M_2, MID_i, M_{GS}, T_2\} \longrightarrow$	
		Checks T_2
		$(R_u \| R_g) = M_2 \oplus h(SID_j \| X_j \| T_2)$
		$M_{GS}^* = h(MID_i \| SID_j \| R_u \| R_g \| X_j \| T_2)$
		checks $M_{GS}^* \stackrel{?}{=} M_{GS}$
		Generates random nonce R_s and timestamp T_3
		$M_3 = R_s \oplus h(R_u \| SID_j \| X_j \| T_3)$
		$M_{SG} = h(R_u \| R_g \| SID_j \| X_j \| T_3)$
		$SK = h(R_u \| R_s)$
		$M_{SU} = h(SK \| R_s \| R_u \| SID_j \| MID_i)$
		$\longleftarrow \{M_3, M_{SG}, M_{SU}, T_3\}$
	Checks T_3	
	$R_s = M_3 \oplus h(R_u \| SID_j \| X_j \| T_3)$	
	$M_{SG}^* = h(R_u \| R_g \| SID_j \| X_j \| T_3)$	
	checks $M_{SG}^* \stackrel{?}{=} M_{SG}$	
	Generates a timestamp T_4	
	$MID_i^{new} = h(ID_i \| h(K_{GWN} \| R_g))$	
	$X_i^{new} = h(MID_i^{new} \| R_g \| K_{GWN})$	
	$M_4 = (MID_i^{new} \| X_i^{new} \| R_s \| R_g) \oplus h(MID_i \| X_i \| T_4)$	
	$M_{GU} = h(R_u \| R_g \| MID_i \| X_i \| T_4)$	
	$\longleftarrow \{M_4, M_{SU}, M_{GU}, T_4\}$	
checks T_4		
$(MID_i^{new} \| X_i^{new} \| R_s \| R_g) = M_4 \oplus h(MID_i \| X_i \| T_4)$		
$M_{GU}^* = h(R_u \| R_g \| MID_i \| X_i \| T_4)$		
Checks $M_{GU}^* \stackrel{?}{=} M_{GU}$		
Computes		
$SK = h(R_u \| R_s)$		
$M_{SU}^* = h(SK \| R_s \| R_u \| SID_j \| MID_i)$		
Checks $M_{SU}^* \stackrel{?}{=} M_{SU}$		
Computes		
$Q_i^{new} = h(MID_i^{MPW} \| MPW_i) \oplus X_i^{new}$		
$W_i^{new} = h(MPW_i \| X_i^{new})$		
$(Q_i, W_i, MID_i) \leftarrow (Q_i^{new}, W_i^{new}, MID_i^{new})$		

3. E utilizes the above message X_i, M_4, T_4 and MID_i to computes $(MID_i^{new} \| X_i^{new} \| R_s \| R_g) = M_4 \oplus h(MID_i \| X_i \| T_4)$. Now E can obtain R_s.
4. At last, E can compute the session key by $SK = h(R_u \| R_s)$.

In a nutshell, if an attacker E obtains transmitted messages over the insecure channel and the session-specific temporary information R_u, E can compute the session key.

4 Conclusion

In this paper, we analyzed Yu et al.'s protocol. Although Yu et al. claimed that their protocol could resist various kinds of attacks, we still find that their protocol is still vulnerable to known session-specific temporary information attacks. For the attack method proposed in this paper, we will try to propose a new scheme in the future.

References

1. Chen, C.M., Fang, W., Wang, K.H., Wu, T.Y.: Comments on "an improved secure and efficient password and chaos-based two-party key agreement protocol". Nonlinear Dyn. **87**(3), 2073–2075 (2017)
2. Chen, C.M., Li, C.T., Liu, S., Wu, T.Y., Pan, J.S.: A provable secure private data delegation scheme for mountaineering events in emergency system. Ieee Access **5**, 3410–3422 (2017)
3. Chen, C.M., Xiang, B., Wu, T.Y., Wang, K.H.: An anonymous mutual authenticated key agreement scheme for wearable sensors in wireless body area networks. Appl. Sci. **8**(7), 1074 (2018)
4. Chen, C.L., Deng, Y.Y., Weng, W., Chen, C.H., Chiu, Y.J., Wu, C.M.: A traceable and privacy-preserving authentication for uav communication control system. Electronics **9**(1), 62 (2020)
5. Du, W., Deng, J., Han, Y.S., Chen, S., Varshney, P.K.: A key management scheme for wireless sensor networks using deployment knowledge. In: IEEE INFOCOM 2004, vol. 1. IEEE (2004)
6. Estrin, D., Girod, L., Pottie, G., Srivastava, M.: Instrumenting the world with wireless sensor networks. In: 2001 IEEE International Conference on Acoustics, Speech, and Signal Processing. Proceedings (Cat. No. 01CH37221). vol. 4, pp. 2033–2036. IEEE (2001)
7. Huang, D., Mehta, M., Medhi, D., Harn, L.: Location-aware key management scheme for wireless sensor networks. In: Proceedings of the 2nd ACM Workshop on Security of ad hoc and Sensor Networks, pp. 29–42 (2004)
8. Hussain, S., Ullah, I., Khattak, H., Khan, M.A., Chen, C.M., Kumari, S.: A lightweight and provable secure identity-based generalized proxy signcryption (ibgps) scheme for industrial internet of things (iiot). J. Inf. Secur. Appl. **58**, 102625 (2021)
9. Khan, M.A., Salah, K.: Iot security: review, blockchain solutions, and open challenges. Futur. Gener. Comput. Syst. **82**, 395–411 (2018)
10. Kumari, S., Chaudhary, P., Chen, C.M., Khan, M.K.: Questioning key compromise attack on ostad-sharif et al.'s authentication and session key generation scheme for healthcare applications. IEEE Access **7**, 39717–39720 (2019)
11. Kwon, D., Yu, S., Lee, J., Son, S., Park, Y.: Wsn-slap: secure and lightweight mutual authentication protocol for wireless sensor networks. Sensors **21**(3), 936 (2021)
12. Lee, I., Lee, K.: The internet of things (iot): applications, investments, and challenges for enterprises. Bus. Horiz. **58**(4), 431–440 (2015)
13. Li, C.T., Lee, C.C., Weng, C.Y., Chen, C.M.: Towards secure authenticating of cache in the reader for rfid-based iot systems. Peer Peer Netw. Appl. **11**(1), 198–208 (2018)
14. Lu, Y., Xu, G., Li, L., Yang, Y.: Anonymous three-factor authenticated key agreement for wireless sensor networks. Wireless Netw. **25**(4), 1461–1475 (2019)
15. Mo, J., Chen, H.: A lightweight secure user authentication and key agreement protocol for wireless sensor networks. Secur. Commun. Netw. 2019 (2019)
16. Moghadam, M.F., Nikooghadam, M., Al Jabban, M.A.B., Alishahi, M., Mortazavi, L., Mohajerzadeh, A.: An efficient authentication and key agreement scheme based on ecdh for wireless sensor network. IEEE Access 8, 73182–73192 (2020)

17. Renuka, K., Kumar, S., Kumari, S., Chen, C.M.: Cryptanalysis and improvement of a privacy-preserving three-factor authentication protocol for wireless sensor networks. Sensors **19**(21), 4625 (2019)
18. Sarma, S.E., Weis, S.A., Engels, D.W.: Rfid systems and security and privacy implications. In: International Workshop on Cryptographic Hardware and Embedded Systems, pp. 454–469. Springer (2002)
19. Shafiq, A., Ayub, M.F., Mahmood, K., Sadiq, M., Kumari, S., Chen, C.M.: An identity-based anonymous three-party authenticated protocol for iot infrastructure. J. Sens. 2020 (2020)
20. Sun, H.M., Wang, K.H., Chen, C.M.: On the security of an efficient time-bound hierarchical key management scheme. IEEE Trans. Depend. Sec. Comput. **6**(2), 159–160 (2009)
21. Wang, C., Xu, G., Sun, J.: An enhanced three-factor user authentication scheme using elliptic curve cryptosystem for wireless sensor networks. Sensors **17**(12), 2946 (2017)
22. Wang, D., Wang, P., Wang, C.: Efficient multi-factor user authentication protocol with forward secrecy for real-time data access in wsns. ACM Trans. Cyber-Phys. Syst. **4**(3), 1–26 (2020)
23. Wang, E.K., Liang, Z., Chen, C.M., Kumari, S., Khan, M.K.: Porx: A reputation incentive scheme for blockchain consensus of iiot. Futur. Gener. Comput. Syst. **102**, 140–151 (2020)
24. Wu, T.Y., Chen, C.M., Wang, K.H., Wu, J.M.T.: Security analysis and enhancement of a certificateless searchable public key encryption scheme for iiot environments. IEEE Access **7**, 49232–49239 (2019)
25. Yahuza, M., Idris, M.Y.I., Wahab, A.W.A., Nandy, T., Ahmedy, I.B., Ramli, R.: An edge assisted secure lightweight authentication technique for safe communication on the internet of drones network. IEEE Access **9**, 31420–31440 (2021)
26. Yu, S., Park, Y.: Slua-wsn: secure and lightweight three-factor-based user authentication protocol for wireless sensor networks. Sensors **20**(15), 4143 (2020)

Design and Optimization of the Seat of New Energy Electric Vehicle

Ya-Zheng Zhao, Yi-Jui Chiu, and Shu-Hao Zhao

Abstract In this paper, through the analysis of the human seat model, the boundary conditions of the seat lumbar support area are set and the objective function is calculated to study the influence of the seat lumbar support parameters on the seat. This paper mainly studies the design parameters of pressure value and its distribution, the contact area between the back surface and the human body, and the supporting force of the backrest obtained by the human body.

1 Introduction

Analyzing the human–machine-environment system shows that the machine refers to the unit that affects the comfort of the car. From the analysis of the whole vehicle, there are many parts that have contact with people and affect the comfort of the car, such as seats, gear levers, steering wheels, brake pedals, and accelerator pedals.

Regarding the simulation experiment of the pressure distribution of the car seat, the finite element analysis method is mainly used. Mohajer et al. [1] proposed a HBM (Human Body Model) for objective estimation of road vehicle driving comfort based on simulation. The results show that the head vibration is highly correlated with the vertical ISO cushion index (about 99.8%) and the longitudinal ISO seat back index (about 99.7%). Vink et al. [2] have the following conclusion: the sensitivity of the area contacting the back seat and the hips of the seat is significantly different, and the body parts contacting the front of the seat chassis are more sensitive. This is consistent with the results of other studies on ideal pressure distribution and pain threshold. By adjusting the shape of the seat, the sensitive area is less stressed, and the insensitive area is slightly increased to improve comfort. Liu et al. [3] studied the vibration performance of passengers sitting on chairs in a dynamic state. Smulders et al. [4] completed the design of the car interior by analyzing the sitting posture of the human body based on his revised human body parameters. Bai et al. [5] proposed

---

Y.-Z. Zhao · Y.-J. Chiu (✉) · S.-H. Zhao
School of Mechanical and Automotive Engineering, Fujian Province, Xiamen University of Technology, No. 600, Ligong Rd, Xiamen 361024, China
e-mail: chiuyijui@xmut.edu.cn

and demonstrated a method of systematically identifying the best configuration or structure of a 4-degree-of-freedom (4DOF) human body vibration model and identifying its parameters. The research of Guo et al. [6] shows that car seats will affect the ride comfort of the driver during long-term driving. They modeled the human body and applied a finite element model to analyze the influence of the seat lumbar support on the human–computer interaction, and obtained the pressure distribution between the body and the seat under the action of gravity, the contact area, the total backrest force and the intervertebral disc stress. The results show that after comprehensive comparison and evaluation, the optimal thickness of the seat lumbar support size is 10 mm for seat comfort.

This paper is mainly aimed at the comfort design requirements of the seat, from the perspective of ergonomics, the seat is designed and optimized. After completing the design, use workbench to analyze the designed seat.

2 Human Body and Human Body Model

2.1 Sitting Position

The stress on the spine in a sitting position is different from that in a standing position. The pelvis will rotate backwards, which will increase the pressure on the intervertebral discs. If you continue to sit, your waist will be sore. Therefore, the seat curve should be as close as possible to the curvature of the lumbar spine when standing to give sufficient support to the waist. The spine of the human body is connected by ligaments and muscles between each vertebra, and the spine is positioned by the force of the back muscles. In addition to a well-designed seat curve, a comfortable backrest and cushion material should be selected to make the contact muscles in a more relaxed state.

2.2 Factors of Seat Comfort

When the human body is in a sitting position, the spine maintains an S-shaped curve through the shape of the back of the chair, and the pressure on the back will be less. A good body pressure distribution should make the pressure on the scapula and lumbar spine larger than other places, and the pressure on the back of the spine is evenly distributed and smaller. The softness and hardness of the cushion makes the ischial tuberosity bear 60% of the torso weight.

Table 1 2015 Jiangsu body size survey

Measurement items	Percentile					
	Male			Female		
Height/mm	1683	1750	1830	1516	1600	1710
Shoulder height/mm	1400	1450	1557	1203	1395	1420
Elbow height/mm	1053	1156	1254	783	1050	1250
Shoulder width/mm	400	430	480	343	370	420
Sitting height/mm	890	950	1410	771	880	1120
Sitting elbow height/mm	226	300	384	193	300	480
Sitting knee height/mm	420	530	580	363	495	550
Hip-knee distance/mm	429	540	600	397	510	570
Sit deep/mm	386	430	537	350	430	520
Sitting elbow width/mm	340	460	557	313	435	480
Sitting hip width/mm	270	350	465	287	350	420

2.3 Body Size

The main purchasers of the electric vehicles designed in this article will be the 20 to 40-year olds with low spending power. Therefore, this article chooses to use the measurement data of 104 boys and 100 girls in a college in 2015 by Zhu as the body size data. The data is as shown in Table 1.

3 Backrest and Cushion Surface Optimization

3.1 Backrest Curve Optimization

3.1.1 Spine Size and Image Preprocessing

Adjust the size of the spine in the 3ds Max software according to the human body size table to meet the design requirements, and then export the side view of the spine into an image. Use Photoshop to mark the spinous processes of the thoracic spine, lumbar spine, and cervical spine (as shown in Fig. 1) for later programming.

3.1.2 Fitted Curve Equation

Input the coordinate result of the point-taking program into the table, and insert it as a matrix into the MATLAB software for curve fitting.

Figure 2 shows the fitting diagram of the center contour line of the male driver

Fig. 1 Side view of 3D model of spine

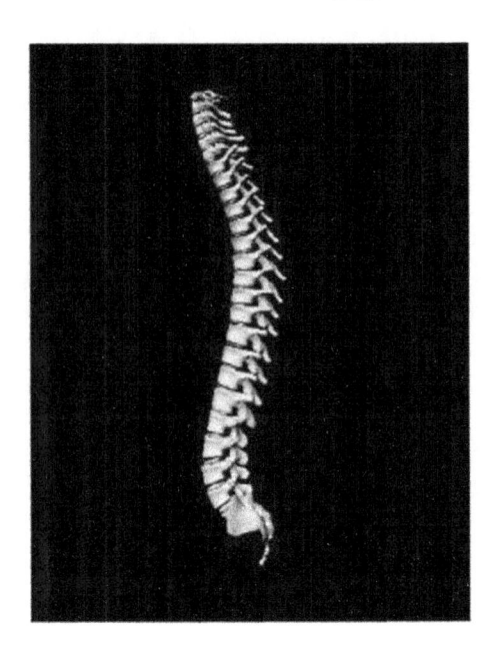

Fig. 2 Male curve equation fitting result

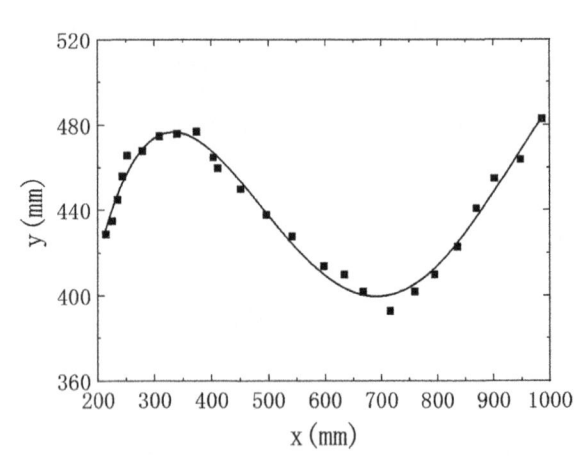

backrest. x is the vertical distance from a point on the male spine to the origin, and y is the horizontal distance from a point on the human spine to the origin.

The male curve equation is

$$y = A_0 + A_1 x + A_2 x^2 + A_3 x^3 + A_4 x^4 + A_5 x^5 \tag{1}$$

Fig. 3 Female curve equation fitting result

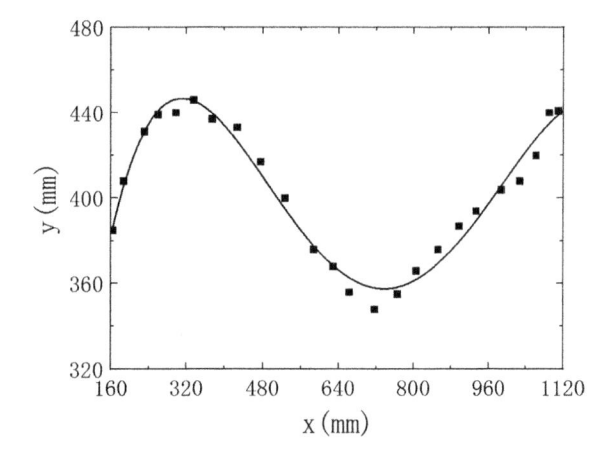

Coefficients:

| A0 = −87.445 | A1 = 4.478 | A2 = −0.012 |
| A3 = 1.314*10⁻⁵ | A4 = −5.141*10⁻⁹ | A5 = 2.756*10⁻¹³ |

Goodness of fit:

R-square: 0.977   Adjusted R-square: 0.970

Figure 3 shows the fitting diagram of the center contour line of the female driver backrest. x is the vertical distance from a point on the female spine to the origin, and y is the horizontal distance from a point on the human spine to the origin.

The female curve equation is

$$y = A_0 + A_1x + A_2x^2 + A_3x^3 + A_4x^4 + A_5x^5 \qquad (2)$$

Coefficients:

| A0 = 71.064 | A1 = 3.112 | A2 = −0.009 |
| A3 = 9.565*10⁻⁶ | A4 = − 4.026*10⁻⁹ | A5 = 4.101*10⁻¹³ |

Goodness of fit:

R-square: 0.974   Adjusted R-square: 0.967

Fig. 4. 3D model of backrest and cushion

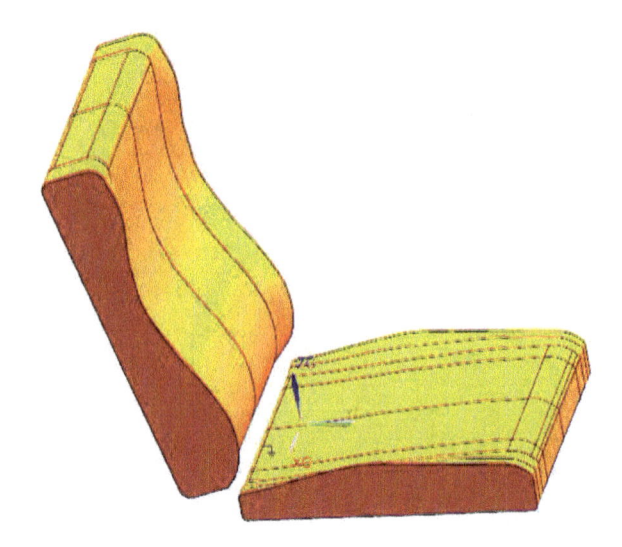

3.2 Seat Structure Design

The length of the seat cushion of this car cannot be adjusted. In order to meet the needs of the 5th to 95th percentile people, the seat cushion cannot be too long, and the seat depth of the 5th percentile is 350 mm. The width is taken as the sitting hip width of the 95th percentile human body is 465 mm. The backrest height is 650 mm from the sitting height of the 50th percentile mannequin minus the head to shoulder height. The seat width uses the 95th percentile mannequin's shoulder width of 480 mm.

Based on the above analysis, the model of the backrest and cushion is shown in Fig. 4.

4 Finite Element Analysis

4.1 Finite Element Model

Since Poser cannot export the solid model, Geomagic Studio software was used to reverse the reconstruction of the .stl file exported by 3ds Max. Because the surfaces of fingers and toes are too complex and are not helpful for finite element analysis and have little effect on the results, no model is built.

Assemble the human body model and the seat model so that the human body is as close to the seat as possible, but does not overlap with the seat. The reconstructed three-dimensional model of the human body is shown in Fig. 5.

Fig. 5 Assembly model

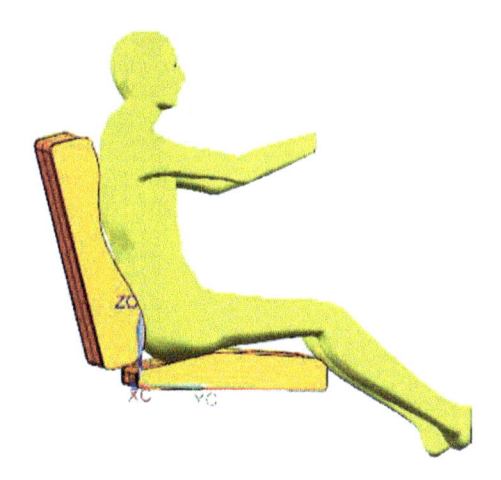

4.2 Finite Element Analysis

4.2.1 Part Contact and Constraint Settings

The driver's hand grasps the steering wheel, so this study restricts his hand. The feet are in frictional contact with the floor, and the friction coefficient is set to 0.2. The human body and the cushion are also in frictional contact, and the friction coefficient is set to 0.3. The joints are arranged as hinges, and the bones and the body are in tied contact.

4.2.2 Meshing and Material

Considering the calculation time, this research divides the grid reasonably. Encrypt the place that is in contact with the human body and is helpful for the analysis. Encrypt the grid in the red area in Fig. 6. The grid size is 5 mm, and the rest is 20 mm. The transition is a slow transition, and the unit is a tetrahedral unit. The total number of seat grids is 23508.

The tissues and organs are set as isotropic linear elastic materials to simplify the material properties. Young's modulus is 0.85 Mpa, Poisson's ratio is 0.46, and density is 1100 kg/m^3. The tissues and organs are set as isotropic linear elastic materials to simplify the material properties. Young's modulus is 0.85 Mpa, Poisson's ratio is 0.46, and density is 1100 kg/m^3. The skeleton is also simplified. After simplification, Young's modulus is 17 Gpa, Poisson's ratio is 0.49, and density is 1700 kg/m^3. The seat material parameters are also simplified. The simplified strength limit is 178.65 KPa, Young's modulus is 39.6 MPa, and Poisson's ratio is 0.42.

Fig. 6 Selection of encrypted mesh surface

4.2.3 Analysis Result

The maximum stress of the cushion appears at the position corresponding to the ischial tuberosity, from which the stress value gradually decreases in a circular distribution, and the stress value gradually decreases from the hip to the thigh. The maximum stress are 9.561, 9.632, and 11.354 kPa. The main pressure on the backrest is concentrated on the waist and shoulder blades, and the stress distribution conforms to the law. The maximum stress value and contact area are shown in Table 2.

According to the results in Fig. 7, it can be seen that the main stress on the back is concentrated on the lumbar support protrusions and then gradually decreases to the upper part, which is consistent with the design purpose of the curved surface to provide sufficient lumbar support. The contact area between the back of the driver A and the seat back is 40356 mm^2, and the maximum stress is 8.963 kPa. The contact area between the back of the driver B and the seat back is 46358 mm^2, and the maximum stress is 7.855 kPa. The contact area between the back of the driver B and the seat back is 63258 mm^2, and the maximum stress is 8.562 kPa.

Table 2 Analysis result

Driver	body parts	Maximum stress/KPa	Contact area/mm^2
Driver A	Backrest	8.963	40,356
	cushion	9.561	71,231
Driver B	Backrest	7.855	46,358
	cushion	9.632	77,500
Driver C	Backrest	8.562	63,258
	cushion	11.354	88,963

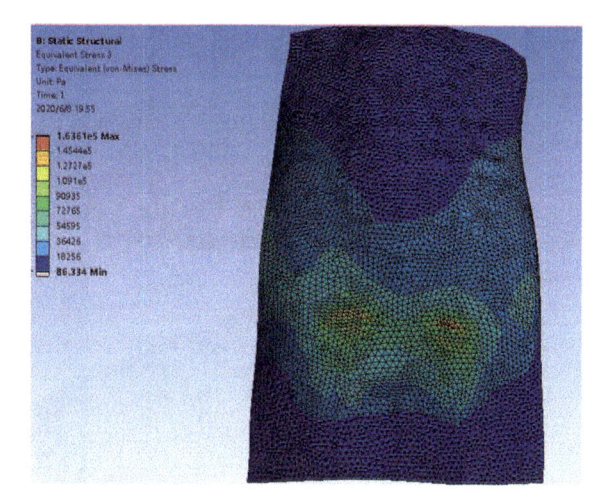

Fig. 7 Back surface stress diagram

From the deformation of the cushion in Fig. 8, it can be seen that the pressure on the seat is mainly from the ischial tuberosity of the body, and then slowly reduced to the front of the thigh.

Figure 9 shows the stress analysis results of the contact surface between the lower limbs and the seat. It can be seen that the main body pressure is distributed on the human buttocks, which is consistent with the original design intention.

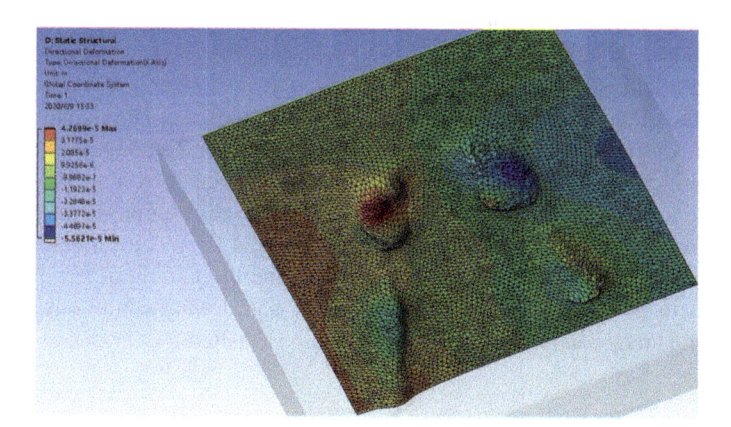

Fig. 8 Deformation diagram of cushion

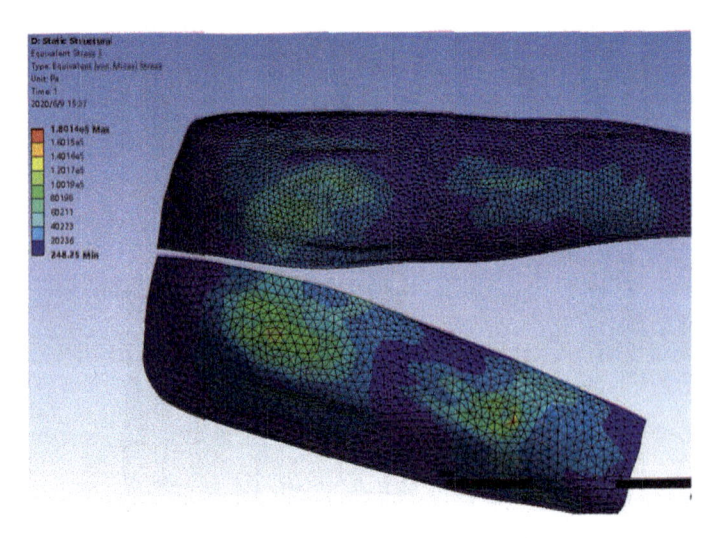

Fig. 9 Stress map of high contact surface

5 Conclusion

This paper designs a new body of the elderly electric vehicle. The backrest height is 650 mm from the sitting height of the 50th percentile mannequin minus the head to shoulder height. The seat width uses the 95th percentile mannequin's shoulder width of 480 mm. The seat depth of the 5th percentile is 350 mm. Backrest curve takes the fitting diagram of the center contour line of the male driver backrest. The curve fitting tool uses MATLAB. Establish 3D model and finite element model, choose appropriate material and mesh size.

The finite element analysis obtained the following results. The maximum stress of the seat back corresponding to driver A is 8.963 kPa, and the maximum stress of the seat cushion is 9.561 kPa. The maximum stress of the backrest and seat cushion corresponding to driver B is 7.855 kPa and 9.632 kPa. C corresponds to the maximum stress of the backrest and cushion is 8.562 and 11.354 kPa. The simulation results show that the top of the elderly electric vehicle and the middle of the chassis are prone to damage. Under normal road conditions and driving conditions, the body will not be damaged and the driving safety will be satisfied.

References

1. Mohajer, N., Abdi, H., Nahavandi, S.: Directional and sectional ride comfort estimation using an integrated human biomechanical-seat foam model. J. Sound Vib. **120**(2), 423–435 (2017)
2. Vink, P., Lips, D.: Sensitivity of the human back and buttocks: the missing link in comfort seat design. J. Appl. Erg. **58**(4), 287–292 (2017)

3. Chi, L., Yi, Q., Michael, J.G.: Dynamic forces over the interface between a seated human body and a rigid seat during vertical whole-body vibration. J. J. Biomech. **61**(5), 176–182 (2017)
4. Smulders, M., Berghman, K., Koenraads, M., Kane, J.A., Krishna, K., Carter, T.K.: Comfort and pressure distribution in a human contour shaped aircraft seat (developed with 3d scans of the human body). J. Work **54**(4), 1–16 (2016)
5. Bai, X.X., Xu, S.X., Cheng, W.: On 4-degree-of-freedom biodynamic models of seated occupants: Lumped-parameter modeling. J. Sound Vib. **402**(2), 122–141 (2017)
6. Guo, L.X., Dong, R.C., Zhang, M.: Effect of lumbar support on seating comfort predicted by a whole human body-seat model. Int. J. Ind. Erg. **53**(1), 319–327 (2016)

Cryptanalysis of an Authentication Protocol for IoT-Enabled Devices in Distributed Cloud Computing Environment

Zhen Li, Lei Yang, Tsu-Yang Wu, and Chien-Ming Chen

Abstract In recent years, more and more authentication protocols combining cloud computing and IoT are proposed for secure communication and access of the Internet of things. In 2020, Kang et al. Proposed a mutual authentication protocol based on distributed cloud computing environment for secure communication. They claim that their agreement can withstand all kinds of attacks. Unfortunately, this paper certificates that Kang et al.'s scheme is very vulnerable to temporary information attack of specific session.

1 Introduction

With the flourishing development of IoT and Internetwork, our life has changed tremendously. The Internet, as a carrier, is further extended and expanded, and the Internet of things is born. IoT is not just a concept; it has been integrated into our lives. Many primary applications based on IoT have been described [5, 6, 13, 14, 16]. However, there are also efficiency related problems, because the sensors it uses have the characteristics of insufficient memory space and weak computing performance. The use of cloud service [3, 8, 19] technology in the Internet of things environment can be effectively improved.

For the sake of providing a safe environment for the IoT, it is necessary to increase the process of authentication and key exchange between participating roles. At present, there are many ake protocols [4, 10–12]. Among them, the multi-server authentication protocol [1, 2, 15, 17, 18, 21] in the Internet of things and cloud environment has attracted more and more attention. In 2018, Amin et al. [2] proved that

---

Z. Li · L. Yang · T.-Y. Wu · C.-M. Chen (✉)
School of Computer Science and Technology, Shandong University of Science and Technology, Shandong, China
e-mail: chienmingchen@ieee.org

Z. Li
e-mail: lz17806171078@163.com

L. Yang
e-mail: yanglei2200@163.com

the two mutual authentication protocols under the multi-server architecture proposed by Chuang [7] and Xue et al. [20] have security vulnerabilities and some hidden dangers. Subsequently, Kang et al. [9] proposed a protocol combining distributed cloud computing and Internet of things, which supports Internet of things devices. They confidently claim that their protocol can resist all possible security attacks and has high security.

Unfortunately, we further pointed out that Kang et al.'s protocol [9] is still fragile to temporary information exposure attacks in this paper. In a temporary information exposure attack, if the attacker intercepts the information on public channels and obtains random numbers generated by a server, this attacker can calculate the private keys of users and a server. Various protocols ([17, 18]) are also vulnerable to temporary information exposure attacks.

Briefly summarize the organizational structure of this paper. In Sect. 2, we need to review the authentication scheme put forward by Kang et al. We describe the security analysis and attack of Kang et al.'s protocol in Sect. 3. Finally, the conclusion is written in Sect. 4.

2 Review of Kang et al.'s Protocol

This section mainly introduces the scheme [9] proposed by Kang et al. The protocol has three entities participating in and communicating with each other: user, cloud server, and control server. Cloud server provides information communication and other services for users. The control server is used for identity registration and authentication of users and cloud servers.

This section mainly introduces the registration phase, login phase, authentication, and key agreement phase. For brevity, the password change and identity update phases are not reviewed. If you need to know the detailed steps of the protocol, please refer to their paper [9]. The important symbols used are shown in Table 1.

2.1 Registration Phase

The registration stage is divided into two situations. The first situation is the registration stage of cloud server S_j, which selects the identity SID_j and generates a random number r_1. Subsequently, Sm sends (SID_j, d) to the CS on the secure channel, and then calculates

$$PSID_j = h(SID_j || r_1),$$

$$BS_j = h(PSID_j || SID_j || y),$$

Table 1 Symbol table

Symbol	Description
CS	The control server
S_j	jth cloud server
SID_j	Identification of the jth cloud server
U_i	ith user
UID_i	Identification of the U_i
PW_i	Password of the U_i
x, y	Clandestine numbers of CS
$h(\cdot)$	Hash
TS	Timestamp
TS_i	The U_i's current timestamp
TS_m	The S_m's current timestamp
$\oplus$	Exclusive OR
$\|$	Concatenation operator

and sends BS_j to S_j through a reliable channel. Note that x and y are two random numbers generated by CS. Once S_j receives BS_j, S_j stores these secret parameters (BS_j, r_1).

In the second case, the user registers with CS and U_i inputs his identity information UID_i and personal password PW_i. Then, it calculates $A_1 = PW_i \oplus h(BIO)$. This is his biological characteristic. Then U_i sends (UID_i, A_1) to CS. After CS receives the message, CS generates a random number r_2 and calculates

$$PID_i = h(UID_i \| r_2),$$

$$A_2 = h(UID_i \| A_1),$$

$$A_3 = h(PID_i \| x),$$

$$A_4 = A_3 \oplus A_1,$$

$$A_5 = h(PID_i \| UID_i \| x),$$

$$A_6 = r_2 \oplus A_1,$$

Store $(A_2, A_6, A_5, A_4, h(\cdot))$ in the smart card and send it to U_i.

2.2 Login Phase

During the login phase, U_i provides UID_i^*, PW_i^*, and BIO^* to a card reader. The card reader calculates

$$A_1^* = PW_i^* \oplus h(BIO^*),$$

$$A_2^* = h(UID_i^* || A_1^*).$$

Then, the card reader verifies if $A_2^* = A_2$ or not. When $A_2^* = A_2$, $A_3^* = A_3$, $PW_i^* = PW_i$, and $BIO^* = BIO$, $A_1^* = A_1$. Then, the card reader chooses a random number R_i and computes

$$r_2 = A_6 \oplus A_1,$$

$$PID_i = h(UID_i || r_2),$$

$$A_3 = A_4 \oplus A_1,$$

$$B_1 = UID_i \oplus A_3,$$

$$B_2 = h(UID_i || SID_j || R_i || TS_i || A_3),$$

$$B_3 = A_5 \oplus R_i,$$

$$B_4 = SID_j \oplus h(A_3 || R_i),$$

where SID_j is the status of S_j chosen by U_i. Then, the card reader transmits the login information $(B_2, B_3, B_4, B_1, PID_i, TS_i)$ to S_j publicly.

2.3 Authentication Key Agreement Phase

The certification phase consists of four steps.
Step 1: Once S_j receives the login information, S_j checks if the condition $TS_j - TS_i < \Delta T$ is valid. If the verification is successful, S_j generates a random number R_j and computes

$$J_i = BS_j \oplus R_j,$$

$$K_i = h(R_j || BS_j || PID_i || B_2 || TS_j).$$

Then, S_j submits $(J_i, K_i, PSID_j, B_2, B_3, B_4, B_1, PID_i, TS_i, TS_j)$ to CS. Here, TS_j is the current timestamp of the cloud server, and ΔT is the acceptable time delay for transmission.

Step 2: When receiving a message from S_j, CS verifies whether $TS_{CS} - TS_j < \Delta T$ is true. If the verification is successful, then CS calculates

$$A_3 = h(PID_i||x),$$

$$UID_i = B_1 \oplus A_3,$$

$$R_i = B_3 \oplus h(PID_i||UID_i||x),$$

$$SID_j = B_4 \oplus h(A_3||R_i),$$

$$B_2^* = h(UID_i||SID_j||R_i||TS_i||A_3).$$

Then, CS checks whether $B_2^* = B_2$ holds or not. If $B_2^* = B_2$, CS considers U_i as a legitimate user. Then, the CS computes

$$BS_j = h(PSID_j||SID_j||y),$$

$$R_j = BS_j \oplus J_i,$$

$$K_i^* = h(R_j||BS_j||PID_i||B_2||TS_j).$$

Then, CS checks the condition $K_i^* = K_i$. If $K_i^* = K_i$, CS believes S_j with real SID_j is legal and chosen by the user U_i. Then, CS produces a random number R_{CS} and computes

$$D_{CS} = R_j \oplus R_{CS} \oplus h(R_i||A_3||B_3),$$

$$E_{CS} = R_i \oplus R_{CS} \oplus h(BS_j||R_j),$$

$$SK_{CS} = h(R_i||R_j||R_{CS}),$$

$$F_{CS} = h((R_j \oplus R_{CS})||SK_{CS}),$$

$$G_{CS} = h((R_i \oplus R_{CS})||SK_{CS}).$$

Then, CS sends $(D_{CS}, E_{CS}, F_{CS}, G_{CS})$ to the S_j publicly.

Step 3: When receiving a reply from CS, S_j computes

$$W_m = h(BS_j||R_j),$$

$$R_i \oplus R_{CS} = E_{CS} \oplus W_m,$$

$$SK_m = h(R_i \oplus R_{CS} \oplus R_j),$$

$$G^*_{CS} = h((R_i \oplus R_{CS}) || SK_m).$$

Then, S_j checks the condition $G^*_{CS} = G_{CS}$. If $G^*_{CS} = G_{CS}$, S_j sends messages (D_{CS}, F_{CS}) to the U_i publicly.

Step 4: On receiving messages from S_j, U_i calculates

$$L_i = h(R_i || A_3 || B_3),$$

$$R_i \oplus R_{CS} = D_{CS} \oplus L_i,$$

$$SK_i = h(R_j \oplus R_{CS} \oplus R_i),$$

$$G^*_{CS} = h((R_j \oplus R_{CS}) || SK_i).$$

Next, U_i checks whether $G^*_{CS} = G_{CS}$. If the verification is successful, U_i believes the authenticity of S_j and CS and shares a session key $SK_i = SK_m$ with the cloud server S_j.

3 Cryptanalysis of Kang et al.'s Scheme

In this section, we mention the adversary model and analyze the security of Kang et al.'s protocol.

3.1 The Adversary Model

We suppose that attacker E has three capabilities.

1. E can intercept, store, forge, and send messages in the message network without being detected by the main protocol.
2. E can obtain session-specific temporary information.
3. E can try to execute security attack by getting information from smart card.

```
      S_j                              CS                              U_i
  Choose R_j
  Computes
   J_i = BS_j ⊕ R_j
 K_i = h(R_j||BS_j||PID_i||B_2||TS_j)
(J_i, K_i, PSID_j, B_2, B_3, B_4, B_1, PID_i, TS_i, TS_j)
 ─────────────────────────────────→
                                  Calculates.
                                  A_3 = h(PID_i||x)
                                  UID_i = B_1 ⊕ A_3
                                  R_i = B_3 ⊕ h(PID_i||UID_i||x)
                                  SID_j = B_4 ⊕ h(A_3||R_i)
                                  B_2* = h(UID_i||SID_j||R_i||TS_i||A_3)
                                  CheckS B_2* = B_2
                                  Computes
                                  BS_j = h(PSID_j||SID_j||y)
                                   R_j = BS_j ⊕ J_i
                                  K_i* = h(R_j||BS_j||PID_i||B_2||TS_j)
                                  Checks K_i* = K_i
                                  Computes
                                  D_CS = R_j ⊕ R_CS ⊕ h(R_i||A_3||B_3)
                                  E_CS = R_i ⊕ R_CS ⊕ h(BS_j||R_j)
                                  SK_CS = h(R_i||R_j||R_CS)
                                  F_CS = h((R_j ⊕ R_CS)||SK_CS)
                                  G_CS = h((R_i ⊕ R_CS)||SK_CS)
                                  (D_CS, E_CS, F_CS, G_CS)
                                 ←─────────────────────────────
  Computes
  W_m = h(BS_j||R_j)
  R_i ⊕ R_CS = E_CS ⊕ W_m
  SK_m = h(R_i ⊕ R_CS ⊕ R_j)
  G_CS* = h((R_i ⊕ R_CS)||SK_m)
  Checks G_CS* = G_CS
  (D_CS, F_CS)  to  U_i
 ─────────────────────────────────→
                                                              Calculates
                                                              L_i = h(R_i||A_3||B_3)
                                                              R_i ⊕ R_CS = P_CS ⊕ L_i
                                                              SK_i = h(R_j ⊕ R_CS ⊕ R_i)
                                                              G_CS* = h((R_j ⊕ R_CS)||SK_i)
                                                              Checks G_CS* = G_CS
```

Fig. 1 Attack analysis

3.2 Temporary Information Attack of Specific Session

Here, we prove that Kang et al.'s protocol can't resist the temporary information attack of specific session. Figure 1 marks the steps required to execute the attack. The specific attack steps are given below.

If E obtains the session key, the protocol will be in an insecure state and may cause great harm. Mentioned in the authentication phase, $SK_m = SK_i = h(R_j \oplus R_{CS} \oplus R_i)$.

1. Since R_j is a session-specific temporary information and J_i, R_{CS} are transmitted over a public channel, E can obtain R_j and intercept J_i, R_{CS}. $BS_j = R_j \oplus J_i$, so E can get BS_j now.
2. $W_m = h(BS_j||R_j)$. Because E knows the BS_j, R_j, so E can get W_m.
3. E can obtain $R_i \oplus R_{CS}$ by $R_i \oplus R_{CS} = R_{CS} \oplus W_m$.

4. As E can obtain the sensitive parameters above, R can obtain session key by $SK_m = SK_i = h(R_j \oplus R_{CS} \oplus R_i)$.

4 Conclusion

In this paper, we analyzed Kang et al.'s protocol. We demonstrate that this protocol is insecure against a temporary information attack of specific session. An attacker can easily calculate the session key with the information he obtained. Since various AKE protocols suffer the same attack, we will try to design a secure AKE protocol in the future.

References

1. Akram, M.A., Ghaffar, Z., Mahmood, K., Kumari, S., Agarwal, K., Chen, C.M.: An anonymous authenticated key-agreement scheme for multi-server infrastructure. HCIS **10**, 1–18 (2020)
2. Amin, R., Kumar, N., Biswas, G., Iqbal, R., Chang, V.: A light weight authentication protocol for iot-enabled devices in distributed cloud computing environment. Futur. Gener. Comput. Syst. **78**, 1005–1019 (2018)
3. Ateniese, G., Burns, R., Curtmola, R., Herring, J., Kissner, L., Peterson, Z., Song, D.: Provable data possession at untrusted stores. In: Proceedings of the 14th ACM Conference on Computer and Communications Security, pp. 598–609 (2007)
4. Chen, C.M., Fang, W., Wang, K.H., Wu, T.Y.: Comments on "an improved secure and efficient password and chaos-based two-party key agreement protocol". Nonlinear Dyn. **87**(3), 2073–2075 (2017)
5. Chen, C.M., Li, C.T., Liu, S., Wu, T.Y., Pan, J.S.: A provable secure private data delegation scheme for mountaineering events in emergency system. Ieee Access **5**, 3410–3422 (2017)
6. Chen, Y.Q., Zhou, B., Zhang, M., Chen, C.M.: Using iot technology for computer-integrated manufacturing systems in the semiconductor industry. Appl. Soft Comput. **89**, 106065 (2020)
7. Chuang, M.C., Chen, M.C.: An anonymous multi-server authenticated key agreement scheme based on trust computing using smart cards and biometrics. Expert Syst. Appl. **41**(4), 1411–1418 (2014)
8. He, B.Z., Chen, C.M., Wu, T.Y., Sun, H.M.: An efficient solution for hierarchical access control problem in cloud environment. Math. Prob. Eng. (2014)
9. Kang, B., Han, Y., Qian, K., Du, J.: Analysis and improvement on an authentication protocol for iot-enabled devices in distributed cloud computing environment. Math. Prob. Eng. (2020)
10. Kumari, S., Chaudhary, P., Chen, C.M., Khan, M.K.: Questioning key compromise attack on ostad-sharif et al.'s authentication and session key generation scheme for healthcare applications. IEEE Access 7, 39717–39720 (2019)
11. Shamshad, S., Mahmood, K., Kumari, S., Chen, C.M., et al.: A secure blockchain-based e-health records storage and sharing scheme. J. Inf. Secur. Appl. **55**, 102590 (2020)
12. Sun, H.M., Wang, K.H., Chen, C.M.: On the security of an efficient time-bound hierarchical key management scheme. IEEE Trans. Depend. Secur. Comput. **6**(2), 159–160 (2009)
13. Wang, K., Chen, C.M., Liang, Z., Hassan, M.M., Sarné, G.M., Fotia, L., Fortino, G.: A trusted consensus fusion scheme for decentralized collaborated learning in massive iot domain. Inf. Fusion **72**, 100–109 (2021)
14. Wang, K., Xu, P., Chen, C.M., Kumari, S., Shojafar, M., Alazab, M.: Neural architecture search for robust networks in 6g-enabled massive iot domain. IEEE Internet Things J. (2020)

15. Wang, K.H., Chen, C.M., Fang, W., Wu, T.Y.: A secure authentication scheme for internet of things. Pervasive Mob. Comput. **42**, 15–26 (2017)
16. Wu, T.Y., Chen, C.M., Wang, K.H., Wu, J.M.T.: Security analysis and enhancement of a certificateless searchable public key encryption scheme for iiot environments. IEEE Access **7**, 49232–49239 (2019)
17. Wu, T.Y., Lee, Z., Obaidat, M.S., Kumari, S., Kumar, S., Chen, C.M.: An authenticated key exchange protocol for multi-server architecture in 5g networks. IEEE Access **8**, 28096–28108 (2020)
18. Wu, T.Y., Yang, L., Lee, Z., Chen, C.M., Pan, J.S., Islam, S.: Improved ecc-based three-factor multiserver authentication scheme. Secur. Commun. Netw. (2021)
19. Xiong, H., Wang, Y., Li, W., Chen, C.M.: Flexible, efficient, and secure access delegation in cloud computing. ACM Trans. Manag. Inf. Syst. (TMIS) **10**(1), 1–20 (2019)
20. Xue, K., Hong, P., Ma, C.: A lightweight dynamic pseudonym identity based authentication and key agreement protocol without verification tables for multi-server architecture. J. Comput. Syst. Sci. **80**(1), 195–206 (2014)
21. Zhou, L., Li, X., Yeh, K.H., Su, C., Chiu, W.: Lightweight iot-based authentication scheme in cloud computing circumstance. Futur. Gener. Comput. Syst. **91**, 244–251 (2019)

Copyright Storage Method of Dance Short Video Based on Blockchain

Yang Yang and Dingguo Yu

Abstract Blockchain technology is widely used in the field of digital rights protection. The traditional digital rights protection scheme is inefficient, highly centralized, and has the risk of being tampered with. At the same time, blockchain cannot save all original files of digital resources due to its storage size limitation. Firstly, this paper proposes a dynamic limb action recognition algorithm based on dance pose prediction (RADPP). This algorithm recognizes the dance action in dance short video based on deep learning. The algorithm extracts the features of dance action accurately and forms a log file that can represent the short dance video. The algorithm stores the dance pose file into the blockchain to the purpose of preservation. On the one hand, the above methods improve the efficiency of short dance video storage, reduce the degree of centralization of storage, and eliminate the risk of copyright information being easily tampered with; at the same time, the log file calculated by deep learning technology for short video not only ensures the privacy of copyright information, but also ensures the feasibility of storing video information into blockchain. Experiments show that the method proposed in this paper is lighter and more efficient than the existing copyright storage methods, and this method can provide technical support for the media resource management department.

1 Introduction

The short video copyright management scheme based on blockchain technology, relying on the new storage architecture concept of low degree of blockchain centralization, provides a feasible and effective way for short video digital rights management [1]. Firstly, the high dynamic dance action is recognized by image preprocessing

---

Y. Yang
1Intelligent Media Technology Research Institute, Communication University of Zhejiang, Hangzhou 310018, China
e-mail: iimer@qq.com

D. Yu (✉)
College of Media Engineering, Communication University of Zhejiang, Hangzhou 310018, China
e-mail: yann@cuz.edu.cn

[2], template feature extraction [3] and template matching algorithm [4]. Then the extracted dancers' action records are stored in the blockchain in the form of "short video action log" [5], which can accurately represent the short dance video, and store it in the blockchain, so as to achieve the purpose of digital copyright storage [6].

2 Related Work

This paper focuses on the short dance videos with a duration of less than 2 min uploaded by users on the Internet short video platform as the research core, and uses RADPP algorithm as the basis to identify and describe the dynamic posture of dancers in the short video [7]. The short dance video is taken as the input object and the dancer's limb data "dance short video dancer action log" as the output object is integrated and stored in the blockchain [8]. Compared with human eye recognition, computer vision has higher computing power and stronger analysis ability for dynamic and complex human body movements [9]. In order to strengthen the relationship between the relative positions of the body, Suolan [10] proposed a new partition strategy for dividing skeleton joint points based on the popular spatiotemporal graph convolution network (ST-GCN) model in recent years.

3 Dynamic Body Movement Recognition

The blockchain-based dance short video copyright storage method is divided into two modules for algorithm design, which are used to analyze the original video file as a storage log file dance short video dancer action recognition and resolution algorithm module, and the storage module used to store the files formed by the above algorithm analysis into the blockchain for copyright information storage.

3.1 Short Dance Video Frame Preprocessing Algorithm

Firstly, each frame in the short video is sampled frame by frame. Then, the sampled image is preprocessed, and then the dancer's dance action is predicted and recognized. Finally, the accuracy of the identified dancer's action is verified.

Frame Image Gray Processing. The short dance video uploaded by users in mainstream short video software is basically color, but the color is redundant data in the dance action recognition proposed in this paper. The schematic diagram of gray level processing technology of frame picture is shown in Fig. 1.

Frame Image Threshold Segmentation. In this paper, we set $gray(x, y)$ as the gray value of the pixel (x, y) in the frame picture, and set $gradient(x, y)$ as the gray gradient

Fig. 1 Schematic diagram of gray processing technology of frame picture

value of the pixel (*x*, *y*) in the frame picture, then the threshold *th* in the frame image threshold segmentation can be expressed as shown in Formula 1.

$$th = th[x, y, gray(x, y), gradient(x, y)] \qquad (1)$$

In Formula 1, there are some mature algorithms for threshold calculation, but the unreasonable threshold determination algorithm will blur cut the dancer subject and the background picture of the dancer in the frame picture, and even lose the cutting target in the frame picture. The number of pixels occupied by the gray value of each level *l* is $point_l$ the resolution of the frame is *length* * *width*, then the distribution function of the gray probability $prob_l$ can be set as shown in Formula 2.

$$prob_l = \frac{point_l}{length * width} \quad (2)$$

If the number of pixels in the background picture of the dancer and the dancer is *boun*, then the distribution function of the gray probability *back(t)* of the background picture can be set as shown in Formula 3.

$$back(t) = \sum_{i=0}^{boun} prob_l \quad (3)$$

The distribution function of the grey probability *dancer(t)* of the dancer can be set as shown in Formula 4.

$$dancer(t) = \sum_{l=boun+1}^{level-1} prob_l \quad (4)$$

Through the above calculation, the information entropy $entropy_{back}$ of the background picture and the information entropy $entropy_{dancer}$ of the dancer can be obtained, as shown in Formula 5 and Formula 6, respectively.

$$entropy_{back} = \sum_{l=0}^{boun} \frac{prob_l}{back(t)} * \log\left(\frac{prob_l}{back(t)}\right) \quad (5)$$

$$entropy_{dancer} = - \sum_{l=boun+1}^{level-1} \frac{prob_l}{dancer(t)} * \log\left(\frac{prob_l}{dancer(t)}\right) \quad (6)$$

Through the above calculation, according to the information entropy of the background picture and the dancer's own information entropy, the total information entropy $entropy_{total}$ of each frame of short dance video can be obtained, as shown in Formula 7.

$$entropy_{total} = entropy_{back} + entropy_{dancer} \quad (7)$$

The threshold value calculated by the above calculation can be used for frame image segmentation. The schematic diagram of "segmenting" dancers from the background in the frame picture is shown in Fig. 2.

Fig. 2 Schematic diagram of threshold segmentation calculation

3.2 Dance Pose Estimation Algorithm for Short Video of Dance

Based on the above-mentioned part of the paper, the dance gesture prediction model is proposed. The specific calculation process is shown in Fig. 3.

Feature extraction algorithm of dancers in frame images. This paper assumes that the number of feature points that can be extracted from a short dance video is f^{eat}, the number of pictures of a dancer's dance pose template is set to *photo*, and two of the first template pictures of a certain dance posture of the dancer The physical distance between the feature points a and b can be set as $dist_{a,b}^{1}$, then the distance matrix *matr* of a certain dance posture can be calculated, as shown in Formula 8.

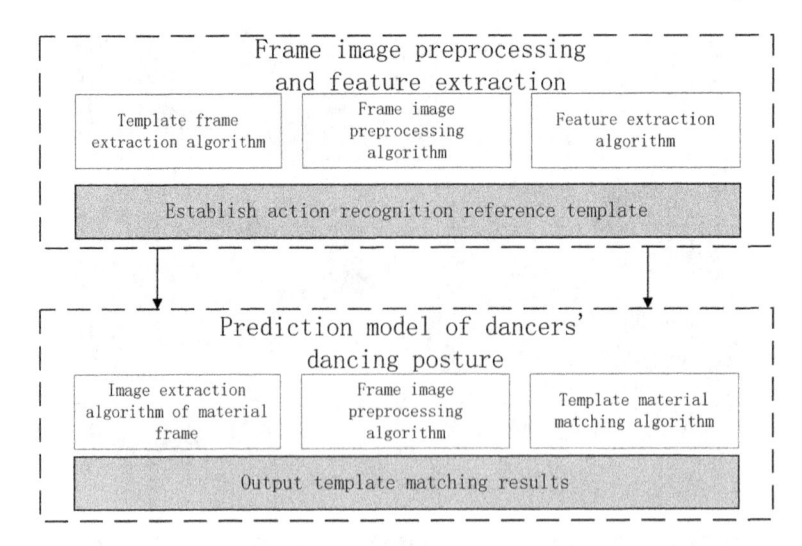

Fig. 3 Calculation process of the prediction model of dancers' movements

$$matr = \begin{pmatrix} dist_{1,2}^1 & \cdots & dist_{feat,(feat-1)}^1 \\ \vdots & \ddots & \vdots \\ dist_{1,2}^{photo} & \cdots & dist_{feat,(feat-1)}^{photo} \end{pmatrix} \qquad (8)$$

4 An Algorithm of Dancing Posture Log File Up Chain

4.1 Data Storage and Architecture Model

If the length of a video file to be stored exceeds the specified duration (5 min), the system will automatically divide it into short video groups with a single duration of 5 min. The elements in the group output logs, respectively, and the total hash value calculated in the form of Merkel tree is written into the block as the Merkel root value of the long video. The schematic diagram of calculating Merkel root value of elements in the group is shown in Fig. 4.

The time-consuming comparison between the traditional storage mode and the paper storage mode is shown in Fig. 5, and the memory consumption comparison between the traditional storage mode and the paper storage mode is shown in Fig. 6. This paper compares the existing traditional copyright storage methods, blockchain copyright storage methods based on POW consensus mechanism and the blockchain version based on RADPP from the aspects of data storage convenience, data capacity, data atomicity (uniqueness), representativeness of stored data, data privacy and security, system operation flexibility and data storage flexibility The results show that the

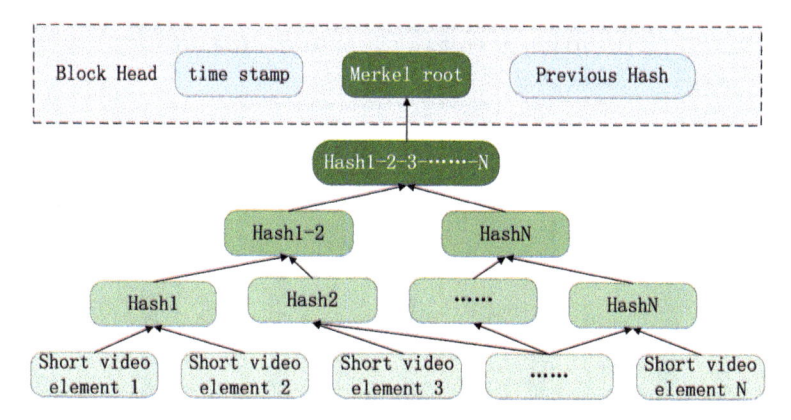

Fig. 4 Schematic diagram of calculating Merkel root value

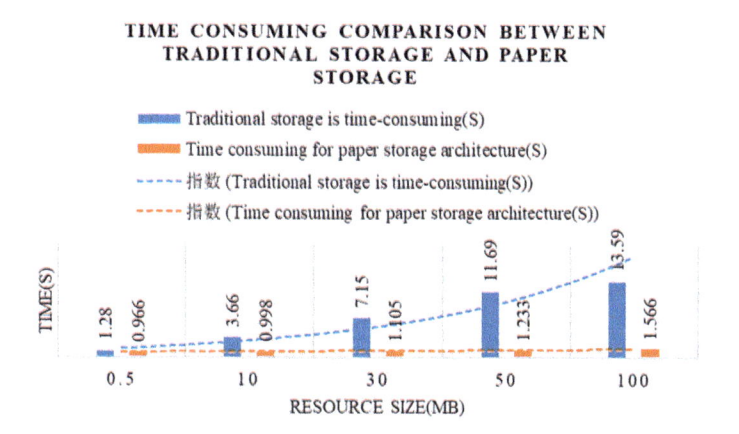

Fig. 5 Time-consuming comparison between traditional storage and paper storage

proposed method is efficient. The detailed comparison dimensions and results are shown in Fig. 7.

5 Conclusion

This paper proposes a dynamic limb action recognition algorithm based on dance pose prediction, which effectively solves the application of short video and other resources copyright storage in blockchain architecture. Next, there are still two directions worthy of breakthrough: how to store larger video such as long video with lower computational burden and whether other deep learning algorithms can be used to calculate and extract frame elements in video.

COMPARISON OF MEMORY CONSUMPTION BETWEEN TRADITIONAL STORAGE AND PAPER STORAGE

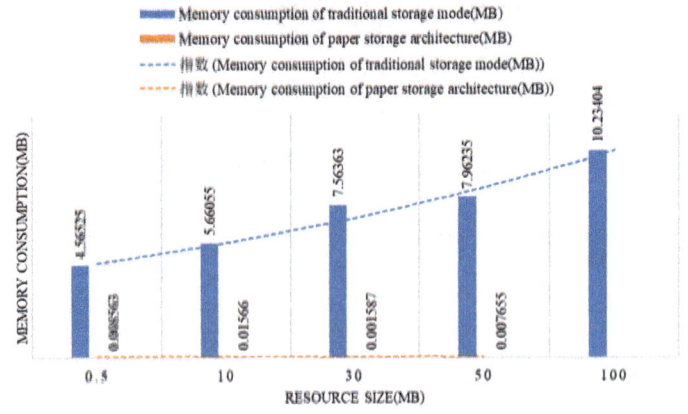

Fig. 6 Comparison of memory consumption between traditional storage and paper storage

Comparison between the existing video copyright storage architecture and the proposed architecture

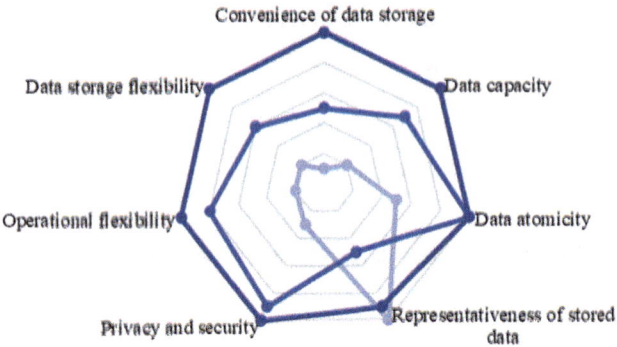

Fig. 7 Comparison between the existing video copyright storage architecture and the architecture proposed

Acknowledgements This paper is supported by the key R&D project of Zhejiang Province, "Research on Key Technologies of all media publishing—Research on Key Technologies of all media press and publication under multi-screen integration environment" (Project No. 2019c03138).

References

1. Qu, C., Tao, M., Zhang, J.: Blockchain based credibility verification method for IoT entities. Secur. Commun. Netw. **2018**(10), 1–11 (2018)
2. Voulodimos, A., Doulamis, N., Doulamis, A.: Deep learning for computer vision: a brief review. Comput. Intell. Neurosci. **2018**(2), 1–13 (2018)
3. Xiaoqiang, Z., Huiping, X.: Image semantic segmentation based on hierarchical feature fusion. Comput. Sci. Explor. **9**(09), 1–8 (2020)
4. Li Changyou, Fu., Huiqi, C.G.: Fast recognition algorithm of improved template matching. J. Hen. Univ. Technol. (Natural Science Edition) **39**(04), 106–111 (2020)
5. Jiang Yingguo, LuLu.: Human motion recognition method based on enhanced spatiotemporal representation and attention model. Comput. Appl. Res. **37**(01), 182–185 (2020)
6. Jingchao, S.: Research on scalable electronic forensics model based on blockchain. Comput. Appl. Res. **09**(09), 1–5 (2019)
7. Kishore, P.V.V., et al.: Indian classical dance action identification and classification with convolutional neural networks. Adv. Multimed. **2018**(01), 1–10 (2018)
8. Zhiguo, M., Jiguan, X., Zhen, L.: Originality of user original short video. J. Dali. Univ. Technol. (Social Science Edition) **12**(01), 1–8 (2019)
9. Yan, X., Zhang, H., Li, H.: Computer vision-based recognition of 3D relationship between construction entities for monitoring struck-by accidents. Comput. Aided Civ. Infrastruct. Eng. **35**(9), 1023–1038 (2020)
10. Liu Suolan, Gu., Jiahui, W.H.: Human behavior recognition based on association partition and st-gcn. Comput. Eng. Appl. **09**(09), 1–11 (2020)

Comments on a Secure and Efficient Three-Factor Authentication Protocol Using Honey List for WSN

Xuanang Lee, Lei Yang, Zhenzhou Zhang, Tsu-Yang Wu, and Chien-Ming Chen

Abstract In 2020, Lee et al. Proposed a secure and efficient three-factor authentication protocol for wireless sensor networks based on a honey list. They claim that their protocol can withstand various attacks. However, this paper still proves that Lee et al. protocol is insecure against known session-specific temporary information exposure attacks.

1 Introduction

In recent years, with the development of science and technology, the Internet of Things (IoT) [7, 8, 13, 14] is becoming more and more important in our life. In a typical IoT environment, various sensors are deployed. These sensors are responsible for detecting target movement, sensing environments, collecting data, and transmitting all necessary messages to gateway servers [5, 7]. Since these messages are often sent through public channels, all the information should be encrypted for data confidentiality. It means that designing a secure and efficient authentication and key agreement protocol is vital [1, 3, 4, 11, 12, 15].

Many authentication and key agreement protocols have been proposed for IoT environments [2, 6, 10, 16, 17]. Recently, Lee et al. [9] Proposed a secure and efficient three-factor authentication protocol for wireless sensor networks based on a honey list. The honey list technology is used for protecting against violence and stolen smart card attacks. This protocol can be applied in many IoT applications

---

X. Lee · L. Yang · Z. Zhang · T.-Y. Wu · C.-M. Chen (✉)
College of Computer and Engineering, Shandong University of Science and Technology, Qingdao 266590, China
e-mail: chienmingchen@ieee.org

X. Lee
e-mail: leonlxaaxl@163.com

L. Yang
e-mail: yanglei2200@163.com

Z. Zhang
e-mail: zhenzhouzhang80@163.com

© The Author(s), under exclusive license to Springer Nature Singapore Pte Ltd. 2022
J.-F. Zhang et al. (eds.), *Advances in Intelligent Systems and Computing*, Smart Innovation, Systems and Technologies 268, https://doi.org/10.1007/978-981-16-8048-9_34

and environments, including intelligent medical, intelligent transportation, smart grid, etc. They also show that their protocol is secure and can withstand a variety of attacks. However, this paper cryptanalysis this protocol [9] and finds that this protocol is still vulnerable to known session-specific temporary information exposure attacks. If an adversary obtains some temporary information in this protocol, he may further calculate the session key.

The other parts of this article are arranged as follows. Section 2 reviews Lee et al.'s protocol. In Sect. 3, we provide cryptanalysis of their protocol. Finally, Sect. 4 concludes this article.

2 Review of Lee et al.'s Protocol

In this section, we review Lee et al.'s protocol [9]. This protocol contains the initialization, registration, login and authentication, and password change phases. In this article, we describe the login and authentication phase. The specific steps of Lee et al.'s protocol can consult in their paper [9]. Table 1 lists the notations used in this paper.

2.1 Registration Phase of Users and Sensors

Assume that a user PU_i and a sensor Sen_j desire to register with the gateway. The following steps are performed:

Table 1 Notations definition

Notations	Descriptions
$Sen_j, SEID_j$	The j-th sensor node and identity
PU_i, PID_i	The i-th user and identity
GWN	Inter network connector
$HIID_i, PIID_j$	Hidden id of i-th user
and j-th sensor O_i, O_G	Rondom numbers of user and inter network connector
y	GWN's secret key
G	Generator
M_{GWN}	Master key
SEK_{ij}	Session key by PU_i, Sen_j
$h(.)$	Hash function
$\|\|$	Conjugation
$\oplus$	Exclusive-or

1. Registration phase of users: PU_i selects his PID_i, PUW_i and $PBIO_i$, generates q_i and then calculates N_i, $H_i = Ge(PBIOi)$, $HIID_i = h(PID_i||q_i)$ and $HPUW_i = h(q_i||PID_i||PUW_i)$. Then PU_i transmits $\{HIID_i, HPUW_i\}$ to GWN through a secure channel. GWN checks if $HIID_i$ is legal. If it is legal, GWN generates F_i and computes $b_i = h(HIID_i||M_GWN||F_i)$, $c_i = b_i \oplus HPUW_i$ and $d_i = h(b_i||HPUW_i)$. After that, GWN stores $HIID_i$ with F_i and $HPUW_i$ and stores values $\{c_i, d_i\}$ into a smartcard. Then, it deliveries this smart card to the user. PU_i calculates $M_i = h(N_i||PUW_i)$ and stores $\{M_i, H_i\}$ into the smartcard.

2. Registration phase of sensors: Sen_j chooses $SEID_j$ and q_j, computes $Sen_1 = SEID_j \oplus h(q_j)$, and sends Sen_1 and q_j to the gateway node. GWN computes $SEID'_j = Sen_1 \oplus h(q_j)$ and $PIID_j = h(SEID_j||r_j)$. After that, GWN generates y and computes $F_j = h(PIID_j||M_GWN||y)$ and stores q_j, $PIID_j$ in its private storage. Finally, GWN sends F_j to Sen_j.

2.2 Login and Authentication Phase

In Fig. 1, we show the authentication process and the detailed steps are given below.

1. PU_i inputs PID_i, PUW_i and $PBIO_i$. Firstly, PU_i calculates $N_i = Re(PBIO_i, H_i)$, $q_i = M_i \oplus h(N_i||HPW_i)$, $HIID_i = h(PID_i||q_i)$ and $HPUW_i = h(q_i||PID_i||PUW_i)$. PU_i gets $b_i = c_i \oplus HPUW_i$ and $d'_i = h(b_i||HPUW_i)$. PU_i calculates $d'_i = h(b_i||HPUW_i)$ and check d'_i and d_i. If they are the same, PU_i generates O_i and calculates $W_1 = h(b_i||SEID_j) \oplus O_i$ and $W_2 = h(b_i||SEID_j||O_i)$. Finally, PU_i sends $HIID_i$, W_1, W_2 to GWN.

2. GWN checks F_i from his database and calculates $b'_i = (HIID_i||M_GWN||F_i)$, $O_i = h(b'_i||SIID_j) \oplus W_1$ and $W'_2 = h(b_i||SIID_j||O_i)$. GWN checks $W_2 \stackrel{?}{=} W'_2$. If it is equal, GWN calculates $F_j = h(h(SEID_j||q_j)||M_GWN||y)$, $W_3 = h(SEID_j||PIID_j||F_j) \oplus O_G$ and $W_4 = h(F_j||PIID_j||O_G)$. Finally, GWN sends W_3, W_4 to Sen_j.

3. Sen_j calculates $O_G = h(PIID_j||F_j) \oplus W_3$, $W'_4 = h(PIID_j||F_j||O_G)$. Sen_j checks legal to compare W_4 with W'_4. If they are legal, Sen_j generates a O_j and computes $SEK_{ij} = h(PIID_j||F_j||O_G)$ and $W_5 = h(SEK_{ij}||F_j||O_G)$. Finally, Sen_j sends W_5 GWN.

4. GWN computes $SEK_{ij} = h(PIID_j||F_j||O_G)$ and $W'_5 = h(SEK_{ij}||F_j||O_G)$. GWN checks $W_5 \stackrel{?}{=} W'_5$. If it is equal, GWN calculates $HIID_{inew} = h(O_G||HIID_i)$, $b_{inew} = h(HIID_{inew}||M_GWN||O_G)$, $W_6 = (O_i||b'_i) \oplus HIID_{inew}$, $W_7 = (O_i||b'_i) \oplus b_{inew}$, $W_8 = (O_i||b'_i) \oplus SEK_{ij}$ and $W_{gu} = (SEK_ij||O_i||b_{inew}||HIID_{inew})$. Finally, GWN sends W_6, W_7, W_8, W_{gu} to PU_i. GWN renews $HIID_i$ to $HIID_{inew}$. If it is not successful, GWN keeps to store $HIID_i$.

5. PU_i calculates $HIID'_{inew} = W_6 \oplus (O_i||b_i)$, $b'_{inew} = W_7 \oplus (O_i||b_i)$, $SEK'_{ij} = W_8 \oplus (O_i||b_i)$ and $W'_{gu} = (SEK'_{ij}||O_i|b_{inew}||HIID_{inew})$. PU_i checks whether

User(PU_i)	Gateway	Sensor(Sen_j)
Inputs PID_i, PUW_i		
Imprints biometric $PBIO_i$		
$N_i = Re(PBIO_i, H_i)$		
$q_i = M_i \oplus h(N_i \| HPW_i)$		
$HIID_i = h(PID_i \| q_i)$		
$HPUW_i = h(q_i \| PID_i \| HPW_i)$		
$b_i = c_i \oplus HPUW_i$		
$d'_i = h(b_i \| HPUW_i)$. Checks $d'_i \stackrel{?}{=} d_i$		
Generates a random number O_i		
$W_1 = h(b_i \| SEID_j) \oplus O_i$		
$W_2 = h(b_i \| SEID_j \| O_i)$		
$\xrightarrow{<HIID_i, W_1, W_2>}$		
	Retrives F_i from a database	
	$b'_i = (HIID_i \| M_{GWN} \| W_i)$	
	$O_i = h(b'_i \| SEID_j) \oplus W1$	
	$W'_2 = h(b'_i \| SEID_j \| O_i)$	
	Checks $W'_2 \stackrel{?}{=} W_2$	
	$if not, b'_i$ inserts into Honey_list	
	Generates a random number O_G	
	Computes	
	$F_j = h(h(SEID_j \| q_j) \| M_{GWN} \| y)$	
	$W_3 = h(SEID_j \| PEID_j \| F_j) \oplus O_G$	
	$W_4 = h(F_j \| PEID_j \| O_G)$	
	$\xrightarrow{<W_3, W_4>}$	
		$O_G = h(PEID_j \| F_j) \oplus W_3$
		$W'_4 = h(PEID_j \| F_j \| O_G)$
		$W'_4 \stackrel{?}{=} W_4$
		Computes
		$SEK_{ij} = h(PEID_j \| F_j \| O_G)$
		$W_5 = h(SEK_{ij} \| F_j \| O_G)$
	$\xleftarrow{<W_5>}$	
	$SEK_{ij} = h(PEID \| F_j \| O_G)$	
	$W'_5 = h(SEK_{ij} \| F_j \| O_G)$	
	$W_5 \stackrel{?}{=} W'_5$	
	$HIID_{inew} = h(O_G \| HIID_i)$	
	$b_{inew} = h(HIID_{inew} \| M_{GWN} \| O_G)$	
	$W_6 = (O_i \| b'_i) \oplus HIID_{inew}$	
	$W_7 = (O_i \| b'_i) \oplus b_{inew}$	
	$W_8 = (O_i \| b'_i) \oplus SEK_{ij}$	
	$W_{gu} = (SEK_{ij} \| O_i \| b_{inew} \| HIID_{inew})$	
	if key agreement is successful	
	replaces $HIID_{inew}$	
	Otherwise, not updated	
$\xleftarrow{<W_6, W_7, W_8, w_{gu}>}$		
$HIID'_{inew} = W_6 \oplus (O_i \| b_i)$		
$b'_{inew} = W_7 \oplus (O_i \| b_i)$		
$SEK'_{ij} = W_8 \oplus (O_i \| b_i)$		
$W'_{gu} = (SEK_{ij} \| O_i \| b_{inew} \| HIID'_{inew})$		
$W'_{gu} \stackrel{?}{=} W_{gu}$		
Computes $c_{inew} = b_{inew} \oplus HPUW_i$		
updates		
$b_{inew}, c_{inew}, HIID_{inew}$		

Fig. 1 Login and authentication phase

W'_{gu} and W_{gu} are equal. PU_i calculates $c_{inew} = b_{inew} \oplus HPUW_i$ and $d_{inew} = h(b_{inew} \| HPUW_i)$ and renews b_{inew}, c_{inew}, d_{inew} and $HIID_{inew}$. Finally, PU_i, GWN and Sen_j have a common session key.

User(PU_i)	Gateway	Sensor(Sen_j)
Inputs PID_i, PUW_i Imprints biometric $PBIO_i$ $N_i = Re(PBIO_i, H_i)$ $q_i = M_i \oplus h(N_i \| HPW_i)$ $HIID_i = h(PID_i \| q_i)$ $HPUW_i = h(q_i \| PID_i \| HPW_i)$ $b_i = c_i \oplus HPUW_i$ $d'_i = h(b_i \| HPUW_i)$. Checks $d'_i \stackrel{?}{=} d_i$ Generates a random number O_i $W_1 = h(b_i \| SEID_j) \oplus O_i$ $W_2 = h(b_i \| SEID_j \| O_i)$ $\xrightarrow{<\boxed{HIID_i}, W_1, W_2>}$		
	Retrives F_i from a database $b'_i = (HIID_i \| M_{GWN} \| W_i)$ $O_i = h(b'_i \| SEID_j) \oplus W1$ $W'_2 = h(b'_i \| SEID_j \| O_i)$ Checks $W'_2 \stackrel{?}{=} W_2$ $if not, b'_i$ inserts into Honey_list Generates a random number $\boxed{O_G}$ Computes $F_j = h(h(SEID_j \| q_j) \| M_{GWN} \| y)$ $W_3 = h(SEID_j \| PEID_j \| F_j) \oplus O_G$ $W_4 = h(F_j \| PEID_j \| O_G)$ $\xrightarrow{<W_3, W_4>}$	
		$O_G = h(PEID_j \| F_j) \oplus W_3$ $W'_4 = h(PEID_j \| F_j \| O_G)$ $W'_4 \stackrel{?}{=} W_4$ Computes $SEK_ij = h(PEID_j \| F_j \| O_G)$ $W_5 = h(SEK_ij \| F_j \| O_G)$ $\xleftarrow{<W_5>}$
	$SEK_{ij} = h(PEID \| F_j \| O_G)$ $W'_5 = h(SEK_{ij} \| F_j \| O_G)$ $W_5 \stackrel{?}{=} W'_5$ $\boxed{HIID_{inew} = h(O_G \| HIID_i)}$ $b_{inew} = h(HIID_{inew} \| M_{GWN} \| O_G)$ $\boxed{W_6 = (O_i \| b'_i) \oplus HIID_{inew}}$ $W_7 = (O_i \| b'_i) \oplus b_{inew}$ $\boxed{W_8 = (O_i \| b'_i) \oplus SEK_{ij}}$ $W_{gu} = (SEK_{ij} \| O_i \| b_{inew} \| HIID_{inew})$ if key agreement is successful replaces $HIID_{inew}$ Otherwise, not updated $\xleftarrow{<W_6, W_7, W_8, W_{gu}>}$	
$HIID'_{inew} = W_6 \oplus (O_i \| b_i)$ $b'_{inew} = W_7 \oplus (O_i \| b_i)$ $SEK'_{ij} = W_8 \oplus (O_i \| b_i)$ $W'_{gu} = (SEK'_{ij} \| O_i \| b_{inew} \| HIID'_{inew})$ $W'_{gu} \stackrel{?}{=} W_{gu}$ Computes $c_{inew} = b_{inew} \oplus HPUW_i$ updates $b_{inew}, c_{inew}, HIID_{inew}$		

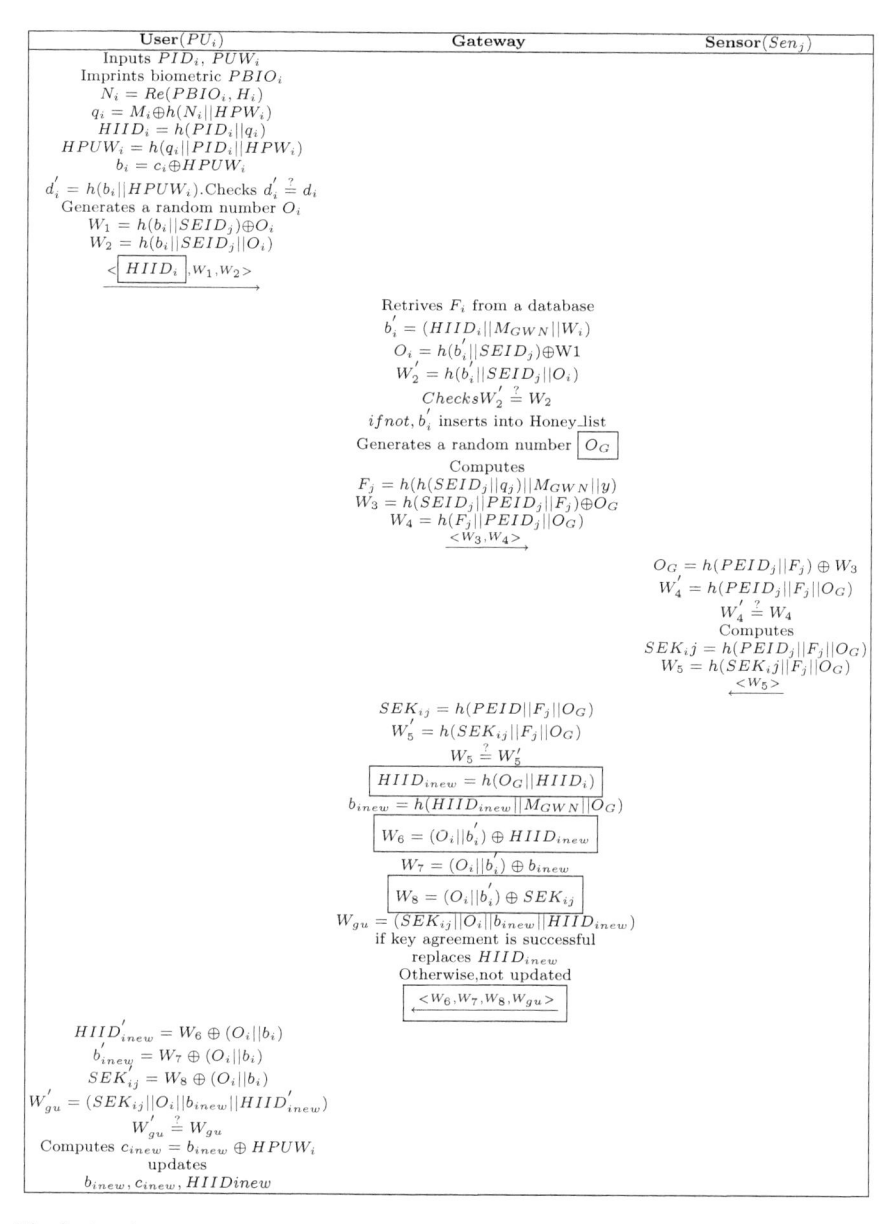

Fig. 2 Attack process

3 Cryptanalysis of Lee et al.'s Protocol

In this section, we demonstrate that Lee et al.'s scheme can't resist a known session-specific temporary information exposure attacks. In Fig. 2, the specific steps are as follows:

3.1 Known Session-Specific Temporary Information Exposure Attacks

If an attacker E obtains a session-specific temporary information N_G, and some parameters are transmitted through an insecure channel, E can calculate the session key. The detailed steps are as follows:

1. HID_{inew} is equal to $h(N_G||HID_i)$, since HID_i is a public message, thus E can get HID_{inew}.
2. $(N_i||a_i')$ is equal to $M_6 \oplus HID_{inew}$, Since M_6 is a public message, thus we can get $(N_i||a_i')$.
3. SK_{ij} is equal to $(N_i||a_i') \oplus M_8$, Since M_8 is a public message, thus we can get SK_{ij}, which means Lee et al.'s protocol does not provide session-specific temporary information secrecy.

4 Conclusion

This paper analyzes the protocol of Lee et al. Although Lee et al.'s claim that their protocol can resist all kinds of attacks, this protocol is still insecure. We show that this protocol is still vulnerable to known session-specific temporary information exposure attacks. In the future, we will try to propose an improved protocol.

References

1. Bojjagani, S., Sastry, V., Chen, C.M., Kumari, S., Khan, M.K.: Systematic survey of mobile payments, protocols, and security infrastructure. J. Ambient Intell. Human. Comput. 1–46 (2021)
2. Chen, C.M., Deng, X., Gan, W., Chen, J., Islam, S.H.: A secure blockchain-based group key agreement protocol for iot. J. Supercomput. 1–23 (2021)
3. Chen, C.M., Fang, W., Wang, K.H., Wu, T.Y.: Comments on "an improved secure and efficient password and chaos-based two-party key agreement protocol". Nonlinear Dyn. 87(3), 2073–2075 (2017)
4. Chen, C.M., Li, C.T., Liu, S., Wu, T.Y., Pan, J.S.: A provable secure private data delegation scheme for mountaineering events in emergency system. Ieee Access 5, 3410–3422 (2017)

5. Estrin, D., Girod, L., Pottie, G., Srivastava, M.: Instrumenting the world with wireless sensor networks. In: 2001 IEEE International Conference on Acoustics, Speech, and Signal Processing. Proceedings (Cat. No. 01CH37221). vol. 4, pp. 2033–2036. IEEE (2001)
6. Hussain, S., Ullah, I., Khattak, H., Khan, M.A., Chen, C.M., Kumari, S.: A lightweight and provable secure identity-based generalized proxy signcryption (ibgps) scheme for industrial internet of things (iiot). J. Inf. Secur. Appl. **58**, 102625 (2021)
7. Khan, M.A., Salah, K.: Iot security: review, blockchain solutions, and open challenges. Futur. Gener. Comput. Syst. **82**, 395–411 (2018)
8. Lee, I., Lee, K.: The internet of things (iot): applications, investments, and challenges for enterprises. Bus. Horiz. **58**(4), 431–440 (2015)
9. Lee, J., Yu, S., Kim, M., Park, Y., Das, A.K.: On the design of secure and efficient three-factor authentication protocol using honey list for wireless sensor networks. IEEE Access **8**, 107046–107062 (2020)
10. Li, C.T., Lee, C.C., Weng, C.Y., Chen, C.M.: Towards secure authenticating of cache in the reader for rfid-based iot systems. Peer Peer Netw. Appl. **11**(1), 198–208 (2018)
11. Shamshad, S., Mahmood, K., Kumari, S., Chen, C.M., et al.: A secure blockchain-based e-health records storage and sharing scheme. J. Inf. Secur. Appl. **55**, 102590 (2020)
12. Sun, H.M., Wang, K.H., Chen, C.M.: On the security of an efficient time-bound hierarchical key management scheme. IEEE Trans. Depend. Secur. Comput. **6**(2), 159–160 (2009)
13. Wang, E.K., Chen, C.M., Hassan, M.M., Almogren, A.: A deep learning based medical image segmentation technique in internet-of-medical-things domain. Futur. Gener. Comput. Syst. **108**, 135–144 (2020)
14. Wang, K., Chen, C.M., Liang, Z., Hassan, M.M., Sarné, G.M., Fotia, L., Fortino, G.: A trusted consensus fusion scheme for decentralized collaborated learning in massive iot domain. Inf. Fusion **72**, 100–109 (2021)
15. Wang, P., Chen, C.M., Kumari, S., Shojafar, M., Tafazolli, R., Liu, Y.N.: Hdma: hybrid d2d message authentication scheme for 5g-enabled vanets. IEEE Trans. Intell. Transp. Syst. (2020)
16. Wu, T.Y., Chen, C.M., Wang, K.H., Wu, J.M.T.: Security analysis and enhancement of a certificateless searchable public key encryption scheme for iiot environments. IEEE Access **7**, 49232–49239 (2019)
17. Yavari, M., Safkhani, M., Kumari, S., Kumar, S., Chen, C.M.: An improved blockchain-based authentication protocol for iot network management. Secur. Commun. Netw. (2020)

Integrating Autonomous Decentralized Communication and Edge Computing for Real-Time Control in IoT System

Masaya Harada, Zhaoyang Du, Celimuge Wu, Tsutomu Yoshinaga, Wugedele Bao, and Yusheng Ji

Abstract With the continuous development of the Internet of Things (IoT) technology, various IoT applications and services are emerging one after another, providing users with a continuous power to enjoy comfortable, efficient, and convenient life, and inject new vitality into the development of society. In this paper, we propose a concept of real-time control in the IoT system by integrating autonomous decentralized communication and the hop-by-hop data transfer method. Furthermore, we prove the effectiveness of the proposed method using Raspberry Pis and laptops.

1 Introduction

In recent years, the number of smartphones and tablet PCs equipped with wireless devices are increasing dramatically. When using these terminals, a communication mode called "infrastructure mode" is used in which communication is performed via an access point such as a wireless LAN [1, 2]. However, in Japan, the public wireless LAN environment is not sufficiently advanced. Therefore, the ad hoc network, in which terminals communicate directly with each other wirelessly without going through an access point, is attracting attention [3, 4]. In a relatively large-scale ad hoc network, multihop communication is performed to enable communication between terminals [5–7].

Since the network topology frequently changes in the ad hoc network, the communication quality is not stable for each hop, and route disconnection is likely to occur [8]. There are problems such as an increase in the network load due to the

---

M. Harada · Z. Du (✉) · C. Wu · T. Yoshinaga
The University of Electro-Communications, 1-5-1, Chofugaoka,
Chofu-shi, Tokyo 182-8585, Japan
e-mail: duzhaoyang@uec.ac.jp

W. Bao
Hohhot Minzu College, Hohhot 010051, China

Y. Ji
National Institute of Informatics, 2-1-2, Hitotsubashi, Chiyoda-ku, Tokyo 101-8430, Japan

route reconstruction and a decrease in communication efficiency [9, 10]. Reference [11] proposed a broadcast routing algorithm that packets could forwarded to multiple intermediate nodes with good link quality and realize simultaneous multipath forwarding autonomously. The packet transmission success rate, throughput, and transmission delay are improved.

5G is expected as a communication system that meets the needs of the next generation. The requirements for 5G include large data capacity, ultra-low latency, and high throughput due to strict QoS (Quality of Service) requirements [12]. Moreover, since 5G is sensitive to delay and always pursues lower delay, 5G data could classify into hard real-time data, soft real-time data, and non-real-time data according to the delay time [13].

In edge computing, the edge server is located close to the client, which reduces delays and is expected to provide services with high real-time requirements such as hard real-time data. Thus, edge computing is expected to realize low latency communication, which is a requirement of 5G [14]. Edge computing is classified according to various characteristics, and further development is expected by the integration of 5G and edge computing. Liu et al. [15] proposed a use case to realize the intelligence edge in the edge computing IoT scenario. The round trip time (RTT) is decreased by reducing the distance between the client and server.

In this paper, we utilize multihop communication in the ad hoc network to transfer data from the terminals to the edge server. The ad hoc network can be constructed between terminals even in a place where no access point exists, and it is possible to establish a route to transfer data by arranging several terminals. Besides, multihop communication makes it possible to construct a network for devices outside the access point range via some relay terminals. However, the traditional multihop communication has the problem that the throughput drops sharply as the number of hops increases. Therefore, we use hop-by-hop transport instead of the traditional end-to-end principle [16] in multihop communication to perform communication, enabling less delay. Besides, an IoT system with real-time control is realized by integrating autonomous decentralized communication using hop-by-hop transport with edge computing. The effectiveness of the system is evaluated by implementing it on Raspberry Pis and laptops.

2 Proposed Scheme

2.1 Hop-by-Hop Data Transfer in Multihop Communication

In the traditional end-to-end method, the radio interference, the hidden terminal problem, and the exposed terminal problem could cause the throughput drops rapidly with the increase of the number of hops. In this section, we focus on the data transmission time and the corresponding ACK time, and try to improve the network throughput.

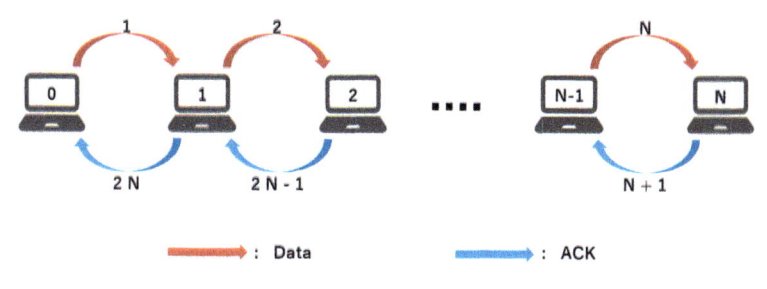

Fig. 1 End-to-end method

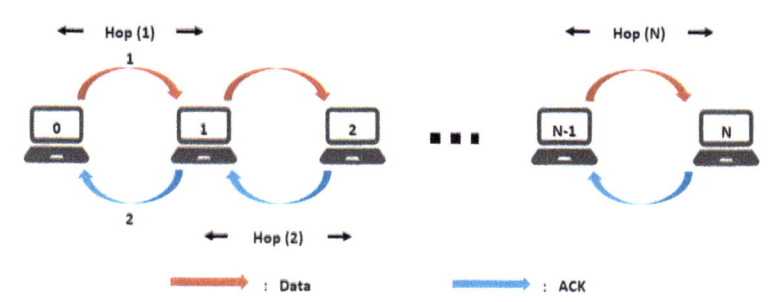

Fig. 2 Hop-by-hop method

In the traditional end-to-end communication, the source terminal sends data to the destination terminal, and then an ACK is sent back. In this method, as shown in Fig. 1, the number of communication time is twice the number of terminals until data transmission is completed.

On the other hand, in the proposed method, multiple one-hop sessions are used. As shown in Fig. 2, since data transmission and ACK are performed in a one-hop session, the number of communication times is the same as the number of terminals until data transmission is completed.

Therefore, the hop-by-hop method could reduce half communication time compared to the end-to-end method, and also, high end-to-end throughput can be realized.

2.2 Integrating Edge Computing and Multihop Communication

In this paper, we propose integrating autonomous decentralized communication and edge computing by using an ad hoc network to enable communication between devices, as shown in Fig. 3. The ad hoc network can be constructed between devices even in a place where no access point exists, so it is possible to establish a route by arranging devices. Besides, due to multihop communication characteristics, the communication between devices in a wide range can be achieved by passing through

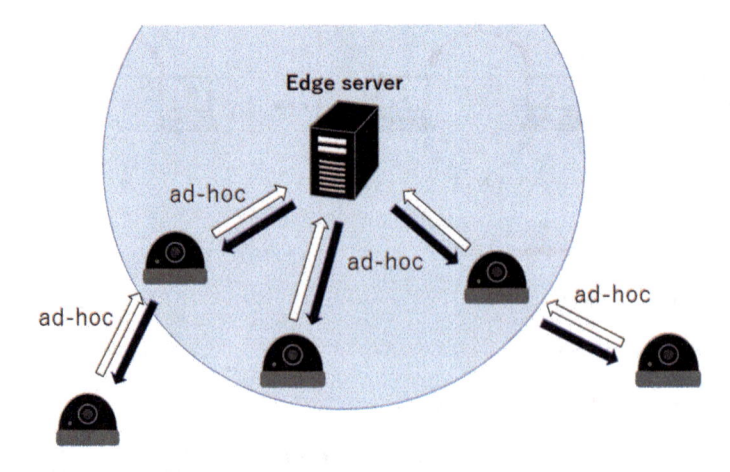

Fig. 3 Integrating edge computing and multihop communication

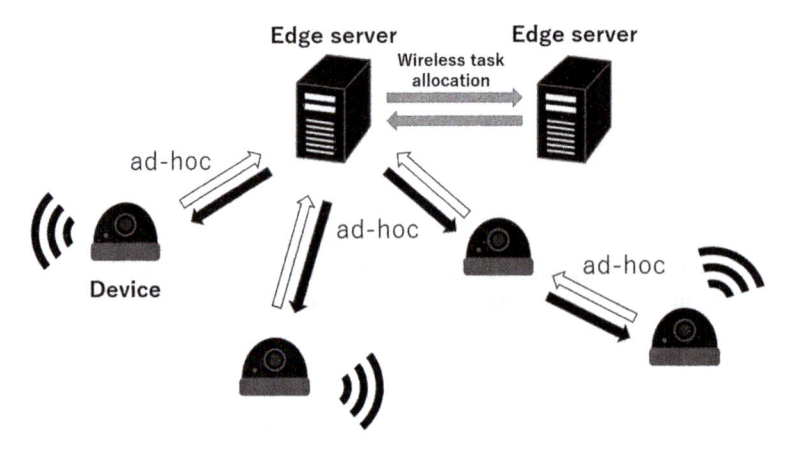

Fig. 4 Load balancing between edge servers

other devices. The real-time control of the IoT system is realized by integrating autonomous decentralized communication and edge computing by adopting hop-by-hop communications.

Load distribution is one of the advantages of edge computing. By arranging multiple edge servers near to clients, the task processing could allocate to other servers. Besides, when arranging multiple edge servers, it is conceivable that some servers will have a large amount of tasks and some will have a small amount of tasks. When the amount of task is large in the server, the processing time inevitably becomes long, so it is reasonable to shorten the processing time by transferring the data to the server with less task. As shown in Fig. 4, the edge server distributes and averages the number of tasks to prevent the concentration of task processing.

3 Evaluation

3.1 Experiment Environment

3.1.1 Computer Environment

The specifications of two PCs and three Raspberry Pis used in this paper are shown in Tables 1 and 2.

3.1.2 Communication Environment

In this paper, each terminal is connected using ad hoc mode. Table 3 shows the wireless environment used in the experiment. In order to improve network throughput by suppressing delays, we adopt OLSR routing protocol [17], which is a proactive routing protocol that does not generate delays during communications.

Table 1 Computer environment

Processor	Intel Core i5-10300H	2.50 GHz × 8
OS	Ubuntu 18.04.5 LTS	
Number of core	2	
GPU	GTX 1650 Ti GDDR6 4 GB	
Memory	DDR4 SO-DIMM 16 GB	
Wireless LAN	Wi-Fi 6AX201NGW	

Table 2 Raspberry Pi environment

Model	Raspberry Pi 3 Model B Plas Rev 1.3
OS	Raspbian GNU Linux 10.6
CPU	Broadcom BCM2835, ARMv7 Processor rev 4
Memory	1GB

Table 3 Communication environment

IEEE	802.11gn
Frequency	2.412GHz
Tx-power	31 dBm
Retry short limit	7
RTS thr	Off
Fragment thr	Off
Power management	Off

Fig. 5 Device configuration of 2-hop model

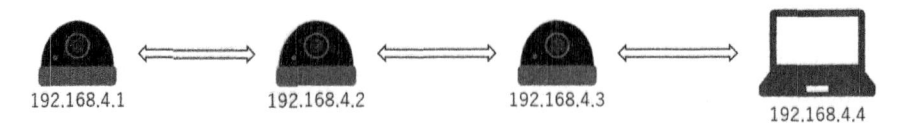

Fig. 6 Device configuration of 3-hop model

3.2 Latency Evaluation of Hop-by-hop Data Transfer

This section shows the comparative evaluation of the proposed hop-by-hop data transfer method with the conventional end-to-end data transfer method. In this experiment, a 2-hop evaluation using three Raspberry Pis, and a 3-hop evaluation using three Raspberry Pis and one laptop are performed. The devices are arranged as shown in Figs. 5 and 6. In order to measure the delay time, an image file is been transferred and the delay time in one frame is measured.

Figure 7a shows the latency comparison in 2-hop, and Fig. 7b shows the latency comparison in 3-hop scenario. The data transfer process is performed ten times for each method, and the average delay time and the dispersion delay time are obtained. In the end-to-end data transport method, data retransmission is performed at the source terminal. On the other hand, in the hop-by-hop case, each relay node is responsible for ensuring that the next node receives the packet. Therefore, as can be seen from the figure, the delay time of the proposed hop-by-hop method is shorter than that of the traditional end-to-end method in both 2-hop and 3-hop scenario.

The network topology in the ad hoc network changes with the movement of terminals, so packet loss often occurs. In the end-to-end communication method, data must be re-sent from the source terminal every time a packet loss occurs, which will cause high delay. In the hop-by-hop communication method, data retransmission can be conducted from intermediate nodes, so the delay time dispersion is low. In summary, these results show that the proposed hop-by-hop data transfer method has a smaller delay than the traditional end-to-end data transfer method. It can be inferred that network throughput also increases since the delay time of the proposed method is stable.

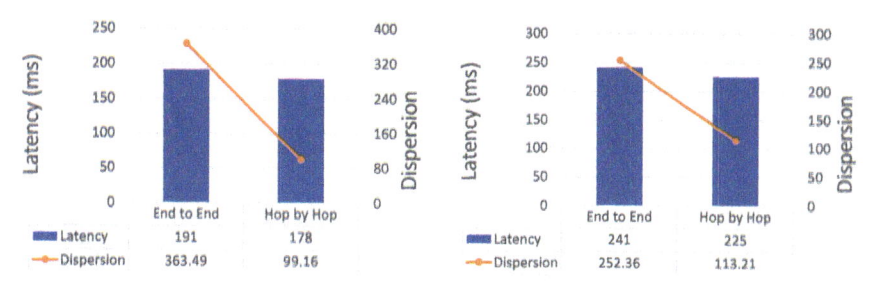

Integrating Autonomous Decentralized Communication ... 373

(a) Delay time in 2-hop scenario.

	End to End	Hop by Hop
Latency	191	178
Dispersion	363.49	99.16

(b) Delay time in 3-hop scenario.

	End to End	Hop by Hop
Latency	241	225
Dispersion	252.36	113.21

Fig. 7 Latency evaluation

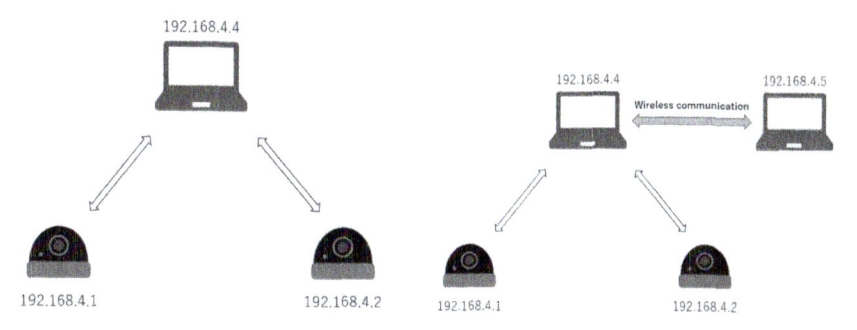

(a) Devices configuration with one edge server.

(b) Devices configuration with two edge servers.

Fig. 8 Devices configuration

3.3 Evaluation of Cooperation Between Edge Servers

In this section, as shown in Fig. 8b, two edge servers are set up, and wireless communication between two edge servers is used. In Fig. 8a, one edge server is set up and uses the traditional method as a baseline.

As shown in Fig. 9, the delay time does not change much in the wired connection scenario, even if the task processing is allocated to two edge servers. In the wireless connection scenario, the delay time when the task processing is allocated to two servers is longer than that in the traditional method. In summary, these results show that when the use of hop-by-hop edge computing is less effective when edge servers are wirelessly connected. Besides, when edge servers are in a wired connection, the proposed method is very effective. Although the delay time of the proposed method is relatively high in this scenario, when the number of tasks is large, it can be expected that the proposed method has a higher processing efficiency.

Fig. 9 Maximum delay time comparison

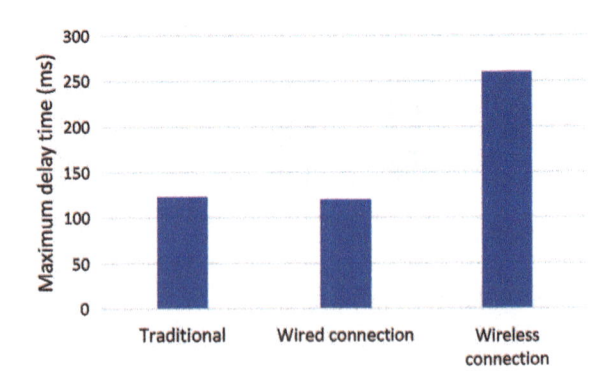

4 Conclusion

In this paper, we first realize the hop-by-hop method instead of the traditional end-to-end method in multihop communications to reduce the communication latency. Then, we propose the integration of edge computing and multihop communications. Finally, we evaluate the effectiveness of the proposal using Raspberry Pis and laptops. The proposed method achieves a shorter delay as compared with the traditional method for data processing in wireless environments by integrating the edge computing and autonomous decentralized communications.

Acknowledgements This research was supported in part by the National Natural Science Foundation Program of China under Grant No. 62062031, in part by the Inner Mongolia natural science foundation grant number 2019MS06035 and Inner Mongolia Science and Technology Major Project, China, and in part by JSPS KAKENHI grant numbers 18KK0279, 19H04093, and 20H00592, Japan.

References

1. Ippisch, A., Graffi, K.: Infrastructure mode based opportunistic networks on android devices. In: 2017 IEEE 31st International Conference on Advanced Information Networking and Applications (AINA), pp. 454–461. Taipei, China (2017)
2. Li, S.: Comparative analysis of infrastructure and Ad-Hoc wireless networks. ITM Web Conf. **25**, 01009 (2019)
3. Keerthana, G., Anandan, P.: A survey on security issues and challenges in mobile Ad-hoc network. EAI Endorsed Trans. Energy Web **7**(20), 1–8 (2018)
4. Olaniyan, O.M., Omodunbi, B.A., Adebimpe, E., Bolanle, W.W., Oyedepo, O.M., Adanigbo, O.O.: Power aware and secured routing protocol in mobile ad-hoc network: a survey. Technology **11**(7), 706–717 (2020)
5. Khanna, G., Chaturvedi, S.K.: A comprehensive survey on multi-hop wireless networks: milestones, changing trends and concomitant challenges. Wireless Pers. Commun. **101**(2), 677–722 (2018)
6. Krishnan, A., Sharma, V.: Distributed control and quality-of-service in multihop wireless networks. In: 2018 IEEE International Conference on Communications (ICC), pp. 1–7. Kansas City, USA (2018)

7. Hassan, N., Yau, K.A., Wu, C.: Edge computing in 5G: a review. IEEE Access **7**, 127276–127289 (2019)
8. Shantaf, A.M., Kurnaz, S., Mohammed, A.H.: Performance evaluation of three mobile Ad-hoc network routing protocols in different environments. In: 2020 International Congress on Human-Computer Interaction. Optimization and Robotic Applications (HORA), pp. 1–6. Ankara, Turkey (2020)
9. Er-rouidi, M., Moudni, H., Mouncif, H., Merbouha, A.: A balanced energy consumption in mobile ad hoc network. Proc. Comput. Sci. **151**, 1182–1187 (2019)
10. Wu, C., Liu, Z., Liu, F., Yoshinaga, T., Ji, Y., Li, J.: Collaborative learning of communication routes in edge-enabled multi-access vehicular environment. IEEE Trans. Cognit. Commun. Netw. **6**(4), 1155–1165 (2020)
11. Yamazaki, T., Yamamoto, R., Miyoshi, T., Asaka, T., Tanaka, Y.: PRIOR: prioritized forwarding for opportunistic routing. IEICE Trans. Commun. **100**(1), 28–41 (2017)
12. Aljiznawi, R.A., Alkhazaali, N.H., Jabbar, S.Q., Kadhim, D.J.: Quality of service (qos) for 5g networks. Int. J. Futur. Comput. Commun. **6**(1), 27 (2017)
13. Dighriri, M., Lee, G.M., Baker, T.: Measurement and classification of smart systems data traffic over 5G mobile networks. In: Technology for Smart Futures, pp. 195–217. Springer, Cham (2018)
14. Taleb, T., Samdanis, K., Mada, B., Flinck, H., Dutta, S., Sabella, D.: On multi-access edge computing: a survey of the emerging 5G network edge cloud architecture and orchestration. IEEE Commun. Surv. Tutor. **19**(3), 1657–1681 (2017)
15. Liu, Y., Peng, M., Shou, G., Chen, Y., Chen, S.: Toward edge intelligence: multiaccess edge computing for 5G and internet of things. IEEE Internet Things J. **7**(8), 6722–6747 (2020)
16. Lee, C., Fumagalli, A.: Internet of things security - multilayered method for end to end data communications over cellular networks. In: 2019 IEEE 5th World Forum on Internet of Things (WF-IoT), pp. 24–28. Limerick, Ireland (2019)
17. Toutouh, J., Garcia-Nieto, J., Alba, E.: Intelligent OLSR routing protocol optimization for VANETs. IEEE Trans. Veh. Technol. **61**(4), 1884–1894 (2012)

Meta-Graph-Based Embedding for Recommendation over Heterogeneous Information Networks

Shiyuan Shuai, Xuewen Shen, Jun Wu, and Zhiqi Xu

Abstract Heterogeneous Information Networks (HINs) are a powerful technique for information expression and have been integrated into recommendation systems as an auxiliary information source. Recently, researchers begin to utilize meta-graphs to capture complex semantics. However, existing recommendation methods based on meta-graphs face two problems. First, they do not learn the representation of meta-graphs. Second, they ignore the dimensional inconsistency of similarities from different meta-graphs, and do not explicitly use similarities. This paper presents MGBE, a meta-graph-based embedding method that learns semantic representations of meta-graphs to generate new features of meta-graphs. Specifically, MGBE utilizes meta-graphs to calculate similarities and standardizes them. It further constructs the embedding vectors of meta-graphs and combines them with similarities to generate new feature vectors. MGBE integrates user features, item features, and meta-graph features to predict the rates of user-item pairs. Extensive experiments on two real-world datasets show the effectiveness of MGBE.

1 Introduction

In the era of knowledge and information explosion, recommender systems have been widely used on commercial platforms such as Amazon or Yelp to alleviate the impact of information overloading and improve user experience. A well-performed recommender system, which provides users with products or services (generally referred to as items), can model users' preferences from feedback.

Traditional recommendation approaches [4, 7, 12] based on collaborative filtering (CF) mainly utilize the interaction information or rates of user-item. These methods

---

S. Shuai
Zhejiang Sci-Tech University, Hangzhou, China

X. Shen (✉) · Z. Xu
Communication University of Zhejiang, Hangzhou, China
e-mail: shenxuewen@cuz.edu.cn

J. Wu
China Mobile Group Zhejjiang Co, Hangzhou, China

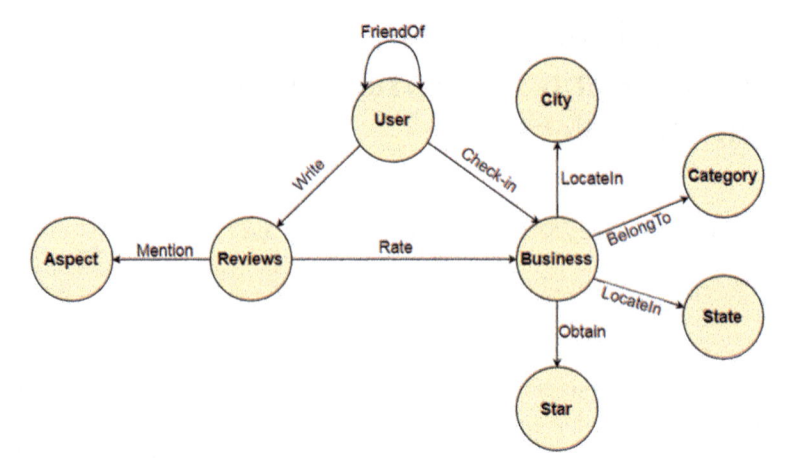

Fig. 1 Example of HIN schema on Yelp dataset

only consider a single relationship type while neglecting the existence of an attribute-rich heterogeneous information network environment. Previous studies show that utilizing additional information about users and items can improve recommendation quality.

Recently, Heterogeneous Information Networks (HINs), which contain multiple types of entities and relations, have been successfully applied to recommender systems [13, 15, 16, 18] due to their powerful ability for information expression. Here, we show an example of the HIN schema from Yelp dataset in Fig. 1. In the network structure of HIN, entity nodes introduce additional information, and relations between entities of different types consist of different semantic relations. So far, there has been a great progress in integrating HINs into recommender systems as an auxiliary information source. Among these approaches, meta-path-based methods [13, 18] have attracted much attention. Predefined meta-paths are utilized to capture the semantic relation between users and items. However, meta-path cannot capture complex semantic information [19]. For example, meta-path cannot capture the semantics between $user_1$ and $user_2$ if they rate the same item and mention the same aspect in their reviews.

Therefore, meta-graph (or meta-structure) [2, 6] has been proposed to capture richer semantic information. Figure 2 shows an example of meta-graph on Yelp. Recently, some work introduces the concept of meta-graph into recommender systems. FMG [19] utilizes the matrix factorization (MF) [8] to factorize user-item similarities from different meta-graphs, and then employs a group lasso regularized factorization machine (FM) [10, 11] to assemble the user and item latent features. To distinguish the influence of different meta-graphs, an attention mechanism is introduced into the feature enhancement model to aggregate user and item latent vectors. MGAR [1] designs a hierarchical feature interaction method to make recommendations. Despite their good performance, the existing methods still have limitations.

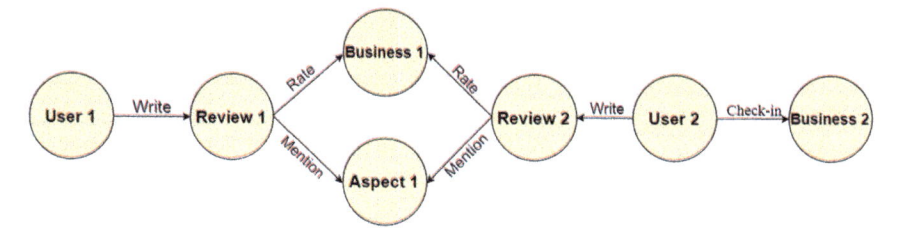

Fig. 2 Example of meta-graph on Yelp dataset

First, they do not learn the representations of meta-graphs. Second, they ignore the dimensional inconsistency of similarities from different meta-graphs, and do not explicitly use similarities. These methods choose to use latent vectors for rating instead of using similarity or representation of meta-graph. In addition, there is a significant difference between the distribution of similarities and latent vectors. For example, the mean value of similarities from the first meta-graph is 0.001 with the standard deviation of 0.224, while the mean value of similarities from the last meta-graph is 410.22 with the standard deviation of 7523.3. Therefore, the above two aspects are not considered, and the meta-graph-based features may not be optimal for recommendation.

To address the above challenges, we propose a new model, MGBE. We first employ matrix factorization (MF) on similarity matrices to obtain a set of user and item latent features. We then utilize the latent features to calculate unobserved similarities of user-item pairs and use the Z-score method to transform similarities. Finally, we take the meta-graph embedding vector as a feature and the similarity as a feature value to generate meta-graph feature vector. After concatenating all vectors, we send them to the recommendation model to make prediction. In order to increase representation power, we integrate an attention mechanism into our model.

The main contributions of our work are summarized as follows:

- We propose a novel approach, MGBE, which combines meta-graphs and an attention mechanism, for recommender systems.
- We learn the representation of meta-graph and dimensionless similarity, which construct a new feature of meta-graph.
- We conduct experiments on two real-world datasets, and the results demonstrate the effectiveness of our model.

2 Problem Definition

In this paper, we consider a recommender system that aims to predict the ratings of unobserved user-item pairs. Given a user set $\mathcal{U} = \{u_1, u_2, \ldots, u_m\}$, an item set $\mathcal{V} = \{v_1, v_2, \ldots, v_n\}$, and a rating matrix $R \in \mathbb{R}^{m \times n}$, our task is essentially to fill the unknown cells of matrix R. In addition, we have an HIN $\mathcal{G} = (\mathcal{V}, \mathcal{E})$, which is composed of additional information. Here, $\mathcal{V}$ and $\mathcal{E}$ denote a node set of different types and a relation set, respectively. Take the HIN on the Yelp dataset as an example. There are eight node types (e.g., business, category, etc.) and nine node connection relationships (e.g., business belongs to category, etc.). Formally, the objective of our system is to learn a prediction function $\hat{r}_{uv} = F(u, v, \mathcal{G}; \Theta)$, where $\hat{r}$ denotes the predicted ratings of user u and item v, and Θ denotes the parameters of the function.

3 Related Work

In recent years, combining heterogeneous information network as an auxiliary information source with recommendation has attracted a lot of attention. HIN- based methods can be divided into two categories: embedding-based methods and path-based methods. In embedding-based methods, they usually learn complex latent embeddings of items and users through auxiliary information and user click history. For example, Wang et al. [16] used the user's history click records and ripple sets to learn the user's preference, and designed a preference propagation technique to update the embedding of user. Sun et al. [15] utilized recurrent network architecture to encode the entire path between user-item pair and both entities in path. However, most of these approaches are lack of explanation, so it is hard to explain the reasons for recommendation.

Path-based methods attempt to mine the semantic information of meta-path for recommendation, which has good recommendation performance and interpretability. In [18], meta-paths like user-item-*-item are used to calculate user preference matrix, which can produce the latent vectors of each meta-path through NMF. In [13], concepts of weighted HIN and weighted meta-path are introduced to calculate user similarity, based on which the model makes recommendations according to similar users. In [5], Convolution Neural Networks are used to learn embedding of path from the embedding matrices composed of user-item path. Ma et al. [9] proposed a joint learning framework to induce rules from item relations, which can achieve better performance and explanation with meta-path. However, meta-path is unable to describe complex network structure and capture rich semantics.

To address the problem, Zhao et al. [19] introduced the concept of meta-graph into recommendation to generate user-item similarity matrix, and adopt FM to capture the interactions between features generated from MF. MGAR [1] employed attention mechanism to enhance useful features, and utilized a hierarchical feature interaction method with factorization machine to capture the interaction of feature for recommendation. As discussed in the introduction, these methods do not consider to learn representations of graphs or explicitly use similarities, thus failing to fully make use of features. In contrast, our approach generates a feature vector of the graph and takes the similarity as feature weight to improve the recommendation effect.

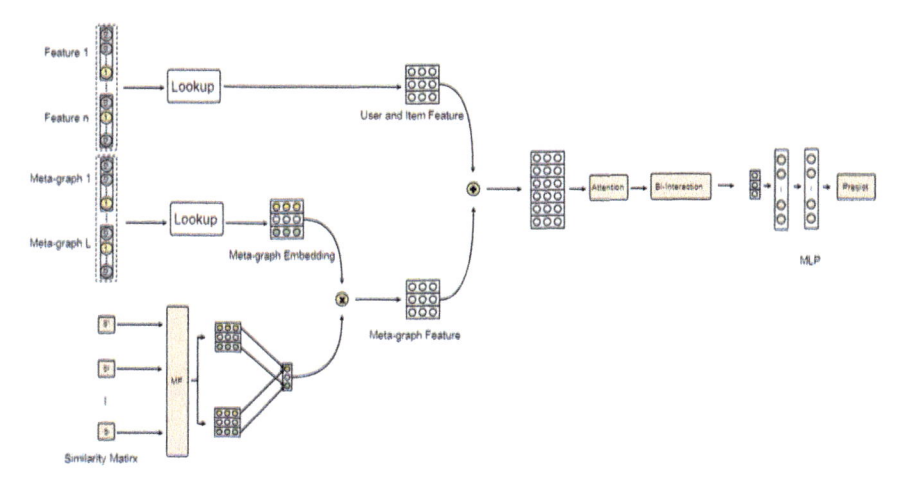

Fig. 3 An overview of MGBE

4 Problem Definition

In this paper, we consider a recommender system that aims to predict the ratings of unobserved user-item pairs. Given a user set $\mathcal{U} = \{u_1, u_2, \ldots, u_m\}$, an item set $\mathcal{V} = \{v_1, v_2, \ldots, u_n\}$, and a rating matrix $R \in \mathbb{R}^{m \times n}$, our task is essentially to fill the unknown cells of matrix R. In addition, we have an HIN $\mathcal{G} = (\mathcal{V}, \mathcal{E})$, which is composed of additional information. Here, $\mathcal{V}$ and $\mathcal{E}$ denote a node set of different types and a relation set, respectively. Take the HIN on Yelp dataset as an example. There are eight node types (e.g., business, category, etc.) and nine node connection relationships (e.g., business belongs to category, etc.). Formally, the objective of our system is to learn a prediction function $\hat{r}_{uv} = F(u, v, \mathcal{G}; \Theta)$, where $\hat{r}$ denotes the predicted ratings of user u and item v, and Θ denotes the parameters of function.

5 Framework Of MGBE

In this section, we describe the structure of MGBE in detail. As shown in Fig. 3, our model consists of three components. (1) Similarity generator: to predict the similarity of user and item, we first utilize matrix factorization to factorize similarity matrices from predefined meta-graphs and then calculate unobserved similarity. (2) Feature generator: to obtain the representations of users, items and meta-graphs, we learn their embeddings by a lookup layer, which transforms one-hot representations into low-dimensional dense vectors. (3) Recommendation model: we utilize a feature interaction layer based on an attention mechanism to predict the rating of user-item pair.

5.1 Similarity Generator

HIN and HIN scheme are first introduced in [14]. Here, meta-graph is defined as follows [19]:

Definition 1 Meta-Graph. A meta-graph $\mathcal{M}$ is a directed acyclic graph (DAG) with a single source node n_s (i.e., with in-degree 0) and a single sink (target) node n_t (i.e., with out-degree 0), defined on an HIN $\mathcal{G} = (\mathcal{V}, \mathcal{E})$ with schema $\mathcal{T_G} = (\mathcal{A}, \mathcal{R})$ where $\mathcal{V}$ is the node set, $\mathcal{E}$ is the edge set, $\mathcal{A}$ is the node type set, and $\mathcal{R}$ is the edge type set. Then, we define a meta-graph as $\mathcal{M} = (\mathcal{V}_M, \mathcal{E}_M, \mathcal{A}_M, \mathcal{R}_M, n_s, n_t)$, where $\mathcal{V}_M \in \mathcal{V}$, $\mathcal{E}_M \in \mathcal{E}$ constrained by $\mathcal{A}_M \in \mathcal{A}$ and $\mathcal{R}_M \in \mathcal{R}$, respectively.

Given the predefined meta-graphs [19], which start with a user node and end with an item node, we first compute the user-item similarity matrix by two operations (Hadamard product and multiplication) and then generate latent features of user and item by MF. Taking Fig. 2 as an example, the similarity matrix of this graph is calculated as

$$S = W_{UR} \cdot ((W_{RI} \cdot W_{RI}^T) \odot (W_{RA} \cdot W_{RA}^T)) \cdot W_{UR}^T \cdot W_{UI}, \tag{1}$$

where $\odot$ is the Hadamard product, W_* is the adjacency matrix between two different types. When there are L meta-graphs, we can obtain L different matrices, denoted by $S^1, \ldots, S^L$. Then, we adopt MF to characterizes items and users from matrix S^t, denotes by low-rank matrices $U^t = [u_1^t, u_2^t, \ldots, u_m^t] \in \mathbb{R}^{m \times d}$ and $I^t = [v_1^t, v_2^t, \ldots, v_n^t] \in \mathbb{R}^{n \times d}$, where d is the length of latent feature. We address only a few known interactions and it is highly prone to overfitting. Thus, we can obtain the U^t and I^t with the Frobenius norm regularization to avoid overfitting. The optimization formula is as follows:

$$min_{U^t, I^t} \frac{1}{2} \|S^t - U^t I^{(t)T}\|_d^2 + \frac{\lambda_{U^t}}{2} \|U^t\|_d^2 + \frac{\lambda_{I^t}}{2} \|I^t\|_d^2, \tag{2}$$

where λ_{U^t} and λ_{I^t} are hyper-parameters of regularization. By repeating the above process for all L similarity matrices, we can obtain L pairs of latent features of users and items, denoted as $U^1, I^1, \ldots, U^L, I^L$. The feature pair (U^t, I^t) can predict the unobserved similarities of the tth graph. As previously discussed, existing methods do not consider the influence of different distributions. Therefore, we use the Z-score method to standardize them. The formalizations are as follows:

$$s_{i,j}^t = u_i^t \cdot (v_j^t)^T, \tag{3}$$

$$\hat{s}_{i,j}^t = f\left(\frac{s_{i,j}^t - \mu^t}{\sigma^t}\right), \tag{4}$$

where i and j denote user i and item j, t denotes the tth meta-graph, μ^t and σ^t are the mean and standard deviation of s^t, $f(\cdot)$ is an activation function, such as Softmax or tanh, and $s_{i,j}^t$ is the similarity between user i and item j in the t meta-graph.

5.2 Feature Generator

Our model contains three features: user features, item features, and meta-graph features. Similar to other recommendation methods, user features (e.g., user ID and last movie rate) and item features (e.g., item ID and average rate) represent user preferences and item attributes, respectively. Different from existing meta-graph-based recommendation models, we consider meta-graph as the feature of (user, item) pairs. For example, when the meta-graph t connects user i and item j at the same time, it means that user i and item j share the semantic connection of graphic structure.

For features of user and item, we employ an embedding layer to learn their representation. Firstly, we encode features into sparse vector $x \in \mathbb{R}^n$ by one-hot encoding, where a feature value $x_t = 1$ means that the t feature exists, and 0, otherwise. Then, we adopt a lookup operation to transform the sparse vectors into a dense vector, denoted by $v \in \mathbb{R}^k$. K is the length of feature vector and is also a hyper-parameter.

To learn the representations of graphs, we embed graph L_t into a low-dimensional vector in the embedding layer, denoted as $e^t \in \mathbb{R}^k$. Note that e^t is shared by all (user-item) pairs, because it only represents the corresponding meta-graph and cannot be used as a feature of a user or item. We then multiply e^t and $\hat{s}_{i,j}^t$ to obtain the feature vector of the tth meta-graph of user i and item j

$$e_{i,j}^t = e^t \cdot \hat{s}_{i,j}^t, \tag{5}$$

where $\hat{s}_{i,j}^t$ measures the degree of correlation between user i and item j in the corresponding meta-graph. For L meta-graphs, we can obtain L graph features for each (user-item) pair, denoted as $e^1, \ldots, e^L$.

5.3 Recommendation Model

After obtaining L graph features, we concatenate them with other feature vectors, and denote it as $V_x = [x_1 v_1, \ldots, x_t v_t, \hat{s}^1 e^1, \ldots, \hat{s}^L e^L]$. For the sake of simplicity, we rephrase V_x as $V_x = [x_1 v_1, \ldots, x_n v_n]$, where v_n and x_n are the nth feature vector and feature value, respectively, and $t + L = n$. To identify the importance of different features, we employ a two-layer structure to implement attention. Specifically, we calculate the attention score as

$$\alpha_i = W_2^T f(W_1^T (x_i v_i) + b_1) + b_2, \tag{6}$$

where W_* and b_* denote the weight and bias of the $*$th layer, and $f(\cdot)$ is ReLU function. Then, we use the Softmax function to normalize the above attention scores

$$h_i = \frac{\exp(\alpha_i)}{\sum_{i \in n} \exp(\alpha_i)}. \tag{7}$$

The value of h_i indicates the importance of the ith feature. Finally, a new feature vector is obtained as

$$\hat{v} = h_i \cdot v_i. \tag{8}$$

Let $\hat{V}_x = [x_1\hat{v}_1, \ldots, x_n\hat{v}_n]$ as the input of Bi-Interaction layer [3], which is a pooling layer with linear time complexity

$$z = \sum_{i=1}^{n} \sum_{j=i+1}^{n} x_i\hat{v}_i \odot x_j\hat{v}_j, \tag{9}$$

where $\odot$ is the Hadamard product, and n is the number of features. The output of the Bi-Interaction layer is a vector with the second-order interaction information. Note that no additional parameters are introduced in this layer, and the calculation can be completed in linear time to reformulate

$$z = \frac{1}{2}\left[\left(\sum_{i=1}^{n} x_i\hat{v}_i\right)^2 - \sum_{i=1}^{n}(x_i\hat{v}_i)^2\right]. \tag{10}$$

Hence, compared with the feature interaction layer of MGAR, our method achieves a faster calculation speed. Then, we use DNN to predict the rating of user-item pair. The formalizations are as follows:

$$h_1 = f(W_1 z + b_1), \tag{11}$$
$$h_2 = f(W_2 h_1 + b_2), \tag{12}$$
$$h_3 = f(W_3 h_2 + b_3), \tag{13}$$

where $f(\cdot)$ is ReLU activation function, and W_* and b_* are the weight matrix and bias of the $*$th layer, respectively.

6 Experiments and Performance Evaluation

6.1 Dataset

Yelp[1] and Amazon,[2] which are provided by [19], are used in the experiments to demonstrate the effectiveness of our model. Yelp dataset consists of the interaction information, such as user ratings and reviews on businesses, as well as network information of business (e.g., category, location). For Amazon dataset, we choose the electronics domain as our second dataset in many Amazon datasets of different domains. Moreover, each dataset has around 200,000 ratings from 1 to 5. In the experiments, we split each dataset at the ratio of 8:2, which means that the training set and the test set are 80% and 20 of the data of the rating matrix, respectively.

6.2 Evaluation Metric

According to the work in [1, 19], we adopt root mean square error (RMSE) as a metric to evaluate the performance of our model

$$RMSE = \sqrt{\frac{\sum_{(i,j) \in \mathcal{R}_{test}} (R_{i,j} - \hat{R}_{i,j})^2}{|\mathcal{R}_{test}|}} \qquad (14)$$

where $R_{i,j}$ and $\hat{R}_{i,j}$ are the real rating and predicted rating of user i to item j, and $\mathcal{R}_{test}$ is the set of user-item in the test data.

6.3 Baseline Models

We compare our model with the following baseline methods:

NeuMF [4]: NeuMF is a neural network model, which adopts GMF and MLP two pathways to model the interaction of users and items. On the first pathway called GMF, it can not only simulate Matrix Factorization (MF), but also extend the non-linearity of MF by using non-linear activation function. On the other pathway, it adopts multi-layer perceptron to learn the interaction between users' and items' potential features.

FMG [19]: FMG is a recommender system for rating prediction, which first combines the concept of meta-graph with HIN for recommendation. It first adopts MF to factorize similarity matrices from different meta-graphs and simply concatenates

---

[1] https://www.yelp.com/dataset.
[2] http://jmcauley.ucsd.edu/data/amazon/.

all the latent features. Then, it utilizes group lasso regularized FM to capture the interactions between features and predicts rate.

AFM [17]: AFM is a neural network model, which combines FM and an attention mechanism. It can reduce noise interactions and enhance useful interactions to distinguish the importance of different feature interactions.

MGAR [1]: MGAR is a meta-graph-based method, which introduces an attention mechanism to distinguish the importance of features from different meta-graphs. The method designs a hierarchical feature interaction method to learn interactions between features.

6.4 Results

6.4.1 Comparison With different Models

The RMSE results of all methods in the two datasets can be found in Table 1. Based on Table 1, we summarize the following observations:

- FMG performs worse than NeuMF and AFM, which may be caused by two reasons. First, FMG does not standardize the features before aggregating them, which affects the fitting of the recommended model. Second, FMG aggregates all features through concatenation, which makes the model unable to recognize the importance of features. Compared with NeuMF, AFM not only uses extra feature information, but also utilizes the attention mechanism to give different weights to different features, hence achieving better performance.
- MGAR performs the best in all baselines. MGAR first designs the attention-based feature enhancement module to consider the importance of different features, and then uses a hierarchical feature interaction method to mine the relationship between features.
- Compared with MGAR, MGBE is still improved in two datasets. This is because we introduce information based on meta-graph and eliminate the influence of dimension by standardization. In addition, we also use the attention mechanism. It is worth noting that the improvement on Amazon dataset is not as obvious as that on Yelp dataset, which may be due to the fact that there are only 5 meta-graphs on Amazon dataset.

6.4.2 Comparison Among MGBE Variants

In order to verify the role of each part, we test the variants of MGBE in the following three aspects: the usage of Z-score method, the usage of attention network and the usage of Bi-Interaction layer. The results are shown in Table 2. Firstly, attention network and standardization can improve the performance of the model, whether used alone or at the same time, which can achieve the best results. The main reason is that

Table 1 RMSE of MGBE and all baseline methods

Methods	Yelp	Amazon
FMG [19]	1.1864(+9.4%)	1.2456(+9.4%)
NeuMF [4]	1.1816(+9.0%)	1.1421(+1.2%)
AFM [17]	1.1680(+7.9%)	1.1360(+0.7%)
MGAR [1]	1.1637(+7.6%)	1.1322(+0.4%)
MGBE	**1.0753**	**1.1281**

Table 2 RMSE results of variants of MGBE

Model	Yelp	Amazon
MGBE without z-score method	1.4647	1.1340
MGBE without attention network	1.1067	1.1552
MGBE without Bi-interaction layer	1.1450	1.1662
MGBE	**1.0753**	**1.1281**

both of them are complementary as the weight of features. In addition, Bi-Interaction layer plays an important role in mining the relationship between features. This proves that the second-order interaction information can improve the performance of the model.

6.5 The Parameter K

In this section, we verify the impact of parameter K on two datasets. K is the length of all feature vectors, including user features, item features and graph features. We set K in the range of [2, 3, 5, 10, 20, 30, 40, 50], and the results of RMSE are plotted in Fig. 4a and b. We observe that with the increase of K, RMSE on the whole presents a downward trend and then an upward trend. For Yelp, the model performs the best when $K = 20$; For Amazon, RMSE first keeps stable and then rises when K is greater than 10. Obviously, $K = 20$ and $K = 10$ are the best parameter settings on Yelp and Amazon datasets, respectively, which achieve better performance and avoid additional computational cost. Therefore, we set $K = 20$ and $K = 10$ for Yelp dataset and Amazon dataset.

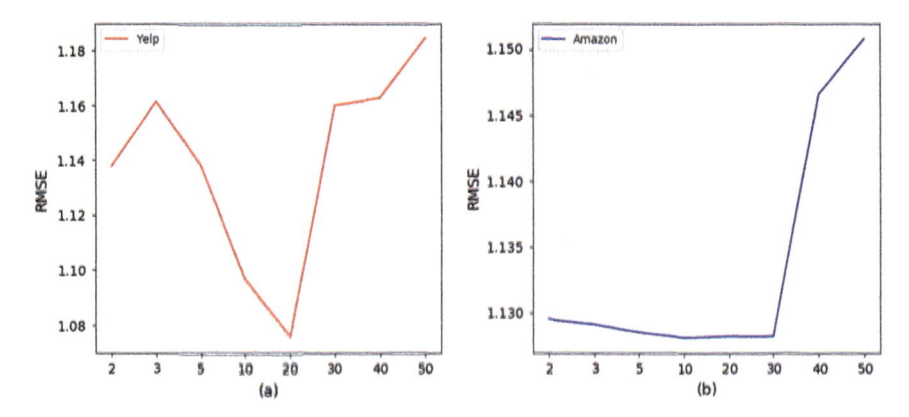

Fig. 4 RMSE with varying K

7 Conclusion

In this paper, we proposed a recommendation method over Heterogeneous Information Networks (HINs), named MGBE. MGBE learns the semantic representations of meta-graphs and combines them with the similarities to generate new features of user-item pairs. Meta-graph features can be well integrated with other features. We also learned that the similarity distributions of different meta-graphs are different, so they are treated with standardization. Finally, we used Bi-interaction layer and attention mechanism to explore the relationship between features. Experimental results on Yelp dataset and Amazon dataset demonstrate that MGBE performs better than several recommendation methods.

Acknowledgements This work was supported by Basic Public Welfare Research Project of Zhejiang Province (LGF19F020002).

References

1. Dai, F., Gu, X., Li, B., Zhang, J., Qian, M., Wang, W.: Meta-graph based attention-aware recommendation over heterogeneous information networks. In: International Conference on Computational Science, pp. 580–594. Springer (2019)
2. Fang, Y., Lin, W., Zheng, V.W., Wu, M., Li, X.L.: Semantic proximity search on graphs with metagraph-based learning. In: IEEE International Conference on Data Engineering (2016)
3. He, X., Chua, T.S.: Neural factorization machines for sparse predictive analytics. In: Proceedings of the 40th International ACM SIGIR Conference on Research and Development in Information Retrieval, pp 355–364 (2017). https://doi.org/10.1145/3077136.3080777
4. He, X., Liao, L., Zhang, H., Nie, L., Chua, T.S.: Neural collaborative filtering. In: the 26th International Conference (2017)
5. Hu, B., Shi, C., Zhao, W.X., Yu, P.S.: Leveraging meta-path based context for top- n recommendation with a neural co-attention model. In: the 24th ACM SIGKDD International Conference (2018)

6. Huang, Z., Zheng, Y., Cheng, R., Sun, Y., Mamoulis, N., Li, X.: Meta structure: Computing relevance in large heterogeneous information networks. In: Proceedings of the 22nd ACM SIGKDD International Conference on Knowledge Discovery and Data Mining, Association for Computing Machinery, New York, KDD '16, pp. 1595–1604 (2016). https://doi.org/10.1145/2939672.2939815
7. Koren, Y.: Factorization meets the neighborhood: A multifaceted collaborative filtering model. In: Proceedings of the 14th ACM SIGKDD International Conference on Knowledge Discovery and Data Mining, Las Vegas, August 24–27 (2008)
8. Koren, Y., Bell, R., Volinsky, C.: Matrix factorization techniques for recommender systems. Computer **42**(8), 30–3 (2009)
9. Ma, W., Min, Z., Cao, Y., Jin, W., Wang, C., Liu, Y., Ma, S., Ren, X.: Jointly learning explainable rules for recommendation with knowledge graph. In: The World Wide Web Conference, pp. 1210–1221 (2019). https://doi.org/10.1145/3308558.3313607
10. Rendle, S.: Factorization machines. In: IEEE International Conference on Data Mining (2010)
11. Rendle, S.: Factorization machines with libfm. Acm Trans. Intell. Syst. Technol. **3**(3), 1–22 (2012)
12. Sarwar, B., Karypis, G., Konstan, J., Riedl, J.: Item-based collaborative filtering recommendation algorithms. In: Proceedings of ACM World Wide Web Conference, vol. 1 (2001). https://doi.org/10.1145/371920.372071
13. Shi, C., Zhang, Z., Luo, P., Yu, P., Yue, Y., Wu, B.: Semantic path based personalized recommendation on weighted heterogeneous information networks. In: Proceedings of the 24th ACM International on Conference on Information and Knowledge Management, pp 453–462 (2015). https://doi.org/10.1145/2806416.2806528
14. Sun, Y., Han, J., Yan, X., Yu, P.S., Wu, T.: Pathsim: meta path-based top-k similarity search in heterogeneous information networks. Proc. Vldb Endow. **4**(11), 992–1003 (2011)
15. Sun, Z., Yang, J., Zhang, J., Bozzon, A., Huang, L.K., Xu, C.: Recurrent knowledge graph embedding for effective recommendation. In: Proceedings of the 12th ACM Conference on Recommender Systems, pp 297–305 (2018). https://doi.org/10.1145/3240323.3240361
16. Wang, H., Zhang, F., Wang, J., Zhao, M., Li, W., Xie, X., Guo, M.: Ripplenet: propagating user preferences on the knowledge graph for recommender systems. In: Proceedings of the 27th ACM International Conference on Information and Knowledge Management, Association for Computing Machinery, New York, CIKM '18, pp. 417–426 (2018). https://doi.org/10.1145/3269206.3271739
17. Xiao, J., Ye, H., He, X., Zhang, H., Wu, F., Chua, T.S.: Attentional factorization machines: learning the weight of feature interactions via attention networks. CoRR (2017)
18. Yu, X., Ren, X., Sun, Y., Gu, Q., Sturt, B., Khandelwal, U., Norick, B., Han, J.: Personalized entity recommendation: a heterogeneous information network approach. In: Proceedings of the 7th ACM International Conference on Web Search and Data Mining, pp 283–292 (2014). https://doi.org/10.1145/2556195.2556259
19. Zhao, H., Yao, Q., Li, J., Song, Y., Lee, D.L.: Meta-graph based recommendation fusion over heterogeneous information networks. In: KDD 2017 (2017)

Comments on a Secure AKA Scheme for Multi-server Environments

Qian Meng, Zhiyuan Lee, Tsu-Yang Wu, Chien-Ming Chen, and Kuan-Han Lu

Abstract In telecare medicine information systems (TMIS), the design of three factors-based schemes using smart cards can be used for remote user authentication. Face on different services, the patient needs clearly to register and login to each server. However, most of the existing schemes in TMIS are usually based on a single server environment, which increases the cost of saving cards and memorizing the passwords for the users. Recently, Ali et al. proposed a three-factor authentication and key agreement (AKA) scheme for multi-server environments. They claimed that their scheme can resist many well-known security attacks. However, Yu et al. have conducted a security analysis on Ali et al.'s scheme. In this paper, we further point out Ali et al.'s AKA scheme exists other flaws which include perfect forward secrecy, temporary value disclosure attacks, and off-line guessing attacks.

1 Introduction

With the rapid development of the Internet of Things (IoT) [23] and cheaper mobile devices, making online medical services have been developed. In an online medical system, registered users can obtain remote medical services from anywhere [17]. By integrating a variety of mobile devices and information technologies, the telecare medicine information system (TMIS) upgrades the traditional medical model to a more convenient and intelligent electronic medical model [14].

Q. Meng · Z. Lee · T.-Y. Wu (✉) · C.-M. Chen
College of Computer Science and Engineering, Shandong University of Science and Technology, Qingdao 266590, China
e-mail: fMengQian@163.com

Z. Lee
e-mail: jlizhiyuan@163.com

K.-H. Lu
Department of Computer Science and Information Management, Soochow University, Taipei 11490, Taiwan

Recently, many scholars have conducted a lot of research on TMIS, which has brought great changes to the electronic medical model. Through the use of TMIS, the patient's medical condition can be accessed and monitored by the doctor at any time [4]. However, due to the openness of this environment, the public channel may be completely controlled by attackers, which may cause the patient's private information to be obtained by attackers. Therefore, it is critical to protect the privacy of patients to safely share electronic medical information.

In many studies, researchers have proposed some authentication schemes [7–10, 19, 20, 22] to improve security and ensure user privacy. In [3, 5, 6, 11, 15, 18, 21, 24], they combined passwords, biometrics, and smart cards to propose a three-factor authentication scheme. In order to ensure user privacy (anonymity), Li et al. [12] proposed an anonymous user authentication scheme based on the IoT medical environment. Pu et al. [16] proposed an authentication scheme using elliptic curve cryptography (ECC). But their scheme has a high demand for computing, communication, and storage. Xu et al. [26] proposed an anonymous, lightweight mutual authentication scheme. However, Alzahrani et al. [2] proved that Xu et al.'s scheme has security issues such as off-line identity-guessing attacks, replay attacks, and key-compromise impersonation (KCI) threat. Liang et al. [13] proposed a PUF-based mutual authentication scheme, but xiao et al. [25] showed that Liang et al.'s protocol will be subject to tracking attacks.

Recently, Ali et al. [1] proposed an improved three-factor symmetric-key authentication scheme called ITSSAKA-MS. In 2020, Yu et al. [27] have demonstrated that Ali et al.'s scheme has some security problems such as masquerade attacks, session key exposure attacks, man-in-the-middle attacks (MIMT), smart card theft attacks, and insecure mutual authentication. However, in this paper, we briefly reviewed the ITSSAKA-MS scheme and then found some other security weaknesses including perfect forward secrecy, temporary value disclosure attacks, and off-line guessing attacks.

2 Review of the Scheme of Ali et al.

In the section, we briefly give an introduction to the scheme of Ali et al. Their scheme consists of three phases: user and sever registration, login and key-establishment phase, and update phase. Messages in the user registration and server registration phases are transmitted on a secure channel. In the other two phases, since the common channel transmission data, the data can be stolen or forged, and modified. Our security analysis of their scheme includes only two phases: registration phase and login and key-establishment phase. The notations used in the paper are shown in Table 1.

Table 1 Notations and their meanings

Notations	Meanings
U_i	The ith user
ID_i	U_i's identity
PW_i	U_i's password
SC_i, PW_i	The Smart card and pseudo-random-password U_i
RC	The Registration center
SK_{RC}	The secret key of RC
S_j	The jth sever
SID_j	S_j's identity
SK_S	The secret key of sever
t, τ_i	Error tolerance threshold, public reproduction parameter
σ_i	Biometric secret key
SK	Session key
T	Timestamp
ΔT	Maximum transmission delay
$h(\cdot)$	A hash function
$\parallel$	Concatenation
$\oplus$	XOR
$Gen(\cdot), Rep(\cdot)$	Fuzzy extractor

2.1 User Registration Phase

1. The users U_i selects ID_i, PW_i, and imprints BIO_i, then computes $VID_i = h(ID_i)$ and sends VID_i to registration center RC over the secure channel.
2. RC selects a random number r_1 and temporary identity TID_i, computes $Auth_i = h(SK_{RC} \parallel VID_i \parallel r_1)$, $P_i = h(VID_i \parallel r_1)$, $Q_i = E_{SK_{RC}}(Auth_i, r_1, VID_i)$, then RC stores $\{P_i, Q_i, TID_i\}$ in smart card SC, then send SC to U_i.
3. U_i computes $Gen(BIO_i) = (\sigma_i, \tau_i)$, $A_i = h(ID_i \parallel PW_i \parallel \sigma_i)$, $Q_i' = Q_i \oplus h(PW_i \parallel \sigma_i)$, $P_i' = P_i \oplus h(PW_i \parallel \sigma_i)$, $TID_i' = TID_i \oplus h(PW_i \parallel \sigma_i)$, then replaces $\{Q_i, P_i, TID_i\}$ with $\{Q_i', P_i', TID_i'\}$ and stores $\{A_i, R_i', P_i', TID_i', h(\cdot), Gen(\cdot), Rep(\cdot), \tau_i, t\}$.

2.2 Server Registration Phase

S_j registers with RC, and the specific steps are as follows:

1. Sever S_j selects its identity SID_j and sends SID_j to RC over the secure channel.
2. RC computes sever's secret key $SK_S = h(SID_j \parallel SK_{RC})$, then sends SK_S to S_j over the secure channel.

3. Finally, sever saves SK_S.

2.3 Login and Key-Establishment Phase

When User U_i wants to access the server S_j, they can establish a session key through RC to ensure secure communication. The detailed steps are as follows:

1. User U_i inserts SC_i, inputs ID_i, PW_i, imprints his/her BIO'_i. U_i computes $Rep(BIO'_i, \tau_i) = \sigma'_i$, then checks $A_i \stackrel{?}{=} h(ID_i \parallel PW_i \parallel \sigma'_i)$. If true U_i chooses a randoom r_2 and temestamp T_1, comeputes $Q_i = Q'_i \oplus h(PW_i \parallel \sigma'_i)$, $P_i = P'_i \oplus h(PW_i \parallel \sigma'_i)$, $TID_i = TID'_i \oplus h(PW_i \parallel \sigma'_i)$, $VID'_i = h(ID_i)$, $r'_2 = r_2 \oplus VID'_i$, $TID''_i = TID_i \oplus VID'_i$, $T'_1 = T_1 \oplus VID'_i$, $W_i = h(VID'_i \parallel P_i \parallel T_1)$, and then sends the $M_1 = \{Q_i, SID_j, r'_2, W_i, TID''_i, T'_1\}$ to RC.
2. Upon receiving M_1, RC computes $(Auth_i, r_1, VID_i) = D_{SK_{RC}}(Q_i)$, $T_1 = T'_1 \oplus VID_i$, checks $|T_1 - T_C| \leq \Delta T$, $W_i \stackrel{?}{=} h(VID_i \parallel h(VID_i \parallel SK_{RC}) \parallel T_1)$ and $Auth_i \stackrel{?}{=} h(SK_{RC} \parallel VID'_i \parallel r_1)$, if both are verified, then RC computes $TID_i = TID''_i \oplus VID_i$, $r_2 = r'_2 \oplus VID_i$. RC generates a timestamp T_2 and computes the $P_j = h(SID_j \parallel SK_{RC})$, $W_{RC} = h(P_j \parallel T_2)$, $Y_{RC} = h(SID_j \parallel VID_i \parallel r_2 \parallel T_1)$, $G_{RC} = E_{P_j}(TID_i, W_{RC}, Y_{RC}, T_1, T_2)$, finally RC sends $M_2 = \{G_{RC}, SID_j\}$ to S_j.
3. Upon receiving M_2, S_j computes $D_{SK_S}(G_{RC}) = (TID_i, W_{RC}, Y_{RC}, T_1, T_2)$, S_j checks condition $|T_2 - T_C| \leq \Delta T$ and $W_{RC} \stackrel{?}{=} h(SK_S \parallel T_2)$, if true S_j generates timestamp T_3 and computes $SK_{ij} = h(Y_{RC} \parallel SID_j \parallel T_3)$, $W_{S_j} = h(SK_{ij} \parallel T_1 \parallel T_3)$, then S_j directly sends $M_3 = \{W_{S_j}, T_3, TID_i\}$ to U_i.
4. Upon receiving M_3, chenks the condition $|T_3 - T_C| \leq \Delta T$. U_i computes $Y_i = h(SID_j \parallel VID'_i \parallel r_2 \parallel T_1)$, $SK'_{ij} = h(Y_i \parallel SID_j \parallel T_3)$, then checks the condition $W_{S_j} \stackrel{?}{=} h(SK'_{ij} \parallel T_1 \parallel T_3)$, if true U_i saves the session key.

3 Statement of the Problem

This section makes a cryptanalysis of the authentication scheme proposed by Ali et al. which is described in the following subsections.

3.1 Perfect Forward Secrecy

It is said that if the master secrecy of any entity covered by the disclosure protocol does not result in the disclosure of the agreed session key for the current/future (previous) session, the authentication will provide forward(backward) secrecy.

In this part, we prove that the scheme of Ali et al. did not satisfy "perfect forward secrecy (PFS)". Assuming that attacker A obtains the private key SK_{RC} of RC, he/she can launch the following attacks:

1. A decrypts $D_{SK_{RC}}(Q_i) = \{AUth_i, r_1, VID_i\}$ to obtain VID_i, and intercepts $M_1\{Q_i, SID_j, r'_2, W_i, TID''_i, T'_1\}$ to obtain $SID_j, r'_2, TID''_i, T'_1$, and then computes $r_2 = r'_2 \oplus VID_i$, $T_1 = T'_1 \oplus VID_i$, $Y_{RC} = h(SID_j \parallel VID_i \parallel r_2 \parallel T_1)$.
2. A intercepts $M_3\{W_{S_j}, T_3, TID_i\}$ to obtain T_3, hence A can successfully computes $SK_{ij} = h(Y_{RC} \parallel SID_j \parallel T_3)$.

3.2 Temporary Value Disclosure Attacks

Temporary value disclosure attack is to point out the process of running an agreement, if the user's temporary secret number was leaked, leading to an attacker can calculate the session key.

In this section, we proved that Ali et al.'s scheme is vulnerable to temporary value disclosure attacks. Assuming that A obtains the random number r_2, then he/she can launch the following attacks:

1. A successfully intercepts $M_1 = \{Q_i, SID_j, r'_2, W_i, TID''_i, T'_1\}$, and then computes $VID_i = r_2 \oplus r'_2$, $T_1 = T'_1 \oplus VID_i$, $Y_{RC} = h(SID_j \parallel VID_i \parallel r_2 \parallel T_1)$.
2. A intercepts $M_3 = \{W_{S_j}, T_3, TID_i\}$ to obtain T_3, hence A can successfully computes $SK_{ij} = h(Y_{RC} \parallel SID_j \parallel T_3)$.

3.3 Off-Line Guessing Attacks

The attacker uses the transmission information intercepted in the protocol to perform off-line guessing.

In this section, we proved that Ali et al.'s scheme is vulnerable to off-line guessing attacks. Assuming that A gets the smart card and obtain the data of $SC = \{Q'_i, P'_i, TID'_i, h(\cdot), Gen(\cdot), Rep(\cdot), \tau_i, t\}$, because the user can chose his/her identity, A can do the following calculations:

1. A intercepts $M_1 = \{Q_i, SID_j, r'_2, W_i, TID''_i, T'_1\}$ to obtain Q_i, then computes $h(PW_i \parallel \sigma'_i) = Q'_i \oplus Q_i$, $P_i = P'_i \oplus h((PW_i \parallel \sigma'_i))$, $W_i = h(VID'_i \parallel P_i \parallel (T'_1 \oplus VID_i))$. Then A computes $VID^*_i = h(ID^*_i)$ to verify $W^*_i \stackrel{?}{=} W_i$ by off-line guessing user's identity ID_i.
2. If A guesses successfully, then A computes $T_1 = T'_1 \oplus VID^*_i$, $r_2 = r'_2 \oplus VID^*_i$, $Y_{RC} = h(SID_j \parallel VID^*_i \parallel r_2 \parallel T_1)$.
3. A intercepts $M_3 = \{W_{S_j}, T_3, TID_i\}$ to obtain T_3, hence A can successfully computes $SK_{ij} = h(Y_{RC} \parallel SID_j \parallel T_3)$.

4 Conclusion

This comment is about an AKA authentication scheme proposed by Ali et al. In this paper, our analysis found that Ali et al.'s protocol was highly sensitive to PFS, temporary value disclosure attacks, and off-line guessing attacks. It is hoped that the research in this paper will provide guidance for the development of better certification technology.

References

1. Ali, Z., Hussain, S., Rehman, R.H.U., Munshi, A., Liaqat, M., Kumar, N., Chaudhry, S.A.: Itssaka-ms: an improved three-factor symmetric-key based secure aka scheme for multi-server environments. IEEE Access **8**, 107993–108003 (2020)
2. Alzahrani, B.A., Irshad, A., Albeshri, A., Alsubhi, K.: A provably secure and lightweight patient-healthcare authentication protocol in wireless body area networks. Wireless Pers. Commun. **117**(1), 47–69 (2021)
3. Arshad, H., Nikooghadam, M.: Three-factor anonymous authentication and key agreement scheme for telecare medicine information systems. J. Med. Syst. **38**(12), 136 (2014)
4. Barman, S., Shum, H.P., Chattopadhyay, S., Samanta, D.: A secure authentication protocol for multi-server-based e-healthcare using a fuzzy commitment scheme. IEEE Access **7**, 12557–12574 (2019)
5. Challa, S., Das, A.K., Odelu, V., Kumar, N., Kumari, S., Khan, M.K., Vasilakos, A.V.: An efficient ecc-based provably secure three-factor user authentication and key agreement protocol for wireless healthcare sensor networks. Comput. Electr. Eng. **69**, 534–554 (2018)
6. Chaudhry, S.A., Shon, T., Al-Turjman, F., Alsharif, M.H.: Correcting design flaws: an improved and cloud assisted key agreement scheme in cyber physical systems. Comput. Commun. **153**, 527–537 (2020)
7. Chen, C.M., Wang, K.H., Fang, W., Wu, T.Y., Wang, E.K.: Reconsidering a lightweight anonymous authentication protocol. J. Chin. Inst. Eng. **42**(1), 9–14 (2019)
8. Chen, C.M., Xiang, B., Wang, K.H., Yeh, K.H., Wu, T.Y.: A robust mutual authentication with a key agreement scheme for session initiation protocol. Appl. Sci. **8**(10), 1789 (2018)
9. Chen, C.M., Xiang, B., Wang, K.H., Zhang, Y., Wu, T.Y.: An efficient and secure smart card based authentication scheme. J. Internet Technol. **20**(4), 1113–1123 (2019)
10. Chen, C.M., Xu, L., Wang, K.H., Liu, S., Wu, T.Y.: Cryptanalysis and improvements on three-party-authenticated key agreement protocols based on chaotic maps. J. Internet Technol. **19**(3), 679–687 (2018)
11. Li, C.T., Hwang, M.S.: An efficient biometrics-based remote user authentication scheme using smart cards. J. Netw. Comput. Appl. **33**(1), 1–5 (2010)
12. Li, C.T., Wu, T.Y., Chen, C.L., Lee, C.C., Chen, C.M.: An efficient user authentication and user anonymity scheme with provably security for iot-based medical care system. Sensors **17**(7), 1482 (2017)
13. Liang, W., Xie, S., Zhang, D., Li, X., Li, K.: A mutual security authentication method for rfid-puf circuit based on deep learning. ACM Trans. Internet Technol. 1–20 (2020)
14. Liu, X., Ma, W., Cao, H.: Mbpa: a medibchain-based privacy-preserving mutual authentication in tmis for mobile medical cloud architecture. IEEE Access **7**, 149282–149298 (2019)
15. Mansoor, K., Ghani, A., Chaudhry, S.A., Shamshirband, S., Ghayyur, S.A.K., Mosavi, A.: Securing iot-based rfid systems: a robust authentication protocol using symmetric cryptography. Sensors **19**(21), 4752 (2019)
16. Pu, Q., Wang, J., Zhao, R.: Strong authentication scheme for telecare medicine information systems. J. Med. Syst. **36**(4), 2609–2619 (2012)

17. Sammoud, A., Chalouf, M.A., Hamdi, O., Montavont, N., Bouallegue, A.: A secure three-factor authentication and biometrics-based key agreement scheme for tmis with user anonymity. In: 2020 International Wireless Communications and Mobile Computing (IWCMC), pp. 1916–1921 (2020)
18. Srivastava, K., Awasthi, A.K., Kaul, S.D., Mittal, R.: A hash based mutual rfid tag authentication protocol in telecare medicine information system. J. Med. Syst. **39**(1), 153 (2015)
19. Wang, Y., Liu, Y., Ma, H., Ma, Q., Ding, Q.: The research of identity authentication based on multiple biometrics fusion in complex interactive environment. J. Netw. Intell. **4**(4), 124–139 (2019)
20. Wu, T.Y., Lee, Y.Q., Chen, C.M., Tian, Y., Al-Nabhan, N.A.: An enhanced pairing-based authentication scheme for smart grid communications. J. Ambient Intell. Human. Comput. (2021). https://doi.org/10.1007/s12652-020-02740-2
21. Wu, T.Y., Lee, Z., Obaidat, M.S., Kumari, S., Kumar, S., Chen, C.M.: An authenticated key exchange protocol for multi-server architecture in 5g networks. IEEE Access **8**, 28096–28108 (2020)
22. Wu, T.Y., Lee, Z., Yang, L., Luo, J.N., Tso, R.: Provably secure authentication key exchange scheme using fog nodes in vehicular ad hoc networks. J. Supercomput. (2021). https://doi.org/10.1007/s11227-020-03548-9
23. Wu, T.Y., Wang, T., Lee, Y.Q., Zheng, W., Kumari, S., Kumar, S.: Improved authenticated key agreement scheme for fog-driven iot healthcare system. Secur. Commun. Netw. **2021**, 6658041 (2021)
24. Wu, T.Y., Yang, L., Lee, Z., Chen, C.M., Pan, J.S., Islam, S.: Improved ecc-based three-factor multiserver authentication scheme. Secur. Commun. Netw. **2021**, 6627956 (2021)
25. Xiao, L., Xie, S., Han, D., Liang, W., Guo, J., Chou, W.K.: A lightweight authentication scheme for telecare medical information system. Connect. Sci. 1–17 (2021)
26. Xu, Z., Xu, C., Chen, H., Yang, F.: A lightweight anonymous mutual authentication and key agreement scheme for wban. Concurrency. Comput.: Pract. Experi. **31**(14) (2019)
27. Yu, S., Park, Y.: Comments on "itssaka-ms: An improved three-factor symmetric-key based secure aka scheme for multi-server environments". IEEE Access **8**, 193375–193379 (2020)

Author Index

A
Ali, Liaqat, 3

B
Bao, Wugedele, 367

C
Chen, Chien-Ming, 25, 319, 339, 359, 391
Chen, Jiawei, 275
Chen, Lili, 319
Chen, Weilong, 221
Chen, Xiu, 43
Chiang, Tsui-Lin, 63
Chiu, Yi-Jui, 211, 327
Chu, Shu-Chuan, 85, 95, 241, 253, 265, 285, 305

D
Das, Swagatam, 157
Du, Zhaoyang, 367

F
Fu, Xuefeng, 129

G
Gao, Min, 241
Gong, Yuxiao, 25
Guo, Yundong, 199

H
Harada, Masaya, 367

He, Jianchuan, 275
Hsu, Chih-Hung, 15, 43, 167, 175
Hsub, Chih-Hung, 35
Hung, Wen-Jye, 63

J
Ji, Yusheng, 367

K
Kong, Lingping, 157, 285, 295

L
Lee, Xuanang, 359
Lee, Zhiyuan, 319, 391
Liang, Anhui, 265
Li, Geng-Chen, 241
Li, Jianpo, 241
Li, Ming-Ge, 35
Li, Peiwu, 129
Li, Ranran, 53
Liu, Baohong, 129
Li, Wenqi, 85
Liu, Chang, 77
Liu, Churan, 105
Liuc, Jun-Wei, 35
Liu, Fei-Fei, 305
Liu, Shuangshuang, 319
Liu, Shuo, 25
Liu, Yang, 105, 275
Liu, Yipeng, 3
Liu, Zhenyu, 199
Li, Zhangyan, 105
Li, Zhen, 339

Li, Zhongcui, 25
Lu, Kuan-Han, 391
Luo, Hao, 199
Luo, Qi, 63

M
Meng, Qian, 391
Ming-Tai Wu, Jimmy, 25, 53
Mo, Xiao-Mei, 229

P
Pan, Jeng-Shyang, 85, 157, 241, 253, 265, 285, 295, 305
Pan, Tien-Szu, 253
Pu, Huijie, 199

Q
Qian, Tao, 105, 275
Qi, Shuhan, 105, 275

R
Ren, Zhenwen, 117

S
Shen, Xuewen, 145, 377
Shi, Shuaijie, 105
Shuai, Shiyuan, 377
Snášel, Václav, 157, 295
Song, Pei-Cheng, 265
Su, Jingyong, 105
Sun, An-Ching, 15
Sun, Lingyun, 25
Sun, Xiao-Xue, 285

T
Tang, Linlin, 105, 275
Tan, Jianrong, 199
Teng, Qian, 53
Tian, Ai-Qing, 253

W
Wang, Ke, 53
Wang, Xiaopeng, 305
Wang, Xinjie, 199
Wang, Ya-Min, 63
Wang, Yanlong, 117
Wei, Min, 53
Wu, Celimuge, 367
Wu, Jun, 377
Wu, Runxiu, 129
Wu, Tsu-Yang, 285, 319, 339, 359, 391

X
Xiao, Xian-Tuo, 175
Xu-He, 175
Xu, Zhiqi, 377

Y
Yang, Lei, 339, 359
Yang, Xue-Hua, 167
Yang, Yang, 349
Yin, Shihao, 129
Yoshinaga, Tsutomu, 367
Yu, Ding-Guo, 77, 229, 349
Yu, Ru-Yue, 15

Z
Zeng, Jun-Yi, 43
Zhang, Jiajia, 105, 275
Zhang, Jing, 221
Zhang, Li-Gang, 95
Zhang, Lipu, 145
Zhang, Zhenzhou, 359
Zhang, Zhonghao, 3
Zhao, Hengling, 3
Zhao, Shu-Hao, 327
Zhao, Ya-Zheng, 211, 327
Zhengcheng, Wang, 183
Zhou, Yijie, 77
Zhu, Ce, 3

Printed by Books on Demand, Germany